Boundary Elements and other Mesh Reduction Methods XLI

WITPRESS
WIT Press publishes leading books in Science and Technology.
Visit our website for the current list of titles.
www.witpress.com

WITeLibrary
Home of the Transactions of the Wessex Institute.
Papers published in this volume are archived in the WIT elibrary in volume 122 of WIT Transactions on Engineering Sciences (ISSN 1743-3533).
The WIT electronic-library provides the international scientific community with immediate and permanent access to individual papers presented at WIT conferences.
http://library.witpress.com

Forty First International Conference on Boundary Elements and other Mesh Reduction Methods

BEM/MRM 2018

Conference Chairmen

A.H.D. Cheng
University of Mississippi, USA

S. Syngellakis
Wessex Institute, UK

International Scientific Advisory Committee

Organised by

University of Mississippi, USA

Wessex Institute, UK

Sponsored by

Elsevier

WIT Transactions on Engineering Sciences

International Journal of Computational Methods

WIT Transactions

Wessex Institute
Ashurst Lodge, Ashurst
Southampton SO40 7AA, UK

Editorial Board

Boundary Elements and other Mesh Reduction Methods XLI

Editors

A.H.D. Cheng
University of Mississippi, USA

S. Syngellakis
Wessex Institute, UK

WIT*PRESS* Southampton, Boston

Editors:

A.H.D. Cheng
University of Mississippi, USA

S. Syngellakis
Wessex Institute, UK

Published by

WIT Press
Ashurst Lodge, Ashurst, Southampton, SO40 7AA, UK
Tel: 44 (0) 238 029 3223; Fax: 44 (0) 238 029 2853
E-Mail: witpress@witpress.com
http://www.witpress.com

For USA, Canada and Mexico

Computational Mechanics International Inc
25 Bridge Street, Billerica, MA 01821, USA
Tel: 978 667 5841; Fax: 978 667 7582
E-Mail: infousa@witpress.com
http://www.witpress.com

British Library Cataloguing-in-Publication Data

A Catalogue record for this book is available
from the British Library

ISBN: 978-1-78466-295-0
eISBN: 978-1-78466-296-7
ISSN: (print) 1746-4471
ISSN: (on-line) 1743-3533

The texts of the papers in this volume were set individually by the authors or under their supervision. Only minor corrections to the text may have been carried out by the publisher.

Preface

This special issue contains papers selected from the 41st International Conference on Boundary Elements and other Mesh Reduction Methods, held in the New Forest, UK, organised by the Wessex Institute.

The annual conference on Boundary Elements and other Mesh Reduction Methods (BEM/MRM) started in 1978 and is now in its 41st version. It is well established as the recognised international forum for the latest advances in those techniques and their application in science and engineering.

The conference series was founded by Professor Carlos Brebbia, with its first meeting held in Southampton. For the next 40 years, scientists and engineers have used this gathering to exchange the progresses made in the field. The continued success of the meeting is a result of the strength of the research on boundary elements and mesh reduction techniques being carried out all over the world. It is also a testimonial of the legacy of the founder of the boundary element method, Professor Brebbia. Unfortunately, earlier this year the world lost a giant: Professor Brebbia passed away on March 3, 2018. He may no longer chair the conference and interact with the participants, but his contributions will continue to benefit the field, and will shine the way for future researchers. This particular conference, the 41st in the series, is dedicated to Professor Carlos Brebbia.

The objective of the research papers presented at the meetings is the further development of techniques that reduce or eliminate the type of meshes required by first generation computational methods, such as finite differences or finite elements.

This has steadily been achieved through the development of BEM as a computational tool and continues through more recent research into advanced techniques, leading to further mesh reduction aiming to produce truly meshless methods.

Also included are papers on the use of BEM and, in particular, the description of new applications.

The Editors would like to thank all authors for the quality of their papers and other colleagues for their help in reviewing the material.

The Editors
New Forest, 2018

Contents

FUNDAMENTAL SOLUTION OF COMPOSITE THICK SHELLS WITH CLAMPED EDGES

JIAO JIN[1], HUI WEI[1], JIANLONG ZHENG[1] & PIHUA WEN[2]
[1]School of Traffic and Transportation Engineering, Changsha University of Science and Technology, China
[2]School of Engineering and Materials Science, Queen Mary University of London, UK

ABSTRACT

In this paper, a rectangular composites double curved shell with four clamped edges is studied under static distributed and concentrated loads. The governing equations for the laminated and functionally graded shells with respect to the middle surface are presented, and the fundamental solutions are obtained. The exact solutions of the laminated and functionally graded shells should serve as the benchmark solutions for any numerical computation methods and can be used in the boundary element method and meshless method.

Keywords: fundamental solution, composite and functionally graded materials, plate and shell, moderate thick plate theory.

1 INTRODUCTION

A number of theories exist for layered anisotropic shells [1] and many of them were developed for thin shells based on the Kirchhoff hypothesis. However, the application of thin plate theory could lead to 30% or more errors to layered anisotropic composite shells [2]. The higher-order theory of Reddy and Liu [3] for composite shells is based on five degrees of freedom. This theory assumes a constant transverse deflection through the thickness and the displacements of the middle surface are expanded, as cubic functions of the thickness coordinate. Exact solutions were derived for simply supported laminated shells under static loads by Reddy [2], using double Fourier transform technique. This paper deals with the analytical solutions of the laminated and orthotropic functionally graded, cross-ply laminated, double curved shells with four clamped edges under static transverse concentrated loads. Analytical solutions for transverse deflection, in-plane tensile force and moment are presented for the laminated and various functionally graded shells. The exact solutions presented herein for cylindrical and spherical functionally graded shells under uniformly distributed load and concentrated force can be used as the benchmark results for any numerical methods, such as the finite element, finite difference, boundary element and meshless methods.

2 GOVERNING EQUATIONS FOR ORTHOTROPIC SHELLS

An uniform thickness shell is shown in Fig. 1, where ξ_1, ξ_2 and ς denote the orthogonal curvilinear coordinates, where ξ_1 and ξ_2 are the lines of curvature on the middle surface $\varsigma = 0$. For cylindrical and spherical shells, the lines of principal curvature coincide with the coordinate axis. With the assumptions of a moderate thick shell theory, one has the following displacement fields in the time domain

$$u_1(\xi_1,\xi_2,\varsigma) = u_0(\xi_1,\xi_2) + \varsigma\varphi_1(\xi_1,\xi_2),$$

$$u_2(\xi_1,\xi_2,\varsigma) = v_0(\xi_1,\xi_2) + \varsigma\varphi_2(\xi_1,\xi_2),$$

$$u_3(\xi_1,\xi_2,\varsigma) = w_0(\xi_1,\xi_2), \tag{1}$$

WIT Transactions on Engineering Sciences, Vol 122, © 2019 WIT Press
www.witpress.com, ISSN 1743-3533 (on-line)
doi:10.2495/BE410011

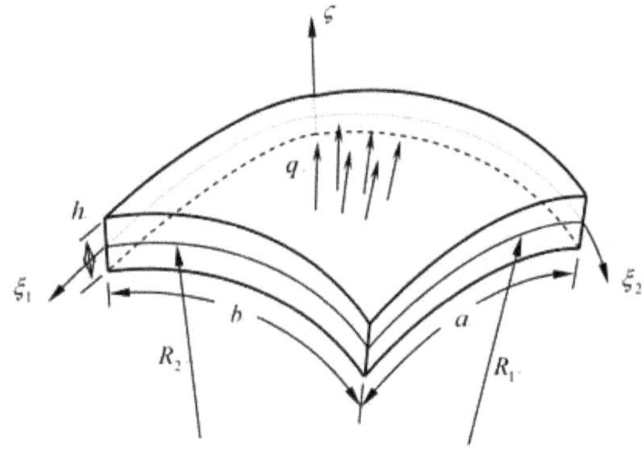

Figure 1: Geometry of a double curved shell.

where (u_0, v_0, w_0) are the displacements of the middle surface and (φ_1, φ_2) the rotations of a normal to the ξ_1 and ξ_2 axes respectively. In the following analysis coordinate axis is (x, y) for convenience. The resultants per unit length can be derived by

$$(N_i, M_i) = \int_{-h/2}^{h/2} \sigma_i(x, y, z)(1, z)\,dz,\ i = 1,2,6;$$
$$Q_1 = \kappa \int_{-h/2}^{h/2} \sigma_4(x, y, z)dz,\ Q_2 = \kappa \int_{-h/2}^{h/2} \sigma_5(x, y, z)dz, \tag{2}$$

where $\sigma_1 = \sigma_x, \sigma_2 = \sigma_y, \sigma_6 = \sigma_{xy}, \sigma_4 = \sigma_{yz}, \sigma_5 = \sigma_{xz}$ and κ denotes the shear correction factor (5/6 for the Reissner's moderate thick plate theory). The stress–strain relations are given, in the shell coordinate system, as

$$\begin{Bmatrix} \sigma_1 \\ \sigma_2 \\ \sigma_4 \\ \sigma_5 \\ \sigma_6 \end{Bmatrix} = \begin{bmatrix} Q_{11} & Q_{12} & 0 & 0 & 0 \\ Q_{12} & Q_{22} & 0 & 0 & 0 \\ 0 & 0 & Q_{44} & 0 & 0 \\ 0 & 0 & 0 & Q_{55} & 0 \\ 0 & 0 & 0 & 0 & Q_{66} \end{bmatrix} \begin{Bmatrix} \varepsilon_1 \\ \varepsilon_2 \\ \varepsilon_4 \\ \varepsilon_5 \\ \varepsilon_6 \end{Bmatrix}, \tag{3}$$

where Q_{ij} are the stiffness of orthotropic shell which can be determined by the engineering constants as

$$Q_{11} = \frac{E_1}{1 - \nu_{12}\nu_{21}},\ Q_{12} = \frac{\nu_{12}E_2}{1 - \nu_{12}\nu_{21}},\ Q_{22} = \frac{E_2}{1 - \nu_{12}\nu_{21}},\ Q_{66} = G_{12},\ Q_{44} = G_{23},\ Q_{55} = G_{13}, \tag{4}$$

in which E, ν and G are Young's moduli, Poisson's ratio and shear moduli respectively. Substituting (3) and (4) into (2) gives the stress resultants, ($N_6 = N_{12}, M_6 = M_{12}$ for convenience), as

$$N_1 = A_{11}\left(\frac{\partial u_0}{\partial x} + \frac{w_0}{R_1}\right) + A_{12}\left(\frac{\partial v_0}{\partial y} + \frac{w_0}{R_2}\right) + B_{11}\frac{\partial \varphi_1}{\partial x} + B_{12}\frac{\partial \varphi_2}{\partial y}$$

$$N_2 = A_{12}\left(\frac{\partial u_0}{\partial x}+\frac{w_0}{R_1}\right)+A_{22}\left(\frac{\partial v_0}{\partial y}+\frac{w_0}{R_2}\right)+B_{12}\frac{\partial \varphi_1}{\partial x}+B_{22}\frac{\partial \varphi_2}{\partial y},$$

$$N_6 = A_{66}\left(\frac{\partial u_0}{\partial y}+\frac{\partial v_0}{\partial x}\right)+B_{66}\left[\frac{\partial \varphi_1}{\partial y}+\frac{\partial \varphi_2}{\partial x}+C_0\left(\frac{\partial u_0}{\partial y}-\frac{\partial v_0}{\partial x}\right)\right],$$

$$M_1 = B_{11}\left(\frac{\partial u_0}{\partial x}+\frac{w_0}{R_1}\right)+B_{12}\left(\frac{\partial v_0}{\partial y}+\frac{w_0}{R_2}\right)+D_{11}\frac{\partial \varphi_1}{\partial x}+D_{12}\frac{\partial \varphi_2}{\partial y},$$

$$M_2 = B_{12}\left(\frac{\partial u_0}{\partial x}+\frac{w_0}{R_1}\right)+B_{22}\left(\frac{\partial v_0}{\partial y}+\frac{w_0}{R_2}\right)+D_{12}\frac{\partial \varphi_1}{\partial x}+D_{22}\frac{\partial \varphi_2}{\partial y},$$

$$M_6 = B_{66}\left(\frac{\partial u_0}{\partial y}+\frac{\partial v_0}{\partial x}\right)+D_{66}\left[\frac{\partial \varphi_1}{\partial y}+\frac{\partial \varphi_2}{\partial x}+C_0\left(\frac{\partial u_0}{\partial y}-\frac{\partial v_0}{\partial x}\right)\right], \tag{5}$$

where the stiffness coefficients are defined as

$$A_{ij} = \int_{-h/2}^{h/2} Q_{ij}dz, \quad B_{ij} = \int_{-h/2}^{h/2} Q_{ij}zdz, \quad D_{ij} = \int_{-h/2}^{h/2} Q_{ij}z^2dz, \quad i,j=1,2,4,5,6. \tag{6}$$

For the laminated shells, we have

$$A_{ij} = \sum_{k=1}^{n} Q_{ij}^{(k)}\left(z_{k+1}-z_k\right), \quad B_{ij} = \frac{1}{2}\sum_{k=1}^{n} Q_{ij}^{(k)}\left(z_{k+1}^2-z_k^2\right), \quad D_{ij} = \frac{1}{3}\sum_{k=1}^{n} Q_{ij}^{(k)}\left(z_{k+1}^3-z_k^3\right), \tag{7}$$

where $Q_{ij}^{(k)}$ denotes the lamina stiffness referred to the material coordinates of the k-th lamina and n is the number of layers in the laminate. For the gradation of material properties along the plate thickness, we assume the variation profile for volume fraction obey [4]

$$P(z) = P_b + (P_t - P_b)\left(z/h+0.5\right)^{\eta}, \tag{8}$$

where P denotes a generic property like modulus, P_t and P_b denote the property on the top and the bottom surfaces respectively, and η is a parameter that dictates the material variation profile through the thickness. Poisson ratios and density of mass are assumed to be uniform. In this case, all material constants, from (7), become

$$A_{ij} = \frac{(\eta+\beta)}{2(\eta+1)}Q_{ij}^{(b)}h, \quad B_{ij} = \frac{\eta(\beta-1)}{2(\eta+1)(\eta+2)}Q_{ij}^{(b)}h^2, \tag{9a}$$

$$D_{ij} = \frac{\eta(\eta^2+3\eta+8)+3\beta(\eta^2+\eta+2)}{12(\eta+1)(\eta+2)(\eta+3)}Q_{ij}^{(b)}h^3, \qquad i,j=1,2,4,5,6, \tag{9b}$$

where $\beta = P_t/P_b$ and $Q_{ij}^{(b)}$ are material stiffness on the bottom surface. Furthermore, we can also assume that the exponential gradation of material properties in thickness direction,

$$P(z) = P_b e^{\alpha(z/h+0.5)}, \tag{10}$$

where $\alpha = \ln(\beta)$ and hence $e^{\alpha} = \beta$. Then material constants can be written as

$$A_{ij} = \frac{(-1+e^{\alpha})}{\alpha} Q_{ij}^{(b)} h, \;\; B_{ij} = \frac{\alpha+2+(\alpha-2)e^{\alpha}}{2\alpha^2} Q_{ij}^{(b)} h^2, \tag{11a}$$

$$D_{ij} = \frac{-\alpha^2-4\alpha-8+(\alpha^2-4\alpha+8)e^{\alpha}}{4\alpha^3} Q_{ij}^{(b)} h^3. \tag{11b}$$

Use of the modified strain-displacements relations in Hamilton's principal yields the following equilibrium equations [5]

$$\frac{\partial N_1}{\partial x} + \frac{\partial}{\partial y}\left(N_6 + C_0 M_6\right) + \frac{Q_1}{R_1} = 0,$$

$$\frac{\partial}{\partial x}\left(N_6 - C_0 M_6\right) + \frac{\partial N_2}{\partial y} + \frac{Q_2}{R_2} = 0,$$

$$\frac{\partial Q_1}{\partial x} + \frac{\partial Q_2}{\partial y} - \left(\frac{N_1}{R_1} + \frac{N_2}{R_2}\right) + q = 0,$$

$$\frac{\partial M_1}{\partial x} + \frac{\partial M_6}{\partial y} - Q_1 = 0,$$

$$\frac{\partial M_6}{\partial x} + \frac{\partial M_2}{\partial y} - Q_2 = 0. \tag{12}$$

The equations of motion can be expressed in terms of the displacement as

$$A_{11}\left(\frac{\partial^2 u_0}{\partial^2 x} + \frac{1}{R_1}\frac{\partial w_0}{\partial x}\right) + A_{12}\left(\frac{\partial^2 v_0}{\partial x \partial y} + \frac{1}{R_2}\frac{\partial w_0}{\partial x}\right) + B_{11}\frac{\partial^2 \varphi_1}{\partial^2 x} + B_{12}\frac{\partial^2 \varphi_2}{\partial x \partial y} + A_{66}\left(\frac{\partial^2 v_0}{\partial x \partial y} + \frac{\partial^2 u_0}{\partial^2 y}\right)$$
$$\left(B_{66} + C_0 D_{66}\right)\left[\frac{\partial^2 \varphi_1}{\partial^2 y} + \frac{\partial^2 \varphi_2}{\partial x \partial y} + C_0\left(\frac{\partial^2 u_0}{\partial^2 y} - \frac{\partial^2 v_0}{\partial x \partial y}\right)\right] + C_0 B_{66}\left(\frac{\partial^2 v_0}{\partial x \partial y} + \frac{\partial^2 u_0}{\partial^2 y}\right)$$
$$+\frac{\kappa A_{55}}{R_1}\left(\frac{\partial w_0}{\partial x} + \varphi_1 - \frac{u_0}{R_1}\right) = 0,$$

$$A_{12}\left(\frac{\partial^2 u_0}{\partial x \partial y} + \frac{1}{R_1}\frac{\partial w_0}{\partial y}\right) + A_{22}\left(\frac{\partial^2 v_0}{\partial^2 y} + \frac{1}{R_2}\frac{\partial w_0}{\partial y}\right) + B_{12}\frac{\partial^2 \varphi_1}{\partial x \partial y} + B_{22}\frac{\partial^2 \varphi_2}{\partial^2 y} + A_{66}\left(\frac{\partial^2 v_0}{\partial^2 x} + \frac{\partial^2 u_0}{\partial x \partial y}\right)$$
$$\left(B_{66} - C_0 D_{66}\right)\left[\frac{\partial^2 \varphi_1}{\partial x \partial y} + \frac{\partial^2 \varphi_2}{\partial^2 x} + C_0\left(\frac{\partial^2 u_0}{\partial x \partial y} - \frac{\partial^2 v_0}{\partial^2 x}\right)\right] - C_0 B_{66}\left(\frac{\partial^2 v_0}{\partial^2 x} + \frac{\partial^2 u_0}{\partial x \partial y}\right)$$
$$+\frac{\kappa A_{44}}{R_1}\left(\frac{\partial w_0}{\partial y} + \varphi_2 - \frac{v_0}{R_2}\right) = 0,$$

$$\kappa A_{55}\left(\frac{\partial^2 w_0}{\partial^2 x}+\frac{\partial \varphi_1}{\partial x}-\frac{1}{R_1}\frac{\partial u_0}{\partial x}\right)+\kappa A_{44}\left(\frac{\partial^2 w_0}{\partial^2 y}+\frac{\partial \varphi_2}{\partial y}-\frac{1}{R_2}\frac{\partial v_0}{\partial y}\right)$$
$$-\frac{1}{R_1}\left[A_{11}\left(\frac{\partial u_0}{\partial x}+\frac{w_0}{R_1}\right)+A_{12}\left(\frac{\partial v_0}{\partial y}+\frac{w_0}{R_2}\right)+B_{11}\frac{\partial \varphi_1}{\partial x}+B_{12}\frac{\partial \varphi_2}{\partial y}\right]$$
$$-\frac{1}{R_2}\left[A_{12}\left(\frac{\partial u_0}{\partial x}+\frac{w_0}{R_1}\right)+A_{22}\left(\frac{\partial v_0}{\partial y}+\frac{w_0}{R_2}\right)+B_{11}\frac{\partial \varphi_1}{\partial x}+B_{12}\frac{\partial \varphi_2}{\partial y}\right]+q=0,$$

$$B_{11}\left(\frac{\partial^2 u_0}{\partial^2 x}+\frac{1}{R_1}\frac{\partial w_0}{\partial x}\right)+B_{12}\left(\frac{\partial^2 v_0}{\partial x\partial y}+\frac{1}{R_2}\frac{\partial w_0}{\partial x}\right)+D_{11}\frac{\partial^2 \varphi_1}{\partial^2 x}+D_{12}\frac{\partial^2 \varphi_2}{\partial x\partial y}+B_{66}\left(\frac{\partial^2 v_0}{\partial x\partial y}+\frac{\partial^2 u_0}{\partial^2 y}\right)$$
$$D_{66}\left[\frac{\partial^2 \varphi_1}{\partial^2 y}+\frac{\partial^2 \varphi_2}{\partial x\partial y}+C_0\left(\frac{\partial^2 u_0}{\partial^2 y}-\frac{\partial^2 v_0}{\partial x\partial y}\right)\right]-\kappa A_{55}\left(\frac{\partial w_0}{\partial x}+\varphi_1-\frac{u_0}{R_1}\right)=0,$$

$$B_{12}\left(\frac{\partial^2 u_0}{\partial x\partial y}+\frac{1}{R_1}\frac{\partial w_0}{\partial y}\right)+B_{22}\left(\frac{\partial^2 v_0}{\partial^2 y}+\frac{1}{R_2}\frac{\partial w_0}{\partial y}\right)+D_{12}\frac{\partial^2 \varphi_1}{\partial x\partial y}+D_{22}\frac{\partial^2 \varphi_2}{\partial^2 y}+B_{66}\left(\frac{\partial^2 v_0}{\partial^2 x}+\frac{\partial^2 u_0}{\partial x\partial y}\right) \quad (13)$$
$$D_{66}\left[\frac{\partial^2 \varphi_1}{\partial x\partial y}+\frac{\partial^2 \varphi_2}{\partial^2 x}+C_0\left(\frac{\partial^2 u_0}{\partial x\partial y}-\frac{\partial^2 v_0}{\partial^2 x}\right)\right]-\kappa A_{44}\left(\frac{\partial w_0}{\partial y}+\varphi_2-\frac{v_0}{R_2}\right)=0.$$

3 ANALYTICAL SOLUTION

3.1 General solution for rectangular shell with two opposite simply supported edges

Consider a rectangular double curved shell of two dimensions $(a\times b)$ with two opposite simply supported edges, a and b are the dimensions of the shell middle surface along the x and y axes. The boundary conditions on these two edges are expressed, called Case (1), as

$$u_0(x,0)=N_2(x,0)=w_0(x,0)=\varphi_1(x,0)=M_2(x,0)=0, \quad (14a)$$

$$u_0(x,b)=N_2(x,b)=w_0(x,b)=\varphi_1(x,b)=M_2(x,b)=0\,. \quad (14b)$$

Then the exact form of the spatial variations of displacement and rotation can be written [6] as

$$u^{(1)}(x,y)=\sum_{n=1}^{\infty}U_n^{(1)}(x)\sin\beta_n ym=,$$

$$v^{(1)}(x,y)=\sum_{n=1}^{\infty}V_n^{(1)}(x)\cos\beta_n y,$$

$$w^{(1)}(x,y)=\sum_{n=1}^{\infty}W_n^{(1)}(x)\sin\beta_n y,$$

$$\varphi_1^{(1)}(x,y)=\sum_{n=1}^{\infty}X_n^{(1)}(x)\sin\beta_n y,$$

$$\varphi_2^{(1)}(x,y)=\sum_{n=1}^{\infty}Y_n^{(1)}(x)\cos\beta_n y, \tag{15}$$

where $\beta_n = n\pi/b$. Substituting (15) into (13) gives a set of differential equations with respect to x, when the applied load $q=0$, as

$$\begin{bmatrix} S_{11}^{(1)} & S_{12}^{(1)} & S_{13}^{(1)} & S_{14}^{(1)} & S_{15}^{(1)} \\ S_{21}^{(1)} & S_{22}^{(1)} & S_{23}^{(1)} & S_{24}^{(1)} & S_{25}^{(1)} \\ S_{31}^{(1)} & S_{32}^{(1)} & S_{33}^{(1)} & S_{34}^{(1)} & S_{35}^{(1)} \\ S_{41}^{(1)} & S_{42}^{(1)} & S_{43}^{(1)} & S_{44}^{(1)} & S_{45}^{(1)} \\ S_{51}^{(1)} & S_{52}^{(1)} & S_{53}^{(1)} & S_{54}^{(1)} & S_{55}^{(1)} \end{bmatrix} \begin{Bmatrix} U_n^{(1)} \\ V_n^{(1)} \\ W_n^{(1)} \\ X_n^{(1)} \\ Y_n^{(1)} \end{Bmatrix} = \begin{Bmatrix} 0 \\ 0 \\ 0 \\ 0 \\ 0 \end{Bmatrix}, \tag{16}$$

where elements in the matrix are presented in [6]. We have the following equation

$$c_0 + c_1\lambda_n^2 + c_2\lambda_n^4 + c_3\lambda_n^6 + c_4\lambda_n^8 + c_5\lambda_n^{10} = 0, \tag{17}$$

where coefficients c_i, $i = 0,1,2,3,4,5$, can be obtained by expansion of determinant of matrix in (16) in the order of λ_n by the mathematical tool such as *Maple*. For example, one of the constants in (17)

$$c_5 = \kappa A_{55}\left(A_{11}A_{66}D_{11}D_{66} + B_{11}^2B_{66}^2 - B_{11}^2A_{66}D_{66} - A_{11}B_{66}^2D_{11}\right). \tag{18}$$

Other coefficients are too lengthy to be listed here. However, we can determine these coefficients easily by selecting arbitrary six values of $\bar{\lambda}_k$, $k = 0,1,2,..5$. Substituting the value of λ into (17) gives

$$c_0 + c_1\bar{\lambda}_k^2 + c_2\bar{\lambda}_k^4 + c_3\bar{\lambda}_k^6 + c_4\bar{\lambda}_k^8 + c_5\bar{\lambda}_k^{10} = \left|S(\bar{\lambda}_k)\right|, \quad k = 0,1,2,\ldots 5. \tag{19}$$

Then, six coefficients c_k can be obtained. Substituting coefficients c_k into (17), one can obtain ten roots $\lambda_{ni}(\beta_n)$, $i = 1,2,\ldots,10$. For each root, λ_{ni}, let $A_{5i}^{(1,n)} = 1$ in (17) and we have four linear equations to determine rest constants $A_{ji}^{(1,n)}$, $j = 1,2,3,4$, by

$$\begin{bmatrix} S_{11}^{(1)} & S_{12}^{(1)} & S_{13}^{(1)} & S_{14}^{(1)} \\ S_{21}^{(1)} & S_{22}^{(1)} & S_{23}^{(1)} & S_{24}^{(1)} \\ S_{31}^{(1)} & S_{32}^{(1)} & S_{33}^{(1)} & S_{34}^{(1)} \\ S_{41}^{(1)} & S_{42}^{(1)} & S_{43}^{(1)} & S_{44}^{(1)} \end{bmatrix} \begin{Bmatrix} A_{1i}^{(1,n)} \\ A_{2i}^{(1,n)} \\ A_{3i}^{(1,n)} \\ A_{4i}^{(1,n)} \end{Bmatrix} = -\begin{Bmatrix} S_{15}^{(1)} \\ S_{25}^{(1)} \\ S_{35}^{(1)} \\ S_{45}^{(1)} \end{Bmatrix}, \quad \text{if } \det \begin{vmatrix} S_{11}^{(1)} & S_{12}^{(1)} & S_{13}^{(1)} & S_{14}^{(1)} \\ S_{21}^{(1)} & S_{22}^{(1)} & S_{23}^{(1)} & S_{24}^{(1)} \\ S_{31}^{(1)} & S_{32}^{(1)} & S_{33}^{(1)} & S_{34}^{(1)} \\ S_{41}^{(1)} & S_{42}^{(1)} & S_{43}^{(1)} & S_{44}^{(1)} \end{vmatrix} \neq 0. \tag{20}$$

Introducing a unit vector of coefficient in complex value as

$$e_{j,i}^{(1,n)}=\frac{A_{ji}^{(1,n)}}{A_i^{(1,n)}},\quad A_i^{(1,n)}=\sqrt{\sum_{j=1}^{5}\left|A_{ji}^{(1,n)}\right|^2}\,,\tag{21}$$

then the general solution can be obtained by

$$\begin{Bmatrix} U_n^{(1)} \\ V_n^{(1)} \\ W_n^{(1)} \\ X_n^{(1)} \\ Y_n^{(1)} \end{Bmatrix}=\begin{bmatrix} e_{1,1}^{(1,n)} & e_{1,2}^{(1,n)} & \cdots & \cdots & \cdots & e_{1,10}^{(1,n)} \\ e_{2,1}^{(1,n)} & e_{2,2}^{(1,n)} & \cdots & \cdots & \cdots & e_{2,10}^{(1,n)} \\ e_{3,1}^{(1,n)} & e_{3,2}^{(1,n)} & \cdots & \cdots & \cdots & e_{3,10}^{(1,n)} \\ e_{4,1}^{(1,n)} & e_{4,2}^{(1,n)} & \cdots & \cdots & \cdots & e_{4,10}^{(1,n)} \\ e_{5,1}^{(1,n)} & e_{5,2}^{(1,n)} & \cdots & \cdots & \cdots & e_{5,10}^{(1,n)} \end{bmatrix}\begin{Bmatrix} B_1^{(n)}e^{\lambda_{n1}x} \\ B_2^{(n)}e^{\lambda_{n2}x} \\ \cdots \\ \cdots \\ B_{10}^{(n)}e^{\lambda_{n10}x} \end{Bmatrix},\tag{22}$$

where coefficients $B_i^{(n)}(\beta_n,s)$, $i=1,2,\ldots,10$, are unknowns in complex value. In the same way, the boundary conditions with other two opposite simply supported edges, defined as Case 2, can be described

$$v(0,y)=N_1(0,y)=w(0,y)=\varphi_2(0,y)=M_1(0,y)=0,\tag{23a}$$

$$v(a,y)=N_1(a,y)=w(a,y)=\varphi_2(a,y)=M_1(a,y)=0.\tag{23b}$$

Then the exact form of the spatial variations of displacement and rotation can be written as

$$u^{(2)}(x,y)=\sum_{m=1}^{\infty}U_m^{(2)}(y)\cos\alpha_m x,$$

$$v^{(2)}(x,y)=\sum_{m=1}^{\infty}V_m^{(2)}(y)\sin\alpha_m x,$$

$$w^{(2)}(x,y)=\sum_{m=1}^{\infty}W_m^{(2)}(y)\sin\alpha_m x,$$

$$\varphi_1^{(2)}(x,y)=\sum_{m=1}^{\infty}X_m^{(2)}(y)\cos\alpha_m x,$$

$$\varphi_2^{(2)}(x,y)=\sum_{m=1}^{\infty}Y_m^{(2)}(y)\sin\alpha_m x,\tag{24}$$

where $\alpha_m=m\pi/a$. Assuming that the general solutions have the following form as

$$\begin{Bmatrix} U_m^{(2)} \\ V_m^{(2)} \\ W_m^{(2)} \\ X_m^{(2)} \\ Y_m^{(2)} \end{Bmatrix}=\begin{Bmatrix} A_1^{(2,m)} \\ A_2^{(2,m)} \\ A_3^{(2,m)} \\ A_4^{(2,m)} \\ A_5^{(2,m)} \end{Bmatrix}e^{\gamma_m(\alpha_m)y},\tag{25}$$

where $A_i^{(2,m)}$ are unknown coefficients. Same as Case 1, we have the following equation

$$d_0+d_1\gamma_m^2+d_2\gamma_m^4+d_3\gamma_m^6+d_4\gamma_m^8+d_5\gamma_m^{10}=0\,,\tag{26}$$

where coefficients d_i, $i = 0,1,2,3,4,5$, can be obtained by expansion of determinant of matrix in (16) in the order of $\gamma_m(\alpha_m)$. The general solution can be obtained by

$$\begin{Bmatrix} U_m^{(2)} \\ V_m^{(2)} \\ W_m^{(2)} \\ X_m^{(2)} \\ Y_m^{(2)} \end{Bmatrix} = \begin{bmatrix} e_{1,1}^{(2,m)} & e_{1,2}^{(2,m)} & \cdots & \cdots & \cdots & e_{1,10}^{(2,m)} \\ e_{2,1}^{(2,m)} & e_{2,2}^{(2,m)} & \cdots & \cdots & \cdots & e_{2,10}^{(2,m)} \\ e_{3,1}^{(2,m)} & e_{3,2}^{(2,m)} & \cdots & \cdots & \cdots & e_{3,10}^{(2,m)} \\ e_{4,1}^{(2,m)} & e_{4,2}^{(2,m)} & \cdots & \cdots & \cdots & e_{4,10}^{(2,m)} \\ e_{5,1}^{(2,m)} & e_{5,2}^{(2,m)} & \cdots & \cdots & \cdots & e_{5,10}^{(2,m)} \end{bmatrix} \begin{Bmatrix} C_1^{(m)} e^{\gamma_{m1} x} \\ C_2^{(m)} e^{\gamma_{m2} x} \\ \cdots \\ \cdots \\ C_{10}^{(m)} e^{\gamma_{m10} x} \end{Bmatrix}, \tag{27}$$

where coefficients $C_i^{(m)}(\alpha_m, s)$, $i = 1,2,\ldots,10$, are unknowns in complex value.

3.2 Particular solution

For general distribution of transverse load, such as a concentrated force, we can use the double Fourier transform method and obtain

$$q(x,y) = \sum_{n=1}^{\infty}\sum_{m=1}^{\infty} q_{mn} \sin\alpha_m x \sin\beta_n y, \quad q_{mn} = \frac{4}{ab}\int_0^a\int_0^b q(x,y)\sin\alpha_m x \sin\beta_n y dx dy. \tag{28}$$

Then the particular solution can be written as

$$u^*(x,y,s) = \sum_{m=1}^{\infty}\sum_{n=1}^{\infty} U_{mn}^* \cos\alpha_m x \sin\beta_n y,$$

$$v^*(x,y,s) = \sum_{m=1}^{\infty}\sum_{n=1}^{\infty} V_{mn}^* \sin\alpha_m x \cos\beta_n y,$$

$$w^*(x,y,s) = \sum_{m=1}^{\infty}\sum_{n=1}^{\infty} W_{mn}^* \sin\alpha_m x \sin\beta_n y,$$

$$\varphi_1^*(x,y,s) = \sum_{m=1}^{\infty}\sum_{n=1}^{\infty} X_{mn}^* \cos\alpha_m x \sin\beta_n y,$$

$$\varphi_2^*(x,y,s) = \sum_{m=1}^{\infty}\sum_{n=1}^{\infty} Y_{mn}^* \sin\alpha_m x \cos\beta_n y. \tag{29}$$

Five coefficients $\{U_{mn}^*, V_{mn}^*, W_{mn}^*, X_{mn}^*, Y_{mn}^*\}$ can be determined directly.

3.3 Analytical solutions with four clamped edges

According to the principal of superposition, the exact form of the spatial variation of displacement and rotation should be in the following form

$$u(x,y) = \sum_{n=1}^{\infty} U_n^{(1)}(x)\sin\beta_n y + \sum_{m=1}^{\infty} U_m^{(2)}(y)\cos\alpha_m x + \sum_{m=1}^{\infty}\sum_{n=1}^{\infty} U_{mn}^* \cos\alpha_m x \sin\beta_n y,$$

$$v(x,y) = \sum_{n=1}^{\infty} V_n^{(1)}(x)\cos\beta_n y + \sum_{m=1}^{\infty} V_m^{(2)}(y)\sin\alpha_m x + \sum_{m=1}^{\infty}\sum_{n=1}^{\infty} V_{mn}^* \sin\alpha_m x \cos\beta_n y,$$

$$w(x,y)=\sum_{n=1}^{\infty}W_n^{(1)}(x)\sin\beta_n y+\sum_{m=1}^{\infty}W_m^{(2)}(y)\sin\alpha_m x+\sum_{m=1}^{\infty}\sum_{n=1}^{\infty}W_{mn}^{*}\sin\alpha_m x\sin\beta_n y,$$

$$\varphi_1(x,y)=\sum_{n=1}^{\infty}X_n^{(1)}(x)\sin\beta_n y+\sum_{m=1}^{\infty}X_m^{(2)}(y)\cos\alpha_m x+\sum_{m=1}^{\infty}\sum_{n=1}^{\infty}X_{mn}^{*}\cos\alpha_m x\sin\beta_n y,$$

$$\varphi_2(x,y)=\sum_{n=1}^{\infty}Y_n^{(1)}(x)\cos\beta_n y+\sum_{m=1}^{\infty}Y_m^{(2)}(y)\sin\alpha_m x+\sum_{m=1}^{\infty}\sum_{n=1}^{\infty}Y_{mn}^{*}\sin\alpha_m x\cos\beta_n y, \tag{30}$$

where coefficients $B_i(\beta_n)$, $C_i(\alpha_m)$, $i=1,2,...,10;\ m=1,2,...,M$ and $n=1,2,...,N$ are unknowns and M and N are the numbers of truncation term of Fourier series. Clamped edge boundary conditions indicate

$$u=v=w=\varphi_1=\varphi_2=0, \tag{31}$$

at all boundaries. Therefore, from (30) and (31), one obtains following linear equations

$$\begin{cases}\sum_{i=1}^{10}e_{1,i}^{(1,n)}B_i^{(n)}+\sum_{m=1}^{M}\sum_{i=1}^{10}a_{mni}e_{1,i}^{(2,m)}C_i^{(m)}+\sum_{n=1}^{L}U_{mn}^{*}=0\\ \sum_{i=1}^{\infty}e_{2,i}^{(1,n)}B_i^{(n)}=0\\ \sum_{i=1}^{10}e_{3,i}^{(1,n)}B_i^{(n)}=0\\ \sum_{i=1}^{10}e_{4,i}^{(1,n)}B_i^{(n)}+\sum_{m=1}^{M}\sum_{i=1}^{10}a_{mni}e_{4,i}^{(2,m)}C_i^{(m)}+\sum_{m=1}^{L}X_{mn}^{*}=0\\ \sum_{i=1}^{10}e_{5,i}^{(1,n)}B_i^{(n)}=0\end{cases}\qquad n=1,2,...,N,$$

for $x=0$, where L is a large number (50 in all numerical examples) and

$$a_{mni}=\frac{2}{b}\frac{\beta_n}{\gamma_{mi}^2+\beta_n^2}\left[1-(-1)^n e^{\gamma_{mi}b}\right].$$

For the second edge, one has

$$\begin{cases}\sum_{i=1}^{\infty}e_{1,i}^{(2,m)}C_i^{(m)}=0\\ \sum_{n=1}^{N}\sum_{i=1}^{10}b_{mni}e_{1,i}^{(1,n)}B_i^{(n)}+\sum_{i=1}^{10}e_{1,i}^{(2,m)}C_i^{(m)}+\sum_{n=1}^{L}V_{mn}^{*}=0\\ \sum_{i=1}^{10}e_{3,i}^{(2,m)}C_i^{(m)}=0\\ \sum_{i=1}^{10}e_{4,i}^{(2,m)}C_i^{(m)}=0\\ \sum_{n=1}^{N}\sum_{i=1}^{10}b_{mni}e_{5,i}^{(1,n)}B_i^{(n)}+\sum_{i=1}^{10}e_{5,i}^{(2,m)}C_i^{(m)}+\sum_{n=1}^{L}Y_{mn}^{*}=0\end{cases}\qquad m=1,2,...,M,$$

on $y=0$, where

$$b_{mni}=\frac{2}{a}\frac{\alpha_m}{\lambda_{ni}^2+\alpha_m^2}\left[1-(-1)^m e^{\lambda_{ni}a}\right].$$

For the third edge, we have

$$\begin{cases}\sum_{i=1}^{10} e_{1,i}^{(1,n)}e^{\lambda_{ni}a}B_i^{(n)}+\sum_{m=1}^{M}\sum_{i=1}^{10}(-1)^m a_{mni}e_{1,i}^{(2,m)}C_i^{(m)}+\sum_{m=1}^{L}(-1)^m U_{mn}^*=0\\ \sum_{i=1}^{\infty} e_{2,i}^{(1,n)}e^{\lambda_{ni}a}B_i^{(n)}=0\\ \sum_{i=1}^{10} e_{3,i}^{(1,n)}e^{\lambda_{ni}a}B_i^{(n)}=0\\ \sum_{i=1}^{10} e_{4,i}^{(1,n)}e^{\lambda_{ni}a}B_i^{(n)}+\sum_{m=1}^{M}\sum_{i=1}^{10}(-1)^m a_{mni}e_{4,i}^{(2,m)}C_i^{(m)}+\sum_{m=1}^{L}(-1)^m X_{mn}^*=0\\ \sum_{i=1}^{10} e_{5,i}^{(1,n)}e^{\lambda_{ni}a}B_i^{(n)}=0\end{cases}\qquad n=1,2,\ldots,N,$$

on $x=a$ and for the last edge

$$\begin{cases}\sum_{i=1}^{\infty} e_{1,i}^{(2,m)}e^{\gamma_{mi}b}C_i^{(m)}=0\\ \sum_{n=1}^{N}\sum_{i=1}^{10}(-1)^n b_{mni}e_{1,i}^{(1,n)}e^{\gamma_{mi}b}B_i^{(n)}+\sum_{i=1}^{10} e_{1,i}^{(2,m)}e^{\gamma_i b}C_i^{(m)}+\sum_{n=1}^{L}(-1)^n V_{mn}^*=0\\ \sum_{i=1}^{10} e_{3,i}^{(2,m)}e^{\gamma_{mi}b}C_i^{(m)}=0\\ \sum_{i=1}^{10} e_{4,i}^{(2,m)}e^{\gamma_{mi}b}C_i^{(m)}=0\\ \sum_{n=1}^{N}\sum_{i=1}^{10}(-1)^n b_{mni}e_{5,i}^{(1,n)}B_i^{(n)}+\sum_{i=1}^{10} e_{5,i}^{(2,m)}e^{\gamma_{mi}b}C_i^{(m)}+\sum_{n=1}^{L}(-1)^n Y_{mn}^*=0\end{cases}\qquad m=1,2,\ldots,M,$$

on $y=b$. We should have $10\times(N+M)$ linear algebraic equations to determine the same number unknown coefficients, i.e. $B_i^{(n)}$ and $C_i^{(m)}$. Therefore, the analytical solutions for the rectangular double covered shells with four clamped edges are obtained.

4 EXACT SOLUTION FOR DISTRIBUTED AND CONENTRATED FORCES

One cross-ply laminates is considered $[0^0/90^0]$ in this example with the same constants of material property in the above example. Truncation number of the Fourier series (M,N) is taken to be 10. Spherical and cylindrical shells are investigated with different boundary constrains under static transverse load q_0. Non-dimensional transverse deflection $\bar{w}_0(=w_0E_2h^3\times10^3/q_0a^4)$, normalised resultants of stresses $\bar{N}_i(=N_i/q_0a)$ and $\bar{M}_i(=M_i/q_0a^2)$ at the centre of the shell are presented in Tables 1 and 2 respectively for various values of radius-to-side (R/a) ratios and two values of thickness-to-side (h/a) ratios. In addition, spherical and cylindrical shells subjected to a concentrated load P_0 are considered. In this case, the non-dimensional transverse deflection $\overline{w}_0(=w_0E_2h^3\times10^3/P_0a^2)$, normalised resultants of stresses $\overline{N}_i(=N_ia/P_0)$ and $\overline{M}_i(=M_i/P_0)$ at the centre of the shell are presented in Table 3 for various values of radius-to-side (R/a) ratios and two values of thickness-to-side (h/a) ratios. As the in-plane force and moment at the point of concentrated force are singular, they divergent with the number of truncation of Fourier series (M,N). However, the resultants of stresses are regular and convergent in the domain of shell elsewhere.

5 CONCLUSIONS

In this paper, the exact solutions for the double curved laminated and functionally graded rectangular shells with four clamped were derived subjected to static distributed and concentrated transverse loads. The governing equations for the laminated and functionally graded shells with respect to the middle surface were presented with the moderate thick plate theory. The fundamental solution can be applied to the boundary element and the method of fundamental solution. In addition, these solutions can be used to assess the degree of accuracy for any numerical methods.

Table 1: Normalised deflection, in-plane force and moment for a laminated spherical shell $[0^0/90^0]$ $(R_1=R_2=R)$.

R/a	$h/a=0.01$					$h/a=0.1$				
	$\overline{w}_0$	$\overline{N}_1$	$\overline{M}_1$	$\overline{N}_2$	$\overline{M}_2$	$\overline{w}_0$	$\overline{N}_1$	$\overline{M}_1$	$\overline{N}_2$	$\overline{M}_2$
1	0.0044	0.5006	0.0011	0.4990	-0.0011	0.5456	0.5072	0.0126	0.5045	-0.0102
2	0.0187	1.0024	0.0023	0.9965	-0.0023	1.9548	0.8283	0.0247	0.8201	-0.0124
3	0.0463	1.5320	0.0035	1.5223	-0.0035	3.2671	0.9047	0.0313	0.8890	-0.0092
4	0.0908	2.0939	0.0048	2.0832	-0.0046	4.2024	0.8684	0.0339	0.8469	-0.0048
5	0.1528	2.6645	0.0063	2.6536	-0.0057	4.8277	0.7977	0.0346	0.7721	-0.0008
10	0.6511	5.0750	0.0144	5.0392	-0.0085	5.9965	0.5013	0.0323	0.4681	0.0104
∞	3.9782	0.1817	0.0239	-0.1817	0.0239	6.5150	0.0183	0.0233	-0.0183	0.0233

Table 2: Normalised deflection, in-plane force and moment for a laminated cylindrical shell $[0^0/90^0]$ $(R_1=\infty, R_2=R)$.

R/a	$h/a=0.01$					$h/a=0.1$				
	$\overline{w}_0$	$\overline{N}_1$	$\overline{M}_1$	$\overline{N}_2$	$\overline{M}_2$	$\overline{w}_0$	$\overline{N}_1$	$\overline{M}_1$	$\overline{N}_2$	$\overline{M}_2$
1	0.0135	0.0269	0.0000	1.0000	-0.0022	1.1762	0.0246	0.0029	0.9075	-0.0167
2	0.0572	0.0555	0.0001	1.9880	-0.0042	3.1310	0.0360	0.0100	1.2210	-0.0170
3	0.1306	0.0854	0.0001	3.0104	-0.0061	4.4235	0.0377	0.0150	1.1500	-0.0108
4	0.2338	0.1163	0.0004	4.0588	-0.0079	5.1545	0.0364	0.0179	1.0026	-0.0049
5	0.3630	0.1468	0.0009	5.0709	-0.0094	5.5775	0.0346	0.0196	0.8652	-0.0003
10	1.1995	0.2653	0.0057	8.4974	-0.0124	6.2562	0.0281	0.0223	0.4767	0.0110
∞	3.9782	0.1817	0.0239	-0.1817	0.0239	6.5150	0.0183	0.0233	-0.0183	0.0233

Table 3: Normalised deflection $\overline{w}_0$ for a laminated spherical shell $[0^0/90^0]$.

	Spherical shell $(R_1 = R_2 = R)$					Cylindrical shell $(R_1 = \infty, R_2 = R)$				
R/a	h/a=0.01	0.05	0.1	0.15	0.2	0.01	0.05	0.1	0.15	0.2
1	0.0044	0.5006	0.0011	0.4990	-0.0011	0.5456	0.5072	0.0126	0.5045	-0.0102
2	0.0187	1.0024	0.0023	0.9965	-0.0023	1.9548	0.8283	0.0247	0.8201	-0.0124
3	0.0463	1.5320	0.0035	1.5223	-0.0035	3.2671	0.9047	0.0313	0.8890	-0.0092
4	0.0908	2.0939	0.0048	2.0832	-0.0046	4.2024	0.8684	0.0339	0.8469	-0.0048
5	0.1528	2.6645	0.0063	2.6536	-0.0057	4.8277	0.7977	0.0346	0.7721	-0.0008
10	0.6511	5.0750	0.0144	5.0392	-0.0085	5.9965	0.5013	0.0323	0.4681	0.0104
∞	3.9782	0.1817	0.0239	- 0.1817	0.0239	6.5150	0.0183	0.0233	-0.0183	0.0233

REFERENCES

[1] Miyamoto, Y. et al., *Functionally Graded Materials: Design, Processing and Applications*, Kluwer: Boston, 1992.

[2] Reddy, J.N., Exact solutions of moderately thick laminated shells. *Journal of Engineering Mechanics,* **110**(5), pp. 749–809, 1984. DOI: 10.1061/(asce)0733-9399 (1984)110:5(794).

[3] Reddy, J.N. & Liu, C.F., A higher-order shear deformation theory of laminated elastic shells. *International Journal of Engineering Science*, **23**(3), pp. 319–30, 1985. DOI: 10.1016/0020-7225(85)90051-5.

[4] Sladek, J., Sladek, V., Zhang, Ch., Krivacek, J. & Wen, PH., Analysis of orthotropic thick plates by meshless local Petrov–Galerkin (MLPG) method. *International Journal for Numerical Methods in Engineering*, **67**(13), pp. 1830–1850, 2006. DOI: 10.1002/nme.1683.

[5] Reddy, J.N., Bending of laminated anisotropic shells by a shear deformable finite element. *Fibre Science and Technology*, **17**(1), pp. 9–24, 1982. DOI: 10.1016/0015-0568(82)90058-6.

[6] Wen, P.H. & Aliabadi, M.H., On the theory of functionally graded moderately thick composite shells. *Journal of Multiscale Modelling*, **5**(3), p. 1350012, 2013. DOI: 10.1142/s1756973713500121.

MULTI-DOMAIN BOUNDARY ELEMENT METHOD FOR AXISYMMETRIC PROBLEMS IN POTENTIAL THEORY AND LINEAR ISOTROPIC ELASTICITY

VASYL V. GNITKO[1], KYRYL G. DEGTYARIOV[1], ARTEM A. KARAIEV[2] & ELENA A. STRELNIKOVA[1,2*]
[1]A.N. Podgorny Institute for Mechanical Engineering Problems of the Ukrainian Academy of Sciences, Ukraine
[2]V.N. Karazin Kharkiv National University, Ukraine

ABSTRACT

The paper presents an approach based on reduced boundary element methods to resolve axisymmetric problems in potential and linear isotropic elasticity theories. The singular integral equations for these problems are received using fundamental solutions. Initially three-dimensional problems expressed in Cartesian coordinates are transformed to cylindrical ones and integrated with respect to the circumference coordinate. So the three-dimensional axisymmetric problems are reduced to systems of singular integral equations requiring the evaluation of linear integrals only. The fundamental solutions and their derivatives are expressed in terms of complete elliptic integrals. The effective algorithm for treatment of the singular integrals is proposed. The multi-domain boundary element method is applied for the numerical simulation. As examples, the following problems are considered: fluid induced vibrations of a compound cylindrical-spherical elastic shell partially filled with an ideal incompressible liquid, and axisymmetric elasticity problems for an isotropic body with rigid or elastic circular cylindrical inclusions.

Keywords: axisymmetric problems, potential theory, linear isotropic elasticity, multi-domain boundary element method, compound cylindrical-spherical shells.

1 INTRODUCTION

The boundary element method (BEM) has been used in potential theory and elasticity problems for decades. In order to solve the large-scale problems, researchers developed the multi-domain boundary element method (MDBEM). The multi-domain collocation strategy was briefly introduced by Brebbia et al. in [1], and then developed by Wang and Gao [2], Kane et al. [3], Zhang et al. [4] and Gao and Yang [5]. In MDBEM, the computational domain is divided into a number of sub-domains and the BEM algebraic equations are formulated for each sub-domain. Then the global system of equations is obtained by assembling results of all sub-domains using the equilibrium conditions in common interface nodes. The global system matrix based on MDBEM has a blocked and sparse character, and therefore the well-developed solvers for sparse systems can be employed to solve it. The use of MDBEM not only improve both the efficiency and computational speed, but also can solve fracture problems by dividing sub-domains along crack surfaces [4], multi-media problems [5], and problems for tanks with baffles [6] by dividing sub-domains along interfaces and baffle surfaces. In many engineering problems both geometry and boundary conditions are axisymmetric about some axis. BEM was initially developed for potential theory problems [1] and later extended to elasticity problems [7] due to analogy between fundamental solutions of both theories. The BEM for axisymmetric elasticity was first applied by Cruse et al. [8], using the fundamental solution obtained by Kermanidis [9]. Further BEM applications to axisymmetric problems in elasticity, thermo-elasticity, and fracture mechanics have been developed by Bakr [10]. But application of multi-domain BEM to axisymmetric problems is still not fully described in the literature.

ORCID: http://orcid.org/0000-0003-0707-7214

www.witpress.com, ISSN 1743-3533 (on-line)
doi:10.2495/BE410021

2 PROBLEM STATEMENTS

In this paper, we describe applications of the MDBEM to solutions of the problems in potential theory and linear isotropic elasticity. The emphasis is on problems in which the region and the boundary conditions are axisymmetric. In this case, the solution is independent of the angular coordinate θ, and the three-dimensional problem is reduced to a two-dimensional problem in the radial coordinate r and the axial coordinate z.

Although boundary integral formulations for different partial differential equations will change due to the presence of different Green's functions, the integral equations of potential theory and theory of elasticity are similar. Moreover, the Green's function behaviour at the singular point $P = Q$ is essentially the same for these problems. This allows us to begin with the simplest boundary integral equation, namely the Laplace equation.

Integral equations for linear potential flow problems are formulated through the application of Green's third identity, that represents a harmonic function as the sum of single- and double-layer potentials. Taking the field point to the boundary, an integral equation relating to only boundary values and normal derivatives of the harmonic function is obtained. Its counterpart in elasticity is Somigliana's identity.

2.1 Liquid induced vibrations of a compound cylindrical-spherical elastic shell

Containment shell structures have been the subject of intense research over the last 50 years, and the literature in this area is substantial. Shells composed of cylindrical and spherical parts in interaction with a fluid have received a little attention in scientific literature in spite of the importance of such thin walled structures in a number of different engineering branches. They are often used as tanks for liquid and gas storage, and as pressure vessels. They also appear as components of aerospace and marine vehicles, such as rocket and balloon skins, and submarine hulls.

The problem of free harmonic vibrations of the fluid-filled elastic shell composed of a cylindrical part bounded by a hemispherical edge is considered. The shell is of uniform thickness h, and height L, made of homogeneous, isotropic material with elasticity modulus E, Poisson's ratio ν and mass density ρ_s. We designate a moistened shell surface by $S=S_1+S_2$, and a free surface by S_0. Surfaces S_1 and S_2 correspond to the spherical and cylindrical parts, respectively (Fig. 1).

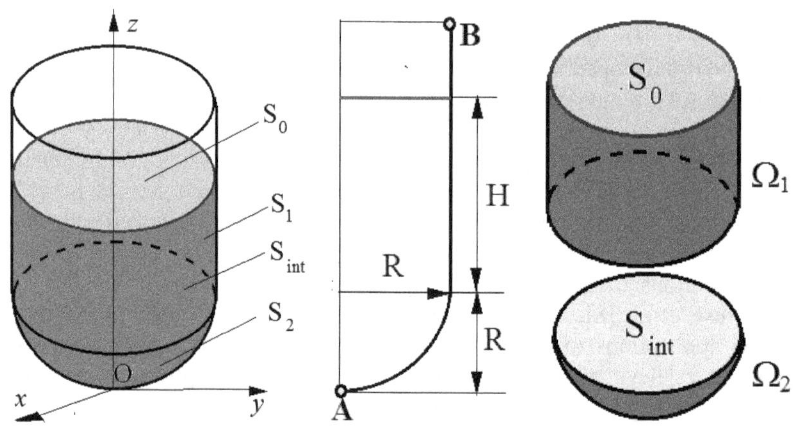

Figure 1: Shell structure, its sketch, and fluid sub-domains.

Denote by R radius of the spherical and cylindrical parts, and by H a height of the wetted cylindrical part. Suppose the Cartesian coordinate system $0xyz$ is connected with the shell, the liquid free surface S_0 coincides with the plane $z=H+R$ at the state of rest (Fig. 1). Let $\mathbf{U}=\left(U_1,U_2,U_3\right)$ denote the vector-function of shell displacements.

Assume that time dependent shell displacements are given by

$$\mathbf{U}=\mathbf{u}\exp(i\Omega t);\quad \mathbf{u}=\left(u_1,u_2,u_3\right).$$

Here Ω is the vibration frequency; the time factor $\exp(i\Omega)$ will be omitted further on. After separation of the time factor, the vibrations of the shell without a liquid are described by a system of three partial differential equations

$$\sum_{i=1}^{3} L_{ij}u_i=\Omega^2 u_j,\quad j=1,2,3,$$

where L_{ij} are linear differential operators of Kirchhoff–Love shell theory (Levitin and Vassiliev [11]). The finite element method was applied by Ravnik et al. in [12] to evaluate the natural frequencies Ω_k and modes $\mathbf{u}_k$, $k=\overline{1,N}$ of the shell of revolution without a liquid. After forming the global stiffness $\mathbf{L}$ and mass $\mathbf{M}$ matrices, the following equation of motion for the shell containing fluid is obtained:

$$\mathbf{LU}+\mathbf{M}\ddot{\mathbf{U}}=p_d\mathbf{n},$$

where $\mathbf{n}$ is an external unit normal to the shell wetted surface, the term $p_d\mathbf{n}$ gives the fluid dynamical pressure upon the shell, normal to its surface.

To model the fluid domain, a mathematical model has been developed based on the following hypotheses: the liquid is incompressible and inviscid one, its motion is irrotational, only small vibrations (linear theory) need to be considered. So a scalar velocity potential $\Phi(x,y,z,t)$ whose gradient represents the fluid velocity can be introduced. The liquid pressure $p=p\left(x,y,z,t\right)$ acting on the wetted shell surface is obtained from the linearized Bernoulli's equation for a potential flow

$$p=-\rho_l\left(\frac{\partial\Phi}{\partial t}+gz\right)+p_0,\quad p_s=-\rho_l gz,\quad p_d=-\rho_l\frac{\partial\Phi}{\partial t},$$

where g is the gravity acceleration, z is the vertical coordinate of a point in the liquid, ρ_l is the liquid density, p_s and p_d are static and dynamic components of the liquid pressure, p_0 is for atmospheric pressure.

At any instant the velocity potential $\Phi(x,y,z,t)$ could be determined from the following boundary value problem:

$$\nabla^2\Phi=\frac{\partial^2\Phi}{\partial x^2}+\frac{\partial^2\Phi}{\partial y^2}+\frac{\partial^2\Phi}{\partial z^2}=0,\quad \left.\frac{\partial\Phi}{\partial\mathbf{n}}\right|_S=\frac{\partial w}{\partial t},\quad \left.\frac{\partial\Phi}{\partial\mathbf{n}}\right|_{S_0}=\frac{\partial\zeta}{\partial t};\quad \left.\frac{\partial\Phi}{\partial t}+g\zeta\right|_{S_0}=0,\qquad (1)$$

where w denotes the normal component of the displacement vector $\mathbf{U}$, namely, $w=\left(\mathbf{U},\mathbf{n}\right)$; an unknown function $\zeta=\zeta\left(x,y,t\right)$ describes the form and position of the free surface. The second equation in (1) is the impermeability condition on the shell wetted surfaces, the third equation here is the kinematics boundary condition, which assumes that a fluid particle of the free surface will stay on this surface at any instance, the fourth equation in (1) is the dynamic boundary condition. It consists in equality of the liquid pressure on the free surface to atmospheric one. So the problem under consideration is reduced to the following system of differential equations:

$$\mathbf{LU} + \mathbf{M}\ddot{\mathbf{U}} = p_d \mathbf{n}\,; \qquad \Delta\Phi = 0 \tag{2}$$

with boundary conditions from (1), relative to Φ, and fixation conditions of the shell, relative to **U**. In this paper clamped-free (C-F) cylindrical-spherical shells are considered. i.e. shells of revolution are clamped at the ends A, and free at the ends B (Fig. 1).

Consider modes of fluid-filled shell vibrations in the form

$$\mathbf{U} = \sum_{k=1}^{N} c_k(t)\mathbf{u}_k\,, \tag{3}$$

where $c_k(t)$ are unknown coefficients, and $\mathbf{u}_k$ are eigenmodes of the empty shell vibrations. Consider the potential Φ as $\Phi = \Phi_1 + \Phi_2$, as it was done in [6]. The series for potential Φ_1 can be written as

$$\Phi_1 = \sum_{k=1}^{N} \dot{c}_k(t)\phi_{1k}\,.$$

Here time-dependant coefficients $c_k(t)$ are defined in eqn (3). To determine functions φ_{1k} the following boundary value problems is formulated:

$$\Delta\phi_{1k} = 0\,, \left.\frac{\partial\phi_{1k}}{\partial\mathbf{n}}\right|_S = w_k\,,\ \phi_{1k}\big|_{S_0} = 0\,, w_k = (\mathbf{u}_k, \mathbf{n})\,,\ k = \overline{1,N}\,. \tag{4}$$

To determine potential Φ_2 we have the problem of liquid vibrations in the rigid shell including gravity effects. Use the expansion

$$\Phi_2 = \sum_{k=1}^{M} \dot{d}_k(t)\phi_{2k}\,,$$

where $d_k(t)$ are unknown coefficients, and functions φ_{2k} are natural modes of the liquid sloshing in the rigid tank. To obtain these modes the following boundary value problems are considered:

$$\Delta\phi_{2k} = 0\,,\ \left.\frac{\partial\phi_{2k}}{\partial\mathbf{n}}\right|_S = 0;\ \left.\frac{\partial\phi_{2k}}{\partial t} + g\zeta\right|_{S_0} = 0;\ \left.\frac{\partial\phi_{2k}}{\partial\mathbf{n}}\right|_{S_0} = \frac{\partial\varsigma}{\partial t},\ k = \overline{1,N}\,. \tag{5}$$

The zero eigenvalue obviously exists for problem (5), but we exclude it with the help of the following orthogonality condition:

$$\iint_{S_0} \frac{\partial\phi_{2k}}{\partial\mathbf{n}} dS_0 = 0\,.$$

Finally, for the sum of potentials $\Phi = \Phi_1 + \Phi_2$ the following expression is valid:

$$\Phi = \sum_{k=1}^{N} \dot{c}_k(t)\phi_{1k} + \sum_{k=1}^{M} \dot{d}_k(t)\phi_{2k}\,. \tag{6}$$

The unknown function ζ takes the form

$$\varsigma = \sum_{k=1}^{N} c_k(t)\frac{\partial\phi_{1k}}{\partial\mathbf{n}} + \sum_{k=1}^{M} d_k(t)\frac{\partial\phi_{2k}}{\partial\mathbf{n}}\,. \tag{7}$$

Both boundary value problems (4) and (5) are reduced to systems of singular integral equations using a boundary element method in its direct formulation, Brebbia et al. [1]. Dropping indexes $1k$ and $2k$ the main relation for determining functions φ_{1k} and φ_{2k} can be written in the form

$$2\pi\phi(P_0) = \iint\limits_{\sigma} q\frac{1}{|P-P_0|}d\sigma - \iint\limits_{\sigma} \phi\frac{\partial}{\partial\mathbf{n}}\frac{1}{|P-P_0|}d\sigma, \tag{8}$$

where $\sigma = S \cup S_0$, the value $r=|P-P_0|$ represents the Cartesian distance between a field point $P(x_i(P),\ i=1,2,3)$ on the boundary S and a collocation point $P_0(x_i(P_0),\ i=1,2,3)$. The function φ, defined on the surface S, presents the pressure on the wetted shell surface, and the function q, defined on the surface S_0, is the flux, $q = \partial\phi / \partial\mathbf{n}$. There are two types of kernels in the integral operators introduced above

$$\mathbf{A}(\sigma_1,\sigma_2)\psi = \iint\limits_{\sigma_1} \psi\frac{\partial}{\partial\mathbf{n}}\frac{1}{|P-P_0|}d\sigma_1;\quad \mathbf{B}(\sigma_1,\sigma_2)\psi = \iint\limits_{\sigma_1} \psi\frac{1}{|P-P_0|}d\sigma_1;\quad P\in\sigma_1; P_0\in\sigma_2, \tag{9}$$

where $\frac{1}{|P-P_0|}$ and $\frac{\partial}{\partial\mathbf{n}}\frac{1}{|P-P_0|}$ are fundamental solution of the Laplace equation. It would be noted that if $\sigma_1 = \sigma_2 = \sigma$, then

$$\mathbf{A}(\sigma,\sigma)\psi = 2\pi\mathbf{I}\psi + \iint\limits_{\sigma} \psi\frac{\partial}{\partial\mathbf{n}}\frac{1}{|P-P_0|}d\sigma,$$

where $\mathbf{I}$ is the identity tensor.

So the problem of free harmonic vibrations of the fluid-filled elastic cylindrical-spherical shell is reduced to defining the unknown potential and flux from relation (8) that can be rewritten as follows

$$\mathbf{A}(\sigma,\sigma)\phi = \mathbf{B}(\sigma,\sigma)q. \tag{10}$$

2.2 Axisymmetric elasticity problems for an isotropic body with an inhomogeneity

Consider an elastic finite cylinder with an inhomogeneity. Suppose that the inhomogeneity has a form of a rounded cylinder (Fig. 2).

The external boundary of the body and its domain are denoted by S and Ω_1, respectively. The external boundary of the inclusion and its domain are denoted by S_{int} and Ω_2. The surface $S = \partial\Omega_1$ represents the boundary surface including inhomogeneity boundary S_{int}; so $S = S_{int} \cup S_1$, where S_1 is the cylinder boundaries, $S_2 = S_{int}$ (Fig. 2).

Suppose that the body is statically loaded along the boundary S by boundary tractions $\mathbf{t}$ and inside the domain by body forces $\mathbf{f}$.

This elastic cylinder with the inhomogeneity may be considered as a representative volume element for studying composites and nano-composites. Further, let Ω_1 be the region of the matrix, and Ω_2 be the region occupied by the inhomogeneity

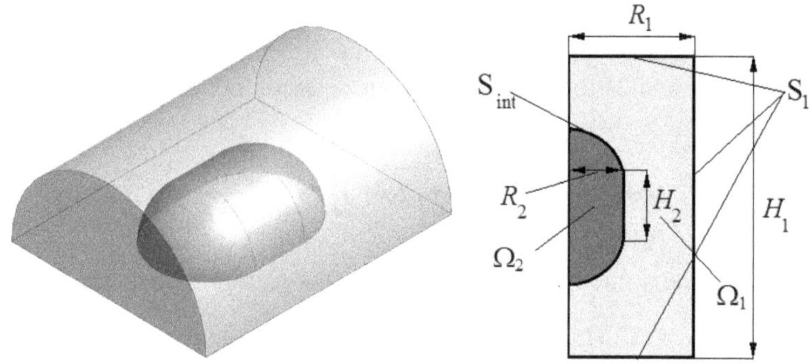

Figure 2: Elastic cylinder with an inclusion, its sketch, and sub-domains.

The three-dimensional isotropic elasticity equations state that the divergence of the stress tensor is zero

$$\nabla \cdot \boldsymbol{\sigma} = \mathbf{f}; \quad \boldsymbol{\sigma} = 2G\left[\nu/(1-2\nu)\operatorname{tr}(\mathbf{e})\mathbf{I} + \mathbf{e}\right]; \quad 2\mathbf{e} = \nabla\mathbf{u} + (\nabla\mathbf{u}); \quad \operatorname{tr}(\mathbf{e}) = e_{ii},$$

where $\mathbf{u}$ is the displacement vector, $\mathbf{e}$ is the infinitesimal strain tensor, $\boldsymbol{\sigma}$ is the stress tensor, $\mathbf{f}$ is the body force, and G=0.5E/(1+ν) is the shear modulus.

Let points P_0, P and Q are the collocation point, the points on the external boundary S and inside the domain, respectively. The relation between the body tractions and its displacements can be expressed by the boundary integral equation (the Somigliana's identity) in the following form [1]:

$$\mathbf{Cu} = \iint_{\sigma} \mathbf{U}(P,P_0)\mathbf{t}(P)\,d\sigma - \iint_{\sigma} \mathbf{T}(P,P_0)\mathbf{u}(P)\,d\sigma + \iiint_{\Omega} \mathbf{U}(P,Q)\mathbf{f}\,d\Omega,$$

where $\mathbf{C}$ is a constant matrix [8], that depends on the position of the point P on the surface σ, $\mathbf{U}(P,P_0) = \{U_{ij}\}$ and $\mathbf{T}(P,P_0) = \{T_{ij}\}$ are the Kelvin fundamental solutions of elastostatics

$$U_{ij}(P,P_0) = \frac{1}{16\pi(1-\nu)Gr}\left[(3-4\nu)\delta_{ij} + r_{,i}r_{,j}\right],$$

$$T_{ij}(P,P_0) = \frac{1}{8\pi(1-\nu)r^2}\left\{\left[(1-2\nu)\delta_{ij} + 3r_{,i}r_{,j}\right]\frac{\partial r}{\partial \mathbf{n}} - (1-2\nu)\left(r_{,i}n_j - r_{,j}n_i\right)\right\}.$$

Here $r = (r_i r_i)^{1/2}$, $r_i = x_i(P) - x_i(P_0)$, $r_{,i} = \partial r / \partial x_i(P) = r_i / r$.

Introducing the matrixes operators as in eqn (9)

$$\mathbf{A}(\sigma_1,\sigma_2)\mathbf{u} = \iint_{\sigma_1} \mathbf{T}(P,P_0)\mathbf{u}(P)\,d\sigma_1, \quad \mathbf{B}(\sigma_1,\sigma_2)\mathbf{t} = \iint_{\sigma_1} \mathbf{U}(P,P_0)\mathbf{t}(P)\,d\sigma_1, \; P_0 \in \sigma_2,$$

and taking into account that

$$\mathbf{A}(\sigma_1,\sigma_1)\mathbf{u} = \mathbf{Cu} + \iint_{\sigma_1} \mathbf{T}(P,P_0)\mathbf{u}(P)\,d\sigma_1,$$

we obtain the main relation in the absence of body forces, in the following form

$$\mathbf{A}(\sigma,\sigma)\mathbf{u} = \mathbf{B}(\sigma,\sigma)\mathbf{t}\,. \tag{11}$$

The integral eqns (10), (11) for considered problems look the same. The important point here is that the singularity order at $r = 0$ is precisely the same for both integral equations.

3 MULTI-DOMAIN APPROACH

Analytic solutions of the singular integral equations are restricted by simple geometrical shapes of the bodies, and by impossibility of studying cracks or inclusions, and thus it is necessary to use computational techniques. The proposed approach is relied on the boundary element method implementation in getting over the continuous models to their discrete analogues. For this purpose we use the multi-domain boundary element method, proposed by Brebbia et al. in [1].

Suppose that the integration domain Ω is divided into two parts Ω_1 and Ω_2 by the interface surface S_{int} as in Figs 1, 2. Here $\partial\Omega_i = S_{\text{int}} \cup S_i$, ($i$=1,2).

Rewrite eqns (10), (11) in the form

$$\mathbf{A}(\sigma,\sigma)\mathbf{X} = \mathbf{B}(\sigma,\sigma)\mathbf{Y}\,,$$

where vector **X** is for values of the potential (or displacements) in BEM mesh nodes, and **Y** is for values of the flux (or tractions). Introducing $\Sigma_1 = S_1$, $\Sigma_2 = S_{\text{int}}$, and $\Sigma_3 = S_2$ allows us to obtain $A_{ij} = \mathbf{A}(\Sigma_i,\Sigma_j)$; $B_{ij} = \mathbf{B}(\Sigma_i,\Sigma_j)$, $i,j = \overline{1,3}$. On the interface surface S_{int} the following compatibility conditions are valid:

$$\mathbf{X}_{S_{\text{int}}\cap\Omega_1} = \mathbf{X}_{S_{\text{int}}\cap\Omega_2}, \quad \mathbf{Y}_{S_{\text{int}}\cap\Omega_1} = -\mathbf{Y}_{S_{\text{int}}\cap\Omega_2}\,. \tag{12}$$

Denote by X_i the unknown values of **X** in nodes of Σ_i. The values of fluxes (tractions) are Y_i (i=1,2,3). By using the multi-domain approach [13] to determine unknown **X** values the next system of integral equations in the operator form is obtained:

$$\begin{aligned}
A_{11}X_1 + A_{12}X_2 &= B_{11}Y_1 + B_{12}Y_2, && P_0 \in \Sigma_1,\\
A_{21}X_1 + A_{22}X_2 &= B_{21}Y_1 + B_{22}Y_2, && P_0 \in S_{\text{int}} \cap \Omega_1,\\
A_{22}X_2 + A_{23}X_3 &= -B_{22}Y_2 + B_{23}Y_3, && P_0 \in S_{\text{int}} \cap \Omega_2,\\
A_{32}X_2 + A_{33}X_3 &= -B_{32}Y_2 + B_{33}Y_3, && P_0 \in \Sigma_2.
\end{aligned} \tag{13}$$

Receiving X_2 from the second and third equations in (13) and substituting it into first and fourth equations respectively, one can obtain the next expressions:

$$X_1 = H_{11}Y_1 + H_{12}Y_2, \quad X_2 = H_{32}Y_2 + H_{33}Y_3,$$

where $H_{ij} = (-1)^{i-j}\left(A_{ii} - A_{i2}A_{22}^{-1}A_{2i}\right)^{-1}\left(B_{ij} - A_{i2}A_{22}^{-1}B_{2i}\right)$, $i = 1,3;\ j = 1,2,3$. Subtracting third equation from the second one in (13) leads to the relation for Y_2

$$(F_1 + F_3)Y_2 = G_1X_1 + G_3X_3,\ G_i = (-1)^{i-2}\left(B_{2i} - A_{2i}H_{ii}\right), F_i = B_{22} + (-1)^{i-2}A_{2i}H_{i2}, i = 1,3\,. \tag{14}$$

Note that system (14) only contains the unknowns Y_2 relative to the interface surface S_{int}. The unknowns for other surfaces are expressed through Y_2 by given formulas.

4 BOUNDARY INTEGRAL EQUATIONS IN AXISYMMETRIC FORMULATION

Accordingly to the features of axisymmetric problems it is useful to transform the Cartesian coordinates system (x,y,z) to a cylindrical one (ρ,θ,z) with following relations:

$$x=\rho\cos\theta;\quad y=\rho\sin\theta;\quad z=z\,.$$

The kernels of integral operators **A** and **B** for both potential and elasticity problems can be represented on terms of complete elliptic integrals [6], [10]. The basic procedure is to start with the standard boundary integral equations, replace Cartesian coordinates with cylindrical ones, and integrate with respect to θ, taking into account that

$$P=(\rho,\theta,z);\quad P=(\rho_0,\theta_0,z_0);\quad |P-P_0|=\sqrt{\rho^2+\rho_0^2+(z-z_0)^2-2\rho\rho_0\cos(\theta-\theta_0)}\,.$$

Then for integral operators of the potential theory problem we have

$$\mathbf{A}(\sigma,\sigma)=2\pi\mathbf{I}+A_1(P,P_0)\mathrm{E}(k)+A_2(P,P_0)\mathrm{K}(k);\quad \mathbf{B}(\sigma,\sigma)=B(P,P_0)\mathrm{K}(k)\,.$$

Here

$$A_1(P,P_0)=\frac{4}{\sqrt{cd}}\left[\frac{\rho^2-\rho_0^2+\hat{z}^2}{2\rho}n_r+\hat{z}n_z\right],\ A_2(P,P_0)=-\frac{2n_r}{\rho\sqrt{c}},\ B(P,P_0)=\frac{4}{\rho\sqrt{c}},$$

$$\hat{z}=z_0-z,\quad a=\rho^2+\rho_0^2+\hat{z}^2,\quad b=2\rho\rho_0,\quad c=a+b,\quad d=a-b,\ k^2=\frac{2b}{a+b},$$

n_r, n_z are components of the external unit normal to the surface σ in the ρ and z directions, respectively, and $\mathrm{E}(k), \mathrm{K}(k)$ are complete elliptic integrals

$$\mathrm{E}(k)=\int_0^{\pi/2}\left(1-k^2\sin^2\psi\right)^{1/2}d\psi,\ \mathrm{K}(k)=\int_0^{\pi/2}\left(1-k^2\sin^2\psi\right)^{-1/2}d\psi\,.$$

The matrix integral operators of axisymmetric elasticity can be expressed in terms of complete elliptic integrals in the similar way

$$\mathbf{A}(\sigma,\sigma)=\mathbf{C}+\{T_{ij}(P,P_0)\},\quad \mathbf{B}(\sigma,\sigma)=\{U_{ij}(P,P_0)\},\quad i,j=\rho,z\,,$$

$$T_{ij}(P,P_0)=T_{ij}^1\mathrm{E}(k)+T_{ij}^2\mathrm{K}(k),\quad U_{ij}(P,P_0)=U_{ij}^1\mathrm{E}(k)+U_{ij}^2\mathrm{K}(k)\,.$$

The displacement kernels here present weak singularities. These kernels are rearranged, so their components are following [10]:

$$U_{11}^1=\frac{-A}{\rho\rho_0\sqrt{c}}\left[(3-4\nu)c+\hat{z}^2\frac{a}{d}\right],U_{11}^2=\frac{A}{\rho\rho_0\sqrt{c}}\left[(3-4\nu)a+\hat{z}^2\right],U_{12}^2=\frac{A\hat{z}}{\rho_0\sqrt{c}},\cdots U_{12}^1=-\frac{U_{12}^2h}{d},$$

$$U_{21}^2=-\frac{A\hat{z}}{\rho\sqrt{c}},\quad U_{21}^1=\frac{U_{21}^2f}{d},\quad U_{22}^2=\frac{2A(3-4\nu)}{\sqrt{c}},\quad U_{22}^1=\frac{2A\hat{z}^2}{d\sqrt{c}},$$

$$A=\left[16\pi^2G(1-\nu)\right]^{-1},\quad h=\rho^2-\rho_0^2+\hat{z}^2,\quad f=\rho_0^2-\rho^2+\hat{z}^2.$$

Similarly, the traction kernels can be expressed as in [10] by following formulas

$$T_{ij}^{k} = 2G\left(T_{ij}^{kr} n_r + T_{ij}^{kz} n_z\right),$$

where

$$T_{11}^{1r} = \frac{A}{\rho_0 \rho^2 \sqrt{cd}}\left(2\nu q + 3\rho^2 f + \left(3m - \frac{2a^2 f}{cd}\right)\hat{z}^2\right); T_{11}^{1z} = \frac{-A}{\rho_0 \rho \sqrt{cd}}\left[(2\nu - 3)a + \left(\frac{4a^2}{cd} - 3\right)\hat{z}^2\right];$$

$$T_{11}^{2r} = \frac{A}{\rho_0 \rho^2 \sqrt{cd}}\left[2\nu m - \frac{3}{2}n + \frac{af\hat{z}^2}{cd}\right]; T_{11}^{2z} = \frac{A\hat{z}}{\rho \rho_0 \sqrt{c}}\left(2\nu - 3 + \frac{a\hat{z}^2}{cd}\right);$$

$$q = \frac{3b^2}{2} + ar^2 - 2a^2; \quad m = 2a^2 - \rho_0^2; \quad n = 2\rho^2 + 3\hat{z}^2;$$

$$T_{12}^{1r} = T_{11}^{1z};\ T_{12}^{1z} = \frac{A}{\rho_0 \sqrt{c}}\left[(2\nu - 1)h + \left(\frac{8f\rho_0^2}{cd} - 1\right)\hat{z}^2\right]; T_{12}^{2r} = T_{11}^{2z}; T_{12}^{2z} = \frac{A\hat{z}}{\rho \rho_0 \sqrt{c}}\left(2\nu - 1 + \frac{h\hat{z}^2}{cd}\right);$$

$$T_{21}^{1r} = \frac{A}{d\sqrt{c}}\left[4(1+\nu) - \frac{a}{\rho^2} + \frac{8a\hat{z}^2}{cd}\right], T_{21}^{1z} = \frac{A}{d\sqrt{c}}\left[(1-2\nu)f + \left(\frac{4fa}{cd} - 3\right)\hat{z}^2\right], T_{21}^{2r} = \frac{A\hat{z}}{\sqrt{c}}\left(\frac{1}{\rho_0^2} + \frac{2\hat{z}^2}{cd}\right);$$

$$T_{21}^{2z} = \frac{A\hat{z}}{\rho_0 \sqrt{c}}\left(2\nu - 1 - \frac{f\hat{z}^2}{cd}\right); T_{22}^{1r} = T_{21}^{1z},\ T_{22}^{1z} = \frac{2A\hat{z}}{d\sqrt{c}}\left(1 - 2\nu + \frac{4a\hat{z}^2}{cd}\right), T_{22}^{2r} = T_{21}^{2z}, T_{22}^{2z} = \frac{2A\hat{z}^3}{\rho_0 \sqrt{c^3}}.$$

The components $U_{ij}^{1,2}$ and T_{ij}^{2} are regular functions. So the only singularity in these components is the logarithmic one due asymptotic behavior of $\mathrm{K}(k)$ for $k \to 1$. The special treatment is applied to evaluating the integrals using the orthogonal polynomials with logarithmic weights [1]. It was shown by Balas et al. [14] that components T_{ij}^{1} contain terms $a_1 = \left[n_r(\rho - \rho_0) + n_z(z - z_0)\right]/d$, $\forall i,j$ and $a_2 = \left[n_z(\rho - \rho_0) - n_\rho(z - z_0)\right]/d$ for $i \neq j$. As in the cylindrical coordinate system there are

$$\rho = \rho(z),\ n_z = -\rho' / \sqrt{1+\rho'^2},\ n_r = 1/\sqrt{1+\rho'^2}\ ,$$

then for $z \to z_0$ one can obtain that $a_1 \to 0$, but $a_2 \to \left[(z - z_0)\sqrt{1+\rho'^2}\right]^{-1}$. Therefore, components T_{11}^{1}, T_{22}^{1} are regular functions, whereas components T_{12}^{1}, T_{21}^{1} have infinite gaps. For their numerical evaluation the method elaborated by Becker [15] is used.

5 NUMERICAL EXAMPLES

The proposed method is applied to solution of axisymmetric problems in potential theory and elastostatics.

5.1 Validation of the proposed axisymmetric BEM

The first step is to define the requisite number of boundary elements for evaluating unknown functions with given accuracy.

To testify the proposed numerical approach to potential theory problems the axisymmetric free liquid vibrations in a rigid cylindrical shell are considered. The circular cylindrical shell with a flat bottom has the following parameters: radius is $R = 1$ m, the fluid filling level $H = 2$ m. The numerical solution is obtained by using the BEM as it is described beforehand. In present numerical simulation we used 60 boundary elements along the bottom, 120 elements along wetted cylindrical parts and 60 elements along the free surface radius. Good agreement of analytical results for frequency parameter obtained by Ibrahim [16] and numerical ones received by proposed axisymmetric BEM is demonstrated in Table 1.

The example for validation in elastostatics is a long cylinder fixed at one edge ($z = 0$). The force F is applied to the other edge ($z = L$). The ratio of radius R to the cylinder length L is 1/10, elasticity modulus is $E = 2{,}11 \cdot 10^6$ MPa. The mesh used here has 120 boundary elements along the cylindrical part and 60 elements along top and bottom radii of the cylinder. This axisymmetric problem is solved using constant elements and integrations with 6 Gauss points. Table 2 shows the normalized displacements $w^* = w\pi ER^2/FL$ obtained at the points $r/R = 0.05, 0.25, 0.5, 0.75, 0.99$ at $z = L$.

The presented BEM and analytical solution $w = Fz / E\pi R^2$ are in very good agreement. The accuracy is decreased at boundary points $r/R = 0.05$ and $r/R = 0.99$. Calculations in these points require special treatment.

5.2 Vibrations of compound cylindrical-spherical elastic shells

Consider the fluid-filled elastic shell composed of a cylindrical part bounded by a hemispherical edge with thickness $h = 0.01$m, radius $R = 1$m, height $L = R+H$=2m, elasticity modulus $E = 2{,}11 \cdot 10^6$ MPa, Poisson's ratio ν =0.3, mass density $\rho_s = 8000$ kg/m^3, and liquid density $\rho_l = 1000$ kg/m^3 (Fig. 1). At first stage the sloshing frequencies are evaluated for angular numbers $\alpha = 0,1,2,4,5,6$. The results are presented in Table 3.

Next stage is to evaluate the frequencies and modes of the elastic empty shell. The shell is supposed to be clamped over its rigid spherical bottom. In Table 3 there are frequencies of the empty cylindrical-spherical shell (CSS) described above and frequencies of an empty cylindrical shell (CS) clamped over its rigid flat bottom with radius $R = 1$m and height $H = 2$m, evaluated for $\alpha = 0,1,2,4,5,6$.

The numerical analysis demonstrates that the lowest frequencies are liquid sloshing ones, and occur for the angular number $\alpha = 1$. Analyzing results of Table 3 one can observe, that the sloshing frequencies of cylindrical and cylindrical-spherical shells with equal heights are very close, whereas frequencies of these both empty and fluid-filled elastic shells with equal heights and radii of free surfaces are differed essentially.

Table 1: Comparison of analytical and numerical results.

Method	Frequency parameter ω^2/g				
	$n = 1$	$n = 2$	$n = 3$	$n = 4$	$n = 5$
BEM	1.833886	5.331447	8.536322	11.706103	14.864072
ANALYTICAL	1.833885	5.331442	8.536316	11.706005	14.863589

Table 2: Normalized static displacements w^* of the elastic cylinder.

Method	Points, r/R				
	0.05	0.25	0.5	0.75	0.99
BEM	1.006	1.0003	1.00001	1.0004	1.02
Analytic	1.0	1.0	1.0	1.0	1.0

Table 3: Frequencies of empty and fluid-filled shells, Hz.

α	m	Sloshing CS	Sloshing CSS	Empty CS	Empty CSS	Fluid-filled CS	Fluid-filled CSS
0	1	6.1278	6.1283	398.132	796.263	145.582	221.553
	2	8.1217	8.2929	610.929	799.048	344.468	417.007
	3	9.9849	9.9871	810.703	817.043	398.132	512.361
1	1	4.3494	4.2392	235.485	473.302	77.780	169.845
	2	7.2283	7.2291	606.710	779.754	348.210	389.965
	3	9.1463	9.1479	730.413	811.617	629.659	492.862
2	1	5.4709	5.4708	117.679	290.119	49.489	136.445
	2	8.1076	8.1077	389.150	671.876	184.712	366.504
	3	9.8843	9.8861	619.164	774.231	319.253	483.943
4	1	7.2188	7.2188	54.491	134.018	28.032	73.046
	2	9.5376	9.5383	186.299	426.793	100.414	251.088
	3	11.1482	11.1494	374.973	654.876	213.825	421.393
5	1	7.9292	7.9286	65.136	110.755	35.401	60.490
	2	10.1535	10.1535	148.954	348.935	83.067	204.221
	3	11.7078	11.7082	300.608	592.120	176.017	379.170
6	1	8.5739	8.5724	88.609	112.813	52.074	68.111
	2	10.7239	10.7227	139.468	300.132	83.090	188.162
	3	12.2322	12.3824	255.945	538.350	158.914	359.072

The lowest frequency of the cylindrical shell caused by its elastic walls vibrations is nearly two times less than that one of the considered CSS. The frequencies of fluid-filled shell vibrations differ drastically from frequencies of empty ones for both shells. But with increasing the circumferential wave number α this difference become gradually smaller. Fig. 3 shows the modes of lowest frequencies.

For liquid sloshing the lowest frequencies corresponds to α= 1 (Fig. 3(a)), and for the elastic CSS vibrations it occurs at α= 5 (Fig. 3(b)).

5.3 Stress-strain state of the elastic cylinder with the inhomogeneity

Consider the cylinder fixed at one edge (z=0.0), loaded uniformly at the other one (z = 5.0) and having the inclusion in the form of rounded cylinder (Fig. 2). Eqn (12) are perfect contact conditions. The ratio of cylinder and inclusion heights is $H_1/H_2 = 5$, and the ratio of their radii is $R_1/R_2 = 2$, elasticity moduli are $E_1 = 2{,}11\cdot 10^6$ MPa, $E_2 = 2{,}11\cdot 10^7$ MPa, and Poisson's ratios are $\nu_1 = \nu_2 = 0.3$. We investigate the disturbance in a stress-strain state in the cylinder caused by the presence of this inhomogeneity. The multi-domain approach is applied here, and the resulting system of linear algebraic equations contains unknown displacements only along the interface surface S_{int} (Fig. 2). Let now Σ_1 and Σ_2 are cylinder surface parts with given displacements and tractions, respectively, $\Sigma_3 = S_{\text{int}}\cap\Omega_1$, and $\Sigma_4 = S_{\text{int}}\cap\Omega_2$. System (13) is rewritten for the case of given displacements at the elastic cylinder edge. Namely, we have

$$-B_{i1}Y_1 + A_{i2}X_2 + A_{i3}X_3 - B_{i3}Y_3 = B_{12}Y_2, \quad P_0 \in \Sigma_i, \quad A_{43}X_3 + B_{43}Y_3 = 0, \quad P_0 \in \Sigma_4 .$$

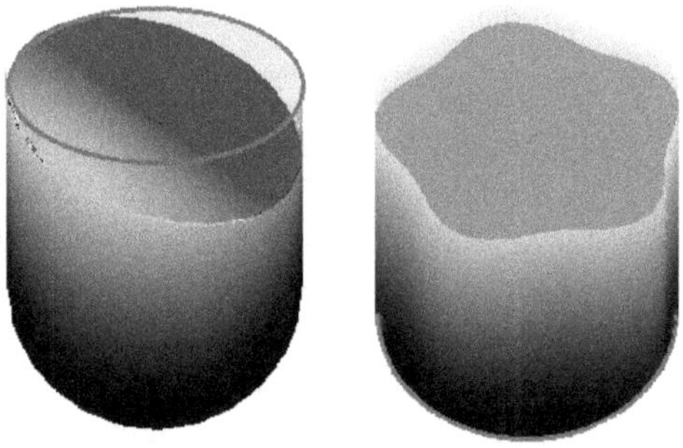

(a) (b)

Figure 3: Modes of lowest frequencies.

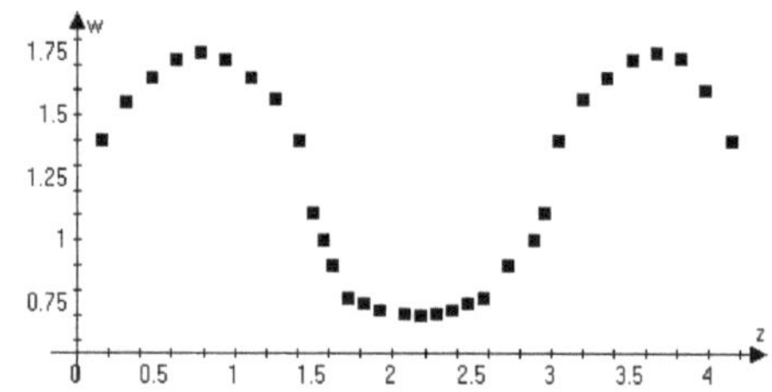

Figure 4: Displacements of interface surface.

So the relation between tractions and displacements of the cylinder on the interface surface is $Y_3 = -B_{41}^{-1} A_{43} X_3$, and we obtain the final system of equations for the elastic cylinder with the inclusion. For numerical simulation we use following numbers N_i of one-dimensional boundary elements along surfaces Σ_I (i=1,2,3) : N_1 =150, N_2 =30, N_3=240. Fig. 4 shows the distributions of displacements in the elastic cylinder along the interface surface. In contrast to homogeneous cylinder the displacements are substantially non-uniform.

6 CONCLUSION

The effective method is elaborated for numerical solution of the boundary integral equations originated in axisymmetric problems of potential theory and elastostatics. Their form is received by using transition to cylindrical coordinates with following integration over angular variable. The approach is relied on the boundary element method implementation in getting from continuous models to their discrete analogues with thorough analysis of displacement and traction kernels. It reveals the similarities of the considered kernels in the singular integral operators that are represented on terms of complete elliptic integrals of first and second kinds both in potential and elasticity theories. Analysis of integrals proved the existence of logarithmic and Cauchy's singularities. The effective procedures for numerical treatment of these integrals taking into account the presence of these singularities are built. To verify the proposed method the numerical examples are provided. The coupled liquid and elastic walls vibrations of composed shells of revolution are considered to testify the proposed numerical approach to axisymmetric problems of potential theory. The long cylinder fixed at one edge and loaded by uniform stress at the other edge is considered for validation of the proposed method. The disturbance in a stress-strain state in the elastic cylinder caused by the presence of the inclusion is investigated with using the multi-domain approach.

ACKNOWLEDGEMENTS

Financial support for the project from STCU under grant No. 6247 is gratefully acknowledged. The authors would also like to thank our foreign collaborator Alexander Cheng, University of Mississippi, USA, for his constant support and interest to our research.

REFERENCES

[1] Brebbia, C.A., Telles, J.C.F. & Wrobel, L.C., *Boundary Element Techniques: Theory and Applications in Engineering*. Springer-Verlag: Berlin and New York, 1984.

[2] Wang, J. & Gao, X.W., Structural multi-scale boundary element method based on element subdivision technique. *Chinese Journal of Computational Mechanics*, **27**(2), pp. 258–263, 2010.

[3] Kane, J.H., Kashava Kumar, B.L. & Saigal, S., An arbitrary condensing, noncondensing solution strategy for large scale, multi-zone boundary element analysis. *Computer Methods in Applied Mechanics and Engineering*, **79**(2), pp. 219–44, 1990. DOI: 10.1016/0045-7825(90)90133-7.

[4] Zhang, Ch., Cui, M., Wang, J., Gao, X.W., Sladek, J. & Sladek, V., 3D crack analysis in functionally graded materials. *Engineering Fracture Mechanics,* **78**, pp. 585–604, 2011.

[5] Gao, X.W. & Yang, K., Thermal stress analysis of functionally graded material structures using boundary element method. *Chinese Journal of Theoretical Chinese Journal of Theoretical and Applied Mechanics*, **43**(1), pp. 136–143, 2011.

[6] Gnitko, V., Degtyariov, K., Naumenko, V. & Strelnikova, E., BEM and FEM analysis of the fluid-structure Interaction in tanks with baffles. *International Journal of Computational Methods and Experimental Measurements*, **5**(3), pp. 317–328, 2017. DOI: 10.2495/cmem-v5-n3-317-328.

[7] Rizzo, F.J. & Shippy, D.J., A boundary integral approach to potential and elasticity problems for axisymmetric bodies with arbitrary boundary conditions. *Mechanics Research Communications,* **6**(2), pp. 99–103, 1979. DOI: 10.1016/0093-6413(79)90022-3.

[8] Cruse, T.A., Snow, D.W. & Wilson, R.B., Numerical solutions in axisymmetric elasticity. *Computers Structures,* **7**(3), pp. 445–451, 1977. DOI: 10.1016/0045-7949(77)90081-5.

[9] Kermanidis, T.A., A numerical solution for axially symmetrical elasticity problems. *International Journal of Solids and Structures,* **11**(4), pp. 493–500, 1975. DOI: 10.1016/0020-7683(75)90084-0.

[10] Bakr, A.A., *The Boundary Integral Equation Method in Axisymmetric Stress Analysis Problems*, Springler Verlag: Berlin Heidelberg, New York, Tokyo, p. 213, 1986.

[11] Levitin, M. & Vassiliev, D., Vibrations of shells contacting fluid: Asymptotic analysis. *Acoustic Interactions with Submerged Elastic Structures,* **5**, pp. 310–332, 1996.

[12] Ravnik, J., Strelnikova, E., Gnitko, V., Degtyarev, K. & Ogorodnyk, U., BEM and FEM analysis of fluid-structure interaction in a double tank. *Engineering Analysis with Boundary Elements*, **67**, pp. 13–25, 2016.

[13] Gnitko, V., Degtyarev, K., Naumenko, V. & Strelnikova, E., Coupled BEM and FEM Analysis of fluid-structure interaction in dual compartment tanks. *International Journal of Computational Methods and Experimental Measurements*, **6**(6), pp. 976–988, 2018.

[14] Balas, J., Sladek, J. & Sladek, V., *Stress Analysis by Boundary Elements. Studies in Applied Mechanics*, Elsevier, 1989.

[15] Becker, A.A., *The Boundary Element Method in Engineering–A Complete Course*, McGraw-Hill, 1992.

[16] Ibrahim, R.A., *Liquid Sloshing Dynamics: Theory and Applications*, University Press: Cambridge, 2005.

MULTIPLE FATIGUE CRACK GROWTH MODELLING IN NONHOMOGENEOUS STRUCTURAL SYSTEMS USING THE DUAL BEM

HEIDER DE CASTRO E ANDRADE & EDSON DENNER LEONEL
São Carlos School of Engineering, Department of Structural Engineering, University of São Paulo, Brazil

ABSTRACT

This study presents a numerical scheme for the fatigue crack growth modelling in plane nonhomogeneous structural systems. The structural domain is assumed as composed of piecewise homogeneous isotropic materials. The high-cycle fatigue case is assumed. Consequently, the linear elastic fracture mechanics theory is utilized. The mechanical behaviour is represented by the dual boundary element method (DBEM), in which singular and hypersingular integral equations are applied. The DBEM is an efficient and robust numerical technique for crack propagation analyses, especially due to the non-requirement of the domain mesh. This aspect enables the accurate representation of the elastic fields around the crack tip and simplifies the remeshing process during the crack propagation. The singular integrals are evaluated through the singularity subtraction method. Third degree polynomial transformation is utilized for improving the accuracy of near-singular integrals. The interface between adjacent materials is assumed as perfectly bonded and the sub-region BEM approach is utilized for representing the nonhomogeneous bodies. The Paris' law is utilized for determining the crack growth rate. The stress intensity factors are assessed by the J-integral and the maximum circumferential stress theory is adopted for defining the crack propagation angle and the range of equivalent stress intensity factor. The structural life is determined by a special scheme, which is based on discrete crack increments. One application is presented to illustrate the efficiency of the proposed model in simulating the fatigue crack growth in multiple cracked structures.

Keywords: fatigue, multiple crack growth, coalescence, dual boundary element method, nonhomogeneous domains.

1 INTRODUCTION

The application of materials with dissimilar mechanical properties in nonhomogeneous structural systems has recently increased in several engineering fields. In this type of typology, the best mechanical feature of each material is explored, resulting in better structural performance. However, the materials contain inherent flaws at some level of its microstructure. Thus, under oscillating stresses, such flaws may join to form a crack, which in turn may grow and lead to the mechanical failure. This phenomenon is known as fatigue and it is a main concern regarding structures subjected to cyclic loading. Fatigue failure may occur even if the stresses along the material are below its static tensile strength. Therefore, this type of collapse is sudden, i.e., without previous notice, and usually catastrophic. Hence, the proposition of models to analyse accurately fatigue crack growth problems and to predict the structural life of nonhomogeneous structures is a priority in engineering design. Moreover, these models must consider the influence of different material properties during the crack propagation through the structure and the interaction between cracks to represent reliably general and complex problems.

The mechanical analysis of cracked nonhomogeneous structures has been performed experimentally [1], [2]. However, the robust mechanical modelling of this problem requires numerical methods because of the complexity involved into the material discontinuities description in addition to the boundary conditions. In spite of domain mesh methods are utilized in this field, such as the finite element method [3] and the extended finite element

WIT Transactions on Engineering Sciences, Vol 122, © 2019 WIT Press
www.witpress.com, ISSN 1743-3533 (on-line)
doi:10.2495/BE410031

method [4], the boundary element method (BEM) is the proper numerical method for handling such mechanical problem. The BEM was accurately utilized for assessing the stress intensity factors (SIF) in cracked nonhomogeneous bodies by [5], [6]. Afterwards, improved BEM models, based on the dual BEM (DBEM) formulation, enabled the simulation of multiple crack propagation and coalescence phenomena [7], [8]. However, the studies considering multiple crack growth in nonhomogeneous bodies are scarce in the literature, which justifies the development of the present research.

In the present study, a numerical model to simulate the fatigue crack growth in multiple cracked nonhomogeneous bodies is presented. The DBEM is utilized to assess the structural mechanical behaviour. The material of each subdomain, which compose the nonhomogeneous structural system, is considered as homogeneous and isotropic. The sub-region formulation of the BEM is applied for coupling the entire domain. The interface between adjacent materials is assumed as perfectly bonded and delamination is not addressed. The proposed approach is capable for simulating the coalescence phenomenon and the crack intersection with the external boundary and with the materials interfaces. The Paris' law is utilized to obtain the crack growth rates and the SIF are evaluated through the J-integral technique. The crack propagation angle is determined by the maximum circumferential stress criterion. An especial methodology is proposed for defining the structural life, which accounts for discrete crack growth increments. One application is utilized to demonstrate the robustness of the proposed numerical approach.

2 THE DUAL BOUNDARY ELEMENT METHOD

Consider a plane, homogeneous and linear elastic domain Ω with boundary Γ subjected to static loads. Using the Betti's reciprocal work theorem and disregarding the body forces, the displacements u_i at an internal point s of this domain are obtained from the displacements u_j and tractions p_j at the boundary as follows:

$$u_i(s) = \int_\Gamma U_{ij}^*(s,f)p_j(f)d\Gamma - \int_\Gamma P_{ij}^*(s,f)u_j(f)d\Gamma, \tag{1}$$

where U_{ij}^* and P_{ij}^* are, respectively, the displacement and traction fundamental solutions, which are functions of the distance r between the source point s and the field point f at the boundary.

By differentiating eqn (1) with respect to the internal point, then using the strain-displacement relation and finally applying the Hooke's law, the equation that define the stress components σ_{jk} at the point s is obtained, which is given by:

$$\sigma_{jk}(s) = \int_\Gamma D_{ijk}^*(s,f)p_i(f)d\Gamma - \int_\Gamma S_{ijk}^*(s,f)u_i(f)d\Gamma, \tag{2}$$

where D_{ijk}^* and S_{ijk}^* are fundamental solutions obtained from the derivatives of U_{ij}^* and P_{ij}^*, respectively, which are also functions of r.

Eqns (1) and (2) contain regular integrals. Moreover, such equations enable the computation of the elastic fields at the internal points because the displacement and traction components at the boundary are known. To define such components, i.e. at the boundary, the source point s is taken to the boundary. By performing this limit process with eqn (1), the displacement boundary integral equation (DBIE) is determined as follows:

$$c_{ij}(s)u_j(s) + c_{ij}(\bar{s})u_j(\bar{s}) + \int_\Gamma P_{ij}^*(s,f)u_j(f)d\Gamma = \int_\Gamma U_{ij}^*(s,f)p_j(f)d\Gamma, \tag{3}$$

where c_{ij} is a coefficient resulting from the limit process on the integral containing P^*_{ij}, which is equal to $\delta_{ij}/2$ for s at the smooth boundaries, being δ_{ij} the Kronecker delta. The point $\overline{s}$ refers to a potential source point at the same position of s but at different surface. This occurs, for instance, for source points at opposite crack surfaces. When s does not have a corresponding point, the $\overline{s}$ term in eqn (3) is nil.

Similar limit process is performed with eqn (2). Then, using the Cauchy's formula, the traction boundary integral equation (TBIE) is obtained. For the source point at smooth boundaries, the TBIE is given by:

$$\frac{1}{2}\left[p_j(s)-p_j(\overline{s})\right]+n_k(s)\int_\Gamma S^*_{ijk}(s,f)u_i(f)d\Gamma = n_k(s)\int_\Gamma D^*_{ijk}(s,f)p_i(f)d\Gamma, \tag{4}$$

where n_k are the components of the normal versor to the boundary at the source point.

Eqns (3) and (4) contain improper integrals because the fundamental solutions present different types of singularity as r approaches zero. The weak singularity, of order $\ln(r)$, occurs in U^*_{ij}; the strong singularity, of order $1/r$, is observed in both P^*_{ij} and D^*_{ijk}; and the hypersingularity, of order $1/r^2$, occurs in S^*_{ijk}. Therefore, the kernels with strong singularity and hypersingularity must be evaluated in the sense of Cauchy principal value and Hadamard finite part, respectively.

In this study, the algebraic system of equations is obtained via DBEM by the numerical evaluation of eqns (3) and (4) [9], [10]. The boundary Γ is subdivided into isoparametric elements with polynomial approximation. High order boundary elements are available. The boundary elements contain the collocation points, which are used as source points of the integral equations. The TBIE is applied to the discretization along one of the crack surfaces, whereas the DBIE is applied to the opposite crack surface and to the external boundary.

The numerical integration is performed with the standard Gauss–Legendre quadrature when the integrated element is far from the source point, because such integrals are regular. The Telles' third degree polynomial transformation [11] is utilized when integrating near singular elements. To integrate the singular elements, i.e. elements that contain the source point, the singularity subtraction method is utilized. Discontinuous or semi-discontinuous elements are utilized for ensuring smooth boundary at collocations points and for enforcing the boundary conditions discontinuity between adjacent elements. In those elements, both or just one of the collocation points at the end are placed inside the element, respectively. Moreover, the existence of the Hadamard finite part in the TBIE requires the continuity of the displacement derivatives at the collocation points, which is guaranteed with discontinuous elements. Therefore, this type of element is also utilized to model the crack surfaces.

In the proposed model, the nonhomogeneous structural system is composed of piecewise homogeneous materials, which are interconnected by interfaces. To model this mechanical condition, the sub-region BEM technique is utilized. Thus, each subdomain is composed exclusively of one material. Hence, the global mechanical problem is assembled by enforcing compatibility of displacements and equilibrium of forces along the adjacent subdomains.

3 LINEAR ELASTIC FRACTURE MECHANICS

The Linear Elastic Fracture Mechanics (LEFM) has been widely applied in fatigue modelling. Particularly, for determining the crack growth rate under oscillatory loading. This theory assumes that the inelastic process zone ahead of the crack tip is small in comparison with the crack length or even with other dimensions of the cracked body. Such behaviour resembles high-cycle fatigue problems, in which the strains developed within the structure are predominantly elastic. According to the LEFM theory, the stress field around the crack tip in a homogeneous material develops a singularity of order $r^{-1/2}$ as r approaches zero. In addition, as for cracks near the interface, the stress field can have others orders of singularity [12]. Therefore, the use of the stress components in this region for analysing crack stability becomes compromised. Alternatively, the so-called stress intensity factors (SIF) are used to assess the crack stability for enabling fully description of the stress field near the crack tip. In plane problems, two values of SIF associated with the basic modes of fracture are defined: K_I, related to opening mode, or mode I, and K_{II}, related to the sliding mode, or mode II.

In the present study, the fatigue model utilized for assessing the structural life is the Paris' law [13]. It states that the crack grows as a function of the amount of load cycle as follows:

$$\frac{da}{dN} = C\left(\Delta K\right)^m, \tag{5}$$

where a is the crack length, N indicates the amount of load cycles, C and m represent the material parameters and ΔK is the range of effective stress intensity factors.

Here, ΔK is assumed as equal to the difference between the maximum and the minimum values of the equivalent SIF, or K_{eq}, during one load cycle. The value of K_{eq} is given by a combination of K_I and K_{II} obtained by a desired criterion, commonly based on stress or energy methods. Several approaches have been introduced in the literature, e.g., the maximum circumferential stress criterion [14], the maximum energy release rate criterion [15] and minimum strain energy density criterion [16]. In this study, the maximum circumferential stress criterion is utilized. This criterion assumes that the crack grows along the direction θ_p, which is perpendicular to the maximum circumferential stress near the crack tip. Thus, θ_p is defined as follows:

$$\theta_p = \sin^{-1}\left(\frac{K_I K_{II} - 3K_{II}\sqrt{8K_{II}^2 + K_I^2}}{9K_{II}^2 + K_I^2}\right). \tag{6}$$

With θ_p value defined, the value of K_{eq} is assessed as follows:

$$K_{eq} = \cos\left(\frac{\theta_p}{2}\right)\left[K_I \cos^2\left(\frac{\theta_p}{2}\right) - \frac{3}{2}K_{II}\sin\left(\theta_p\right)\right]. \tag{7}$$

In this study, the SIF values for modes I and II are obtained with the path-independent J-integral proposed by Rice and Sih [17]. This integral is evaluated along a path Γ_J enclosing the crack tip being expressed by:

$$J = \int_{\Gamma_J}\left(Wn_1 - p_j u_{j,1}\right)d\Gamma, \tag{8}$$

where W is the strain energy density given by $\int_0^{\varepsilon_{ij}} \sigma_{ij} d\varepsilon_{ij}$, p_j are the tractions along Γ_J obtained by $\sigma_{ij} n_i$, u_j are the displacements along Γ_J and n_i are the components of the normal versor to the path.

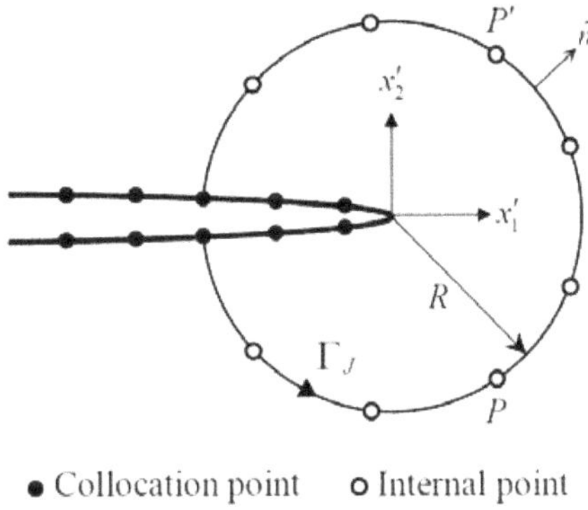

Figure 1: Path used to evaluate the J-integral.

In the proposed model, the path Γ_J is assumed as circular and centred at the crack tip. It starts at a collocation point of the BEM mesh on one crack surface and finishes at the corresponding point on the opposite surface as illustrated in Fig. 1. Moreover, the integration path must be positioned inside the piecewise homogeneous material and cannot cross any other crack tip. To satisfy such conditions, a simpler automatic scheme is utilized to adjust the length of the path radius, which accounts for the intersection distances.

To evaluate the J-integral numerically in the context of the DBEM, a set of internal points is symmetrically positioned at the crack axis along Γ_J. Because the problem is analysed via DBEM, the displacements and stresses of the internal points are obtained into a post-processing phase. For this purpose, the discretized forms of eqns (1) and (2) are applied. Then, the displacement vector and the stress state of the internal points are rotated to the coordinate system oriented according to the crack tip, as shown in Fig. 1. Here, to calculate K_I and K_{II} separately from the J-integral in a mixed-mode problem, the mode decoupling technique proposed by [18] is utilized. In this approach, the displacements and stresses of two symmetric points P and P' are decomposed into their symmetric and antisymmetric parts. By adopting this procedure, the J-integral presented in eqn (8) is also decoupled into a sum of two terms, J^I and J^{II}, with general expression given by:

$$J^M = \int_{\Gamma_J} \left(W^M n_1 - p_j^M u_{j,1}^M \right) d\Gamma, \tag{9}$$

where $M = I, II$ represents the fracture mode. The mode I or mode II are considered if the symmetric or antisymmetric parts of the elastic fields is accounted in eqn (9), respectively.

By discretizing Γ_J into a set of elements that contain the points defined along the path and by adopting a polynomial approximation for displacements and stresses over the element, the integral in eqn (9) can be evaluated numerically using the Gauss–Legendre quadrature. With J^I and J^{II}, the SIF values are determined by:

$$K_I = \sqrt{E^* J^I}, \tag{10}$$

$$|K_{II}| = \sqrt{E^* J^{II}}, \tag{11}$$

where $E^* = E$ for plane stress or $E^* = E/(1-\nu^2)$ for plane strain, with E and ν being the Young's modulus and Poisson's ratio, respectively.

It is worth mentioning that the value of K_I is always positive and is assessed directly from eqn (10). On the other hand, the K_{II} value from eqn (11) gives only its magnitude, because both positive and negative values are allowed. The sign is positive if the displacement about x_1' for the top crack face is higher than for the bottom crack face, and it is negative otherwise.

Another relevant quantity to determine in fatigue problems is the structural life, i.e., the amount of load cycles until the failure. The failure is usually characterized by a high crack growth rate or by a macro crack that develops within the body, compromising its structural integrity. In the present study, the amount of load cycles ΔN_{i+1} required to a crack tip to grow from a certain length a_i to a length a_{i+1} is obtained from the integration of the Paris' law as follows:

$$\Delta N_{i+1} = \int_{a_i}^{a_{i+1}} \frac{da}{C(\Delta K)^m}. \tag{12}$$

The ΔK value depends on the crack length a. Assuming sufficient small crack length increments $\Delta a = a_{i+1} - a_i$, ΔK can be written as a linear function of a according to:

$$\Delta K(a) = (\Delta K_{i+1} - \Delta K_i)\frac{(a - a_i)}{\Delta a} + \Delta K_i, \tag{13}$$

where ΔK_{i+1} and ΔK_i correspond to the values of ΔK at the crack tip evaluated to the lengths a_{i+1} and a_i, respectively.

Substituting eqn (13) into eqn (12) and developing it algebraically, the expression to evaluate the increment in the amount of load cycles is obtained as follows:

$$\Delta N_{i+1} = \frac{\Delta a\left(\Delta K_i^{1-m} - \Delta K_{i+1}^{1-m}\right)}{C(m-1)(\Delta K_{i+1} - \Delta K_i)}. \tag{14}$$

When the crack tip crosses the structural boundary or another crack at the configuration $i+1$, it becomes inactive and the value of ΔK_{i+1} cannot be computed. In this case, only the value of ΔK_i is used to estimate ΔN_{i+1} by applying the discrete form of Paris' law as follows:

$$\Delta N_{i+1} = \frac{\Delta a}{C(\Delta K_i)^m}. \tag{15}$$

Moreover, eqn (14) is only valid if the crack tip remains in the same subdomain. Otherwise ΔN_{i+1} is evaluated by using the discrete form of Paris' law considering each crack increment along a subdomain separately.

Suppose the situation described in Fig. 2(b), where a crack tip initially at material 1 grows Δa_1 in material 1 and Δa_2 in material 2. Taking the corresponding value of ΔK as the one obtained when the tip is at same material, ΔN_{i+1} is given by:

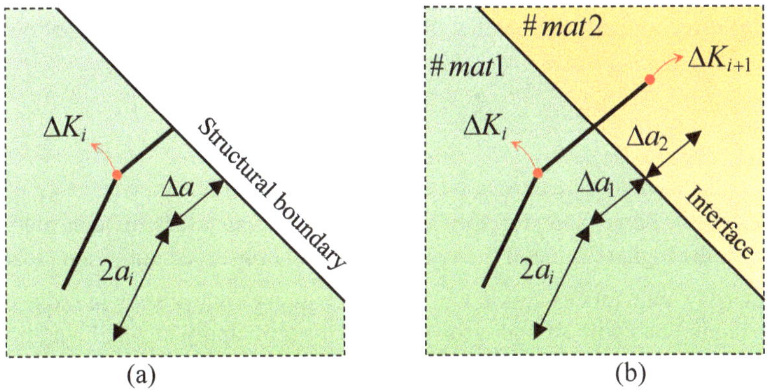

Figure 2: Crack growth. (a) To the structural boundary; and (b) Between subdomains.

$$\Delta N_{i+1} = \frac{\Delta a_1}{C_1\left(\Delta K_i\right)^{m_1}} + \frac{\Delta a_2}{C_2\left(\Delta K_{i+1}\right)^{m_2}}. \tag{16}$$

In a problem with multiple cracks, the value of ΔN_{i+1} is evaluated with eqns (14), (15) or (16) considering the crack tip with the highest growth rate at the configuration *i*.

4 THE CRACK GROWTH SCHEME

The solution process of the fatigue crack growth problem in a nonhomogeneous domain starts with the evaluation of the classical **H** and **G** matrices for the entire boundary element discretization. Such matrices are composed of the kernels presented in eqns (3) and (4). Thus, the DBEM and the sub-region BEM technique are utilized. Then, for each load phase that constitute one load cycle, one response to the boundary values is obtained by applying the corresponding boundary conditions and by solving the algebraic system of equations. The SIF values are also computed for the crack tips at each load phase in a post-processing stage using the J-integral technique. After all load phases are analysed, ΔK is evaluated for each tip by the difference between the maximum and minimum values of K_{eq} determined into one load cycle. A given crack tip grow if the respective value of ΔK is higher than the threshold SIF range ΔK_{th} defined for the material. The propagation angle of the growing crack is evaluated as the weighted average of the values of θ_p defined for all load phases adopting the corresponding K_{eq} as weights, i.e.:

$$\theta_p = \frac{\sum_{j}^{nlp} \theta_p^{(j)} K_{eq}^{(j)}}{\sum_{j}^{nlp} K_{eq}^{(j)}}, \tag{17}$$

in which *nlp* is the amount of load phases during one load cycle and $\theta_p^{(j)}$ and $K_{eq}^{(j)}$ are, respectively, the propagation angle and the equivalent SIF for the *j*-th load phase defined by eqns (6) and (7).

Along this direction, an increment at the crack length ΔL_c is defined as follows:

$$\Delta L_c = \Delta L \frac{C_{sd} \Delta K^{m_{sd}}}{\left[da/dN \right]_{\max}}. \tag{18}$$

where ΔL is the standard crack length increment defined in the analysis, C_{sd} and m_{sd} are the parameters of Paris' law for the subdomain that contains the considered tip and $\left[da/dN \right]_{\max}$ is the highest value of the crack growth rate observed during one load cycle.

From eqn (18), one notices that in a problem involving multiple cracks, some cracks can grow faster than others with smaller growth rate. This may imply in small length increments for the slower growing cracks. In this situation, some collocation points may be very close to other, compromising the accuracy of the DBEM solution. To avoid numerical issues, a virtual crack extension scheme similar to that proposed by [8] is applied. In this strategy, a crack tip propagates only if the following conditions are satisfied: a specified ratio between the lengths of adjacent elements must not be exceeded; and the crack length increment must be higher than a given ratio of ΔL. If these conditions are not observed, the crack tip is not extended and the corresponding ΔL_c defined at that configuration is accumulated as a virtual length. In the course of the analysis, if the total virtual length exceeds both growing condition, then the crack grows with the accumulated length. This scheme certainly introduces approximations in the analysis. However, because the lengths involved in this procedure are small, it does not affect significantly the solution accuracy.

When the crack tips propagate, new collocation points are positioned along the path to discretize the new crack faces as illustrated in Fig. 3. To each collocation point, two additional degrees of freedom in displacements and tractions are created. Therefore, the corresponding influence coefficients are added at **H** and **G** matrices using the DBEM formulation. In addition, the new crack increments may intersect a pre-existing element in the boundary mesh. In this case, the intersected element is modified and a new element is created, Fig. 3(b). The influence coefficients related to the modified element are changed, whereas the influence coefficients related to the new element are added at **H** and **G** matrices. The Fig. 3(c) represents the changes of the BEM matrices with additions and modifications of elements on the mesh.

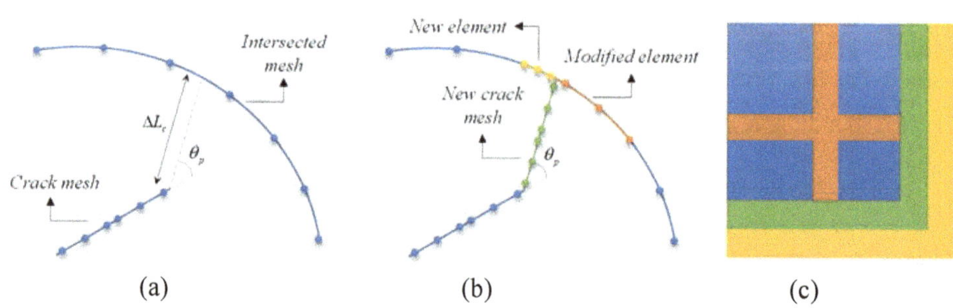

Figure 3: Representation of: (a) Initial mesh composed of quadratic elements; (b) The remeshing procedure during the crack growth; (c) The changes in H and G matrices when elements are created and modified during crack propagation.

After a new crack configuration is defined, the solution of the structure is repeated for each load phase. The amount of load cycles to obtain the new configuration from a previous one is evaluated with eqns (14), (15) or (16), according to the current position of the critical crack tip.

The fatigue failure of the structure is achieved when the growth rate for a given crack tip becomes large, usually in the order of millimetres per cycle. Moreover, if the maximum value of K_{eq} during one load cycle exceeds the fracture toughness of the material K_{Ic}, it is considered that unstable crack propagation occur and the structure fails by fracture.

5 APPLICATION

The nonhomogeneous structure analysed in this application is presented in Fig. 4. The left structural end is fixed along both directions and an oscillatory load *p*, with uniform magnitude ranging from 0 to 10 MPa, was applied along a half of the hole, as illustrated in Fig. 4. Ten cracks with average length equal to 10 mm were randomly distributed at each subdomain and two other cracks with fixed length equal to 10 mm were disposed at the hole boundary along the diagonal perpendicular to load direction. The following dimensions were considered: $L = 400\,mm$, $h = 100\,mm$, $r = 40\,mm$, $R = 100\,mm$ and $\alpha = \omega = 45°$ .Plane stress conditions were assumed in this analysis. The Poisson's rates, ν, and the Paris' law parameters were assumed the same for both materials, with $\nu_1 = \nu_2 = 0.2$, $C_1 = C_2 = 1.0 \cdot 10^{-10}$ and $m_1 = m_2 = 2.75$, in which da/dN is given in $mm/cycle$ and ΔK_{eq} in $MPa \cdot \sqrt{mm}$. Three different Young's moduli ratios were assumed. Such ratio is defined as $\beta = E_1/E_2$ and the following scenarios were considered: (a) $\beta = 1$ (homogeneous domain); (b) $\beta = 0.25$ and (c) $\beta = 4$. For all these scenarios, E_1 was assumed equals to 10000 MPa.

The boundary element mesh was composed of quadratic elements. The structural boundary was discretized into 124 elements, whereas each crack was discretized into 4 elements. During the crack propagation, the maximum and minimum crack length increments were 5 mm and 1 mm, respectively, and the maximum ratio between the lengths of adjacent crack elements was defined equal to 1.5. The Fig. 5 presents the deformed shapes for the final configuration accounting for the three scenarios analysed. The Fig. 5(a) illustrates the final configuration for $\beta = 1$, which represents the homogeneous case. For the initial configuration, the cracks (1) and (2) were critical and developed the highest growth rates. Afterwards, the crack (3) became dominant, because it was at a region with the high stress level. This crack first grew towards and intersected the top edge of the structure. Then, it

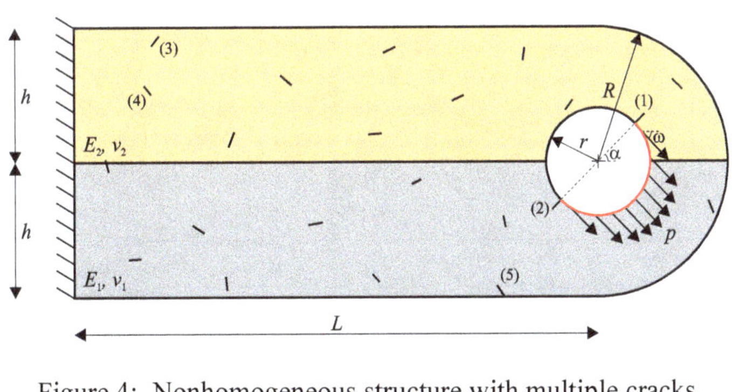

Figure 4: Nonhomogeneous structure with multiple cracks.

propagated downwards, coalesced with crack (4) and crossed the interface between the materials. Finally, it intersected the bottom edge, finishing the analysis. The final deformed shape for $\beta = 0.25$ is presented in Fig. 5(b). In this case, material 2 was subjected to the highest stress level, because it has the higher Young's modulus. Therefore, just the crack (1) was dominant in the beginning. In the course of the analysis, the crack (3) developed the highest growth rates and propagated through a path similar to that observed for $\beta = 1$. Finally, the scenario of $\beta = 4$ is considered and illustrated in Fig. 5(c). Because of material 1 is the stiffest, it developed higher stress levels than in the previous cases. In this scenario, crack (2) was dominant throughout the simulation and intersected the crack (5) in the end of the analysis.

The Fig. 6(a) presents the variation of ΔK_{eq} for the crack tip with the highest growth rate at each configuration with the cumulative crack length increment prescribed for the critical tips during the analysis. The curves for $\beta = 1$ and $\beta = 0.25$ show similar behaviours, because the critical crack tips at both models were the same. At the initial configuration, the maximum value of ΔK_{eq} for $\beta = 0.25$ was higher than that determined for $\beta = 1$ because of the higher stress level developed in material 2, as discussed before. A sudden decrease of ΔK_{eq} is observed when $\beta = 0.25$ for a total crack length increment around 90 mm. This represents the subdomain change of the critical tip to the material with lower stiffness. After the critical crack penetrated into material 1, the values of ΔK_{eq} observed for both curves were similar, because the remained configurations were also similar.

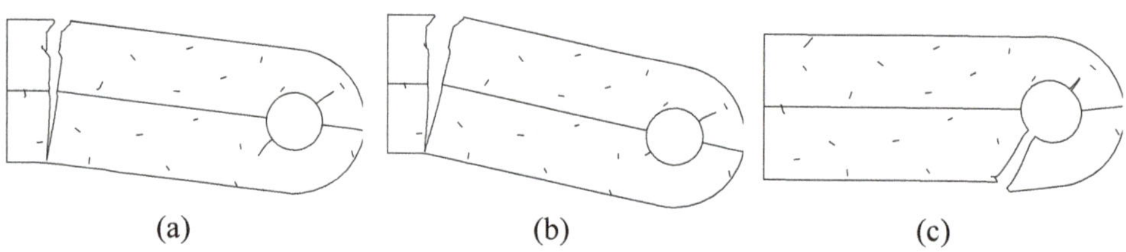

Figure 5: Deformed shapes of the final configurations. (a) $\beta = 1$; (b) $\beta = 0.25$; and (c) $\beta = 4$. Nonhomogeneous structure with multiple cracks.

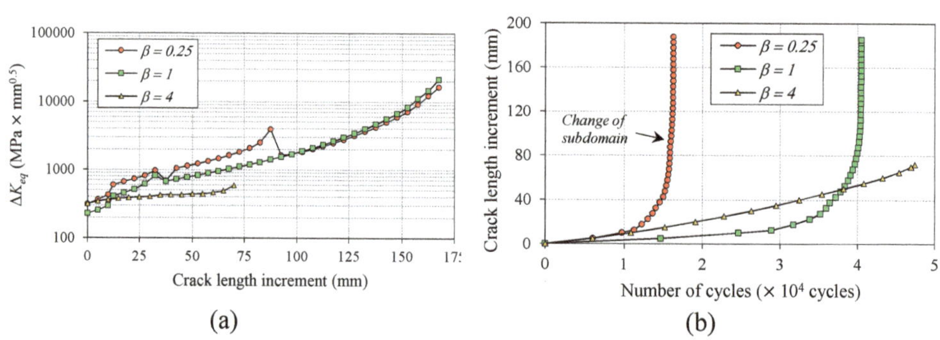

Figure 6: (a) ΔK_{eq} variation; and (b) Structural life.

Other sudden changes in ΔK_{eq} is observed for both $\beta = 0.25$ and $\beta = 1$ cases. The first, around 10 mm, indicates the intersection of one of the tips of crack (3) with the top edge of the structure, resulting into a raise of ΔK_{eq} value for its other tip at the following configuration. The second, around 40 mm, represents the intersection between cracks (3) and (4) for $\beta = 1$ and the shielding effect between the same cracks for $\beta = 0.25$. In the latter situation, the cracks only coalesced in the next configuration. For $\beta = 4$, the response for the initial configuration was the same as for $\beta = 0.25$, because the critical crack in both cases was at the hole edge and in the stiffest material. Afterwards, the ΔK_{eq} values monotonically increased during the analysis, but remained lower than the previous cases.

The structural life predictions for the analysed cases are shown in Fig. 6(b). Because the same Paris' law parameters were assumed for all materials and cases, the curves behaviours are directly related to the ΔK_{eq} variation presented in Fig. 6(a). Therefore, it can be noticed that the higher the value of ΔK_{eq} the higher the crack growth rate and, consequently, the lower the structural life. The following structural lives were determined for the three cases: 16,000, 40,000 and 47,000 load cycles for $\beta = 0.25$, $\beta = 1$ and $\beta = 4$, respectively.

This application demonstrated the robustness of the proposed model in handling fatigue crack propagation problems in multiple cracked nonhomogeneous domains. The simulations were able to efficiently represent the intersection of cracks with the external boundary, the interface between materials and other cracks. Moreover, the responses were stable and in accordance with the expected results.

6 CONCLUSIONS

In this study, a DBEM model was proposed to simulate the fatigue crack growth in plane nonhomogeneous domains considering the LEFM theory. The structures were assumed as composed of piecewise homogeneous and isotropic materials. Moreover, the interfaces between materials were assumed perfectly bonded.

The proposed model was capable to successfully represent the interaction between cracks in problems involving multiple cracked domains, including coalescence. The structural life was also efficiently predicted by the presented scheme. The DBEM model can be further extended to deal with delamination problems by considering the strength of the interface. In addition, anisotropic materials can be accounted to analyse other cases of practical importance.

ACKNOWLEDGEMENT

Sponsorship of this research project by the São Paulo State Foundation for Research (FAPESP), project number 2016/23649-0, is greatly appreciated.

REFERENCES

[1] Jiang, F., Kang, W., Zhao, K., Yan, C., Ye, L. & Mai, Y.M., Experimental investigation on fatigue crack growth normal to a bimaterial interface. *Structural Integrity and Fracture*, eds A. Dyskin, H. Xiaozhi & E. Sahouryeh, pp. 137–140, 2002.

[2] Kolednik, O., Predan, J. & Fischer, F., Cracks in inhomogeneous materials: Comprehensive assessment using the configurational forces concept. *Engineering Fracture Mechanics,* 77(14), pp. 2698–2711, 2010. DOI: 10.1016/j.engfracmech.2010.07.002.

[3] Dias, M., *Crack Propagation on Highly Heterogeneous Composite Materials*. Technische Universiteit Eindhoven: Eindhoven, 2008.

[4] Yu, H., Sumigawa, T., Wu, L. & Kitamura, T., Generalized domain-independent interaction integral for solving the stress intensity factors of nonhomogeneous materials. *International Journal of Solids and Structures*, **67–68**, pp. 151–168, 2015. DOI: 10.1016/j.ijsolstr.2015.03.035.

[5] Ryoji, Y. & Cho, S.B., Efficient boundary element analysis of stress intensity factors for interface cracks in dissimilar materials. *Engineering Fracture Mechanics*, **34**(1), pp. 179–188, 1989. DOI: 10.1016/0013-7944(89)90251-8.

[6] Tullock, D., Reimanis, I., Graham, A. & Petrocic, J., Deflection and penetration of cracks at an interface between two dissimilar materials. *Acta Metallurgica et Materialia*, **42**(9), pp. 3245–3252, 1994. DOI: 10.1016/0956-7151(94)90422-7.

[7] Leonel, E. & Venturini, W., Multiple random crack propagation using a boundary element formulation. *Engineering Fracture Mechanics*, **78**(6), pp. 1077–1090, 2011. DOI: 10.1016/j.engfracmech.2010.11.012.

[8] Price, R. & Trevelyan, J., Boundary element simulation of fatigue crack growth in multi-site damage. *Engineering Analysis with Boundary Elements*, **43**, pp. 67–75, 2014. DOI: 10.1016/j.enganabound.2014.03.002.

[9] Hong, H.K. & Chen, J.T., Derivations of integral equations of elasticity. *Journal of Engineering Mechanics*, **114**(6), pp. 1028–1044, 1988. DOI: 10.1061/(asce)0733-9399(1988)114:6(1028).

[10] Portela, A., Aliabadi, M. & Rooke, D., The dual boundary element method: Effective implementation for crack problems. *International Journal for Numerical Methods in Engineering*, **33**(6), pp. 1269–1287, 1992. DOI: 10.1002/nme.1620330611.

[11] Telles, J., A self-adaptive co-ordinate transformation for efficient numerical evaluation of general boundary element integrals. *International Journal for Numerical Methods in Engineering*, **24**(5), pp. 959–973, 1987. DOI: 10.1002/nme.1620240509.

[12] Zak, A. & Williams, M., Crack point stress singularities at a bi-material interface. *Journal of Applied Mechanics*, **30**(1), pp. 142–143, 1963. DOI: 10.1115/1.3630064.

[13] Paris, P., Gomez, M. & Anderson, W., A rational analytic theory of fatigue. *The Trend in Engineering*, **13**, pp. 9–14, 1961.

[14] Erdogan, F., Stress distribution in a nonhomogeneous elastic plane with cracks. *Journal of Applied Mechanics*, **30**(2), pp. 232–236, 1963.

[15] Hussain, M., Pu, S. & Underwood, J., Strain energy release rate for a crack under combined mode I and mode II. *National Symposium on Fracture Mechanics*, pp. 2–28, 1974.

[16] Sih, G., Strain-energy-density factor applied to mixed mode crack problems. *International Journal of Fracture*, **10**(3), pp. 305–321, 1974. DOI: 10.1007/bf00035493.

[17] Rice, J. & Sih, G., Plane problems of cracks in dissimilar media. *Journal of Applied Mechanics*, **32**(2), pp. 418–423, 1965. DOI: 10.1115/1.3625816.

[18] Ishikawa, H., Kitagawa, H. & Okamura, H., J integral of a mixed mode crack and its application. *Mechanical Behaviour of Materials*, pp. 447–455, 1980. DOI: 10.1016/b978-1-4832-8414-9.50137-1.

ANALYSIS OF THE CRACK PROPAGATION IN FIBER-REINFORCED CONCRETE SPECIMENS USING THE BEM

GABRIELA M. S. BRISOLA[1], LEANDRO PALERMO JR.[2] & LUIZ C. ALMEIDA[2]
[1]FACENS – Sorocaba Engineering College, Sorocaba, Brazil
[2]School Civ. Eng. Arch. Urban Design, University of Campinas, Brazil

ABSTRACT

The mechanical properties of plain concretes are improved with the introduction of steel fibers like the material toughness along with limited crack widths. Numerical investigations on fiber reinforced concrete specimens use models for cohesive crack propagation in quasi-brittle materials. The present crack propagation analysis considered the single-edge notched beam under the three-point bending test. The material softening modeling used the cohesive law with two straight lines within a pure mode I for the crack growth. The dual boundary element method employed quadratic elements and the tangential differential operator in the traction boundary integral equation. The introduction of the constitutive law in the system of equations allowed the direct computation of the cohesive forces at each incremental loading step. The boundary element mesh employed continuous elements along the crack surface and the results obtained are compared with those available in the literature.
Keywords: crack analysis, cohesive model, fiber reinforced concrete, dual boundary element model, tangential differential operator.

1 INTRODUCTION

The improvement of the mechanical properties in plain concrete due to the introduction of steel fibers has motivated the increase of structural applications using this solution. Furthermore, the combination of different types of fibers has extended the range of the applications with fiber reinforced concrete beyond the massive elements. The subsequent researches along the years have demonstrated the fictitious crack model is efficient to describe the tensile behavior of fiber-reinforced concrete (FRC) in the numerical implementations [1]–[3]. The stress concentration in front of the notch tip in concrete specimens with an initial crack develops the damage zone with micro-cracks. In spite of the crack surfaces are not completely separated in this region, the partly destroyed material is still able to transfer stress across the crack trace behind the fictitious crack tip, which is called the fictitious crack or the cohesive zone, as shown in Fig. 1. The fictitious crack model considers the linear elastic behavior before the crack initiation and the stresses along the cohesive zone are taken to be a function of the crack opening. The linear function was employed in the beginning for the relation stress-crack opening to describe the softening in plain concrete specimens within a pure mode I for the crack growth. Peterson was the first to suggest using the function with two straight lines [4] for analysis using plain concrete specimens. Nevertheless, the relation for stress-crack opening function representing the tensile behavior of FRC requires at least a function using two straight lines. The bending failure in concrete beams can be modeled with the development of the fictitious crack in an elastic layer with the thickness proportional to the beam depth [5]. An assessment of the flexural strengths of fiber-reinforced composites and the influence of the fiber bridging on the overall structural behavior were studied in [6], [7], Sousa and Gettu [8], and Slowik et al. [9] determined piecewise linear cohesive models for FRC by minimizing the difference between numerical simulation and experimental test results.

WIT Transactions on Engineering Sciences, Vol 122, © 2019 WIT Press
www.witpress.com, ISSN 1743-3533 (on-line)
doi:10.2495/BE410041

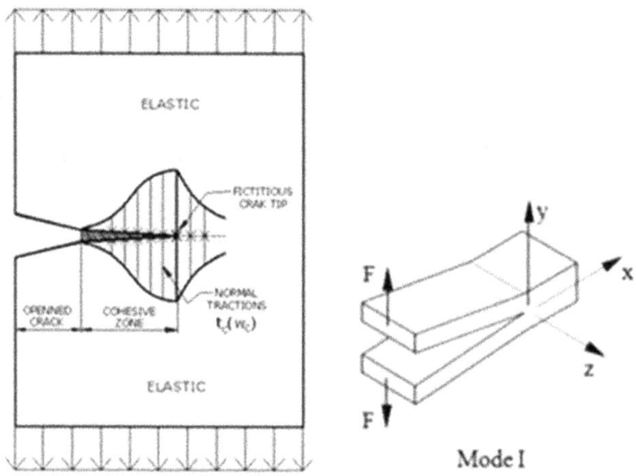

Figure 1: Crack in cohesive material: Stress distribution and fracture mode I.

Several numerical methods besides the FEM have been used to perform crack analyses. The dual boundary element method (DBEM) is one of the most widely used due to its accuracy in computation of the stress intensity factors and the simplicity of adding more elements for crack propagation [10]. The crack analyses for cohesive materials were studied with the DBEM by Saleh and Aliabadi [11], [12], with the Galerkin multizone BEM by Chen et al. [13] and with the displacement discontinuity BEM by Gospodinov [14]. Karlis et al. employed the two dimensional gradient elasticity for crack analyses with the BEM in [15], and Leonel and Venturini used the tangent operator in the non-linear analyses with the BEM for crack propagation in quasi-brittle materials in [16]. Tavara et al. performed a BEM analysis of crack propagation along the fiber–matrix interface of laminates using a linear elastic–brittle interface model in [17]. Trinh et al. employed the fast multipole symmetric Galerkin BEM to solve multizone and multicrack problems in [18], and Benedetti and Aliabadi employed the BEM in multiscale modeling for polycrystalline materials to analyze material degradation and fracture in [19].

The DBEM is adopted here with the strong singularity of the traction boundary integral equation (BIE) reduced with the tangential differential operator (TDO). The TDO, in conjunction with integration by parts, is a technique to reduce the order of singularity in the integrand kernels of BIEs using Kelvin-type fundamental solutions, which is usually referred as the hypersingular kernel. The TDO was first used by Kupradze [20], and by Sladek and Sladek [21] in the solution for a curved crack. Regularized formulations with TDO for gradients in potential problems, and in stress BIEs for elasticity problems were shown by Bonnet in [22]. The non-conformal interpolations require an additional care in the numerical implementations for the TDO, as explained to the traction BIEs of elasticity for two- and three-dimensional problems in [23], [24], and in the BIE for stresses of the plate bending using Reissner and Mindlin models [25].

2 DUAL BOUNDARY INTEGRAL EQUATIONS FOR COHESIVE MATERIALS

A single domain is considered in the solution of general mixed-mode crack problems with the DBEM. The displacement and traction BIEs are used and respectively applied in each one of the crack surfaces. Although the integration path is still the same for coincident points on crack surfaces, the respective boundary integral equations are distinct. The collocation point needed to perform the traction BIE and the strategy used to treat improper integrals are the key features of this formulation. The displacement and traction BIEs for the DBEM are next written:

$$\frac{1}{2}u_i(x') + \int_\Gamma T_{ij}(x',x)u_j(x)d\Gamma(x) = \int_\Gamma U_{ij}(x',x)t_j(x)d\Gamma(x), \tag{1}$$

$$\frac{1}{2}t_k(x') = n_a(x')\left[C_{akim}\int_\Gamma \sigma_{ibj}(x',x)D_{mb}[u_j(x)]d\Gamma(x) - \int_\Gamma \sigma_{jak}(x',x)t_j(x)d\Gamma(x)\right], \tag{2}$$

with

$$C_{akim} = \mu\left(\frac{2v}{1-2v}\delta_{ak}\delta_{im} + \delta_{ai}\delta_{km} + \delta_{am}\delta_{ki}\right),$$

$$D_{bm}[f(x)] = n_b(x)f_{,m}(x) - n_m(x)f_{,b}(x),$$

where $U_{ij}(x', x)$ and $T_{ij}(x', x)$ are the displacement and traction, respectively, in the direction j at boundary point x due to a singular load in the direction i at the collocation point x′ based on the Kelvin solution for two-dimensional problems, $u_i(x)$ and $t_i(x)$ are the displacement and traction at boundary point x, respectively, $n_a(x')$ is the direction cosine of the outward normal at collocation point x′. C_{akim} is the Hooke tensor for an isotropic material, σ_{ibj} is the stress σ_{bj} at boundary point x due to a singular load in the direction i at collocation point x′, μ is the shear modulus, ν is the Poisson ratio, δ_{ij} is the Kronecker delta and $D_{bm}(\,)$ is the tangential operator.

The displacement BIE, eqn (1), has a 1/r singularity and a logarithmic singularity in the integrand kernels at the right and left members, respectively, when the field point approaches the collocation point. The introduction of TDO in the first integral of the right member of the traction BIE, eqn (2), reduced the strong singularity and 1/r singularities result in both integrand kernels. The continuity requirements should be satisfied at the positions of the collocation points for both BIEs. Positions in the interior of the boundary element satisfy the continuity requirements for both BIEs whereas positions at the ends of continuous elements can only be used for the displacement BIE.

The surfaces of the crack region arc connected by the cohesive material in the cohesive zone (Fig. 1). Subsequent loadings increase the separation between these points and eventually lead to cracking. The stress-crack opening curve for the cohesive zone to represent the characteristics of the tensile behavior of FRC uses the function with two straight lines in this study. This function is shown in Fig. 2, where f_t is the tensile strength, and w_c is the maximum widening of the fracture zone when it is still able to transfer stress. The breaking point value is defined in terms of the opening value (w_1) at the corresponding value for the tensile strength (f_1) of the normal traction:

$$\sigma(w) = \begin{cases} 0 \le w \le w_1 \Rightarrow \sigma(w) = f_t\left(1 - \frac{w}{w_c'}\right), \\ w_1 \le w \le w_c \Rightarrow \sigma(w) = f_t'\left(1 - \frac{w}{w_c}\right), \\ w > w_c \Rightarrow \sigma(w) = 0\,. \end{cases} \tag{3}$$

A finite domain with the boundary line Γ_e and containing a crack with the cohesive zone is shown in Fig. 3. The region without the cohesive material is represented by Γ_1 and Γ_4 on

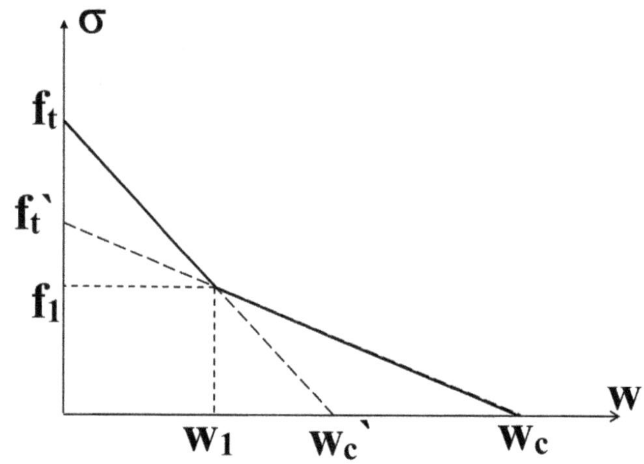

Figure 2: Stress-crack opening function using with two straight lines.

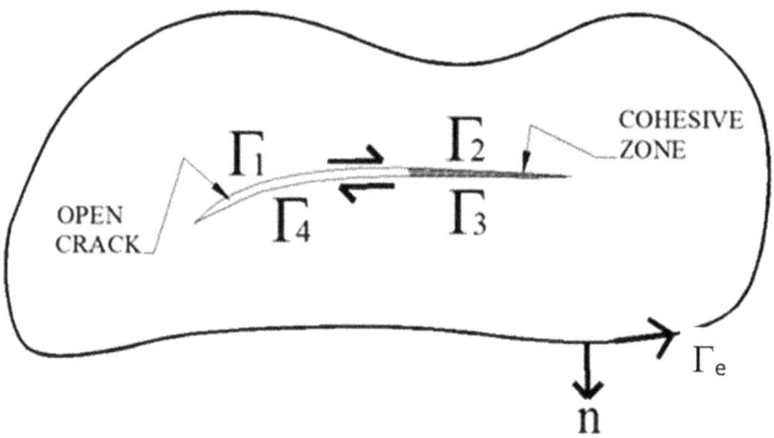

Figure 3: Finite domain containing a crack.

each crack surface whereas Γ_2 and Γ_3 are on each crack surface of the region containing cohesive material (cohesive zone). The normal tractions in the cohesive zone connect the crack surfaces and have opposite signs with equal magnitude, i.e. t^3_j on the crack surface Γ_3 is equal to $-t^2_j$ on the crack surface Γ_2. This condition is used to simplify the equations of DBEM for this problem.

$$\frac{1}{2}u_i(x') + \int_{\Gamma} T_{ij}(x',x)u_j(x)d\Gamma(x) = \int_{\Gamma_e} U_{ij}(x',x)t_j(x)d\Gamma(x)$$

$$+ \int_{\Gamma_2} U_{ij}(x',x)t_j^2(x)d\Gamma(x) - \int_{\Gamma_3} U_{ij}(x',x)t_j^2(x)d\Gamma(x), \quad (4)$$

$$\frac{1}{2}t_k(x') = n_a(x')\left[C_{akim}\int_{\Gamma}\sigma_{ibj}(x',x)D_{mb}[u_j(x)]d\Gamma(x) - \int_{\Gamma_e}\sigma_{jak}(x',x)t_j(x)d\Gamma(x)\right]$$

$$-n_a(x')\int_{\Gamma_2}\sigma_{jak}(x',x)t_j^2(x)d\Gamma(x) + n_a(x')\int_{\Gamma_3}\sigma_{jak}(x',x)t_j^2(x)d\Gamma(x). \qquad (5)$$

The traction in the direction j on the crack surface Γ_2 within the cohesive zone (t^2_j) appeared in both integrals performed on crack surfaces Γ_2 and Γ_3 in eqns (4) and (5) because the traction t^3_j is equal to $-t^2_j$ in the cohesive zone. The numerical implementation in the present study employed eqns (4) and (5), which is different to the formulation used in [26], [27].

The crack propagation is directly related to the maximum principal stress, which has to be always normal to the crack surfaces, and the extension is such that the maximum principal stress at the new tip position is equal to the critical f_t value during continued loading. The beam in the three-point bending test is considered as a generalized plane stress problem and the maximum principal stress is obtained according to following expression:

$$\sigma_{max} = \frac{1}{2}\left(\sigma_n + \sigma_t + \sqrt{(\sigma_n - \sigma_t)^2 + 4{\tau_{nt}}^2}\right). \qquad (6)$$

3 NUMERICAL IMPLEMENTATION

The system of equations obtained from eqns (4) and (5) is given next with the unknown values including the tractions in the cohesive zone (t^2). The additional equation in the system of equation required to the solution corresponds to the cohesive law. The tractions in the cohesive zone are directly obtained at each incremental loading step with this strategy:

$$\begin{bmatrix} H_{ee} & H_{eo} & H_{ez} & -G_{e2} \\ H_{oe} & H_{oo} & H_{oz} & -G_{o2} \\ H_{ze} & H_{zo} & H_{zz} & -G_{z2} \\ 0 & 0 & A & B \end{bmatrix}\begin{Bmatrix} u^e \\ u^o \\ u^z \\ t^2 \end{Bmatrix} = \begin{bmatrix} G_{ee} & 0 \\ G_{oe} & 0 \\ G_{ze} & 0 \\ 0 & C \end{bmatrix}\begin{Bmatrix} t^e \\ w_f \end{Bmatrix}. \qquad (7)$$

Eqn (7) summarizes the system of equations where the dual eqns (4) and (5) are converted to submatrices H_{ij} and G_{ij}, and the cohesive law is converted to submatrices A, B and C. The indices e, o and z are related to the boundary portions and correspond to the external boundary, the open crack boundary and the cohesive zone, respectively. Submatrices A and B relate the displacements and tractions in the normal and tangent directions, respectively, on the crack surfaces in the cohesive zone according to the constitutive law. The numerical algorithm is summarized next, considering the boundary portions shown in Fig. 3 and the constitutive law shown in Fig. 2:

a) Phase 1, $t^2 < f_t$. The traction in the normal direction is less the tensile strength (f_t), the crack surfaces remain jointed, and without relative displacements. Submatrices B and C are zero, whereas submatrix A contains direction cosines relating the displacements in the directions x_i, i.e., the openings in the normal and tangent directions are zero:

$$u_i^2 n_i^2 + u_i^3 n_i^3 = 0, \qquad (8)$$

$$u_i^2 s_i^2 + u_i^3 s_i^3 = 0. \qquad (9)$$

b) Phase 2, $t^2 = f_t$. The traction in the normal direction reaches f_t and openings will appear in the cohesive zone for subsequent loads. The cohesive law is introduced in lines of submatrices B and C to relate openings and tractions in the normal direction. The eqn (8) is modified with the first equation of the constitutive law. No tractions appear in the tangent direction, i.e. the eqn (9) is replaced by the condition of tractions in the tangent direction equal to zero. The following equations replaces eqns (8) and (9), respectively:

$$(u_i^2 n_i^2 + u_i^3 n_i^3) - \frac{w_c'}{f_t} n_i^2 t_i^2 = -w_c', \tag{10}$$

$$s_i^2 t_i^2 = 0. \tag{11}$$

c) Phase 3, $t^2 = f_l$. The breaking point is reached when the traction in the normal direction is equal to f_l. Further openings in the cohesive zone due to subsequent loads will be according to the second equation of the constitutive law show in eqn (3). The condition for tractions in the tangent direction remains unchanged. The following equation replaces eqn (10):

$$(u_i^2 n_i^2 + u_i^3 n_i^3) - \frac{w_c}{f_t'} n_i^2 t_i^2 = -w_c. \tag{12}$$

d) No tractions appear in in the cohesive zone when the critical opening (w_c) is reached. It means that the tractions (t^2_i) at that point on the crack surfaces must be eliminated from the system of equations. Eqns (11) and (12) in the corresponding lines of matrices A and B are replaced by:

$$t_1^2 = 0, \tag{13}$$

$$t_2^2 = 0. \tag{14}$$

There was built a boundary element code using quadratic boundary elements and based in eqns (4) and (5). Quadratic shape functions were employed for the isoparametric boundary elements, with collocation points always placed on the boundary. The same mapping function was used for conformal and non-conformal interpolations, i.e., nodes at the ends of the quadratic elements remained at the ends when discontinuous elements were employed. The collocation points for the displacement BIE were positioned at element nodes in conformal interpolations (ξ'= -1 and ξ'= 0) and shifted to the interior of boundary elements in the case of non-conformal interpolations (ξ'= ±0.67, and ξ'= 0). In contrast, to satisfy the continuity requirements for the traction BIE, collocation points for the traction BIE were always positioned inside boundary elements for both interpolation types, i.e. (ξ'= ±0.67) for conformal interpolations and (ξ'= ±0.67, and ξ'= 0) for non-conformal interpolation. This strategy for the positions of collocation points in the DBEM was discussed in [28]. The singularity subtraction [29] and the transformation of variable technique [30] were employed for the Cauchy and weak-type singularity, respectively, when integrations were performed on elements containing the collocation points. The standard Gauss–Legendre scheme was employed for integrations on elements not containing the collocation points.

4 NUMERICAL RESULTS

Experimental and numerical analyses for plain concrete were performed in [4] using the notched beam under the three-point bending test shown in Fig. 4. This case was studied in [27] and it is repeated here because the formulation and the computer code were changed. The depth of the beam (d) is 0.2 m, the width is 0.05 m, and the length (l) is 2.0 m. The adopted ratio between notched depth and beam depth (a/d) was 0.5. The Poisson's ratio (v) was 0.2 and the Table 1 presents the properties used in the analysis of the crack propagation. Peterson [4] presented results for crack propagation obtained experimentally for two values

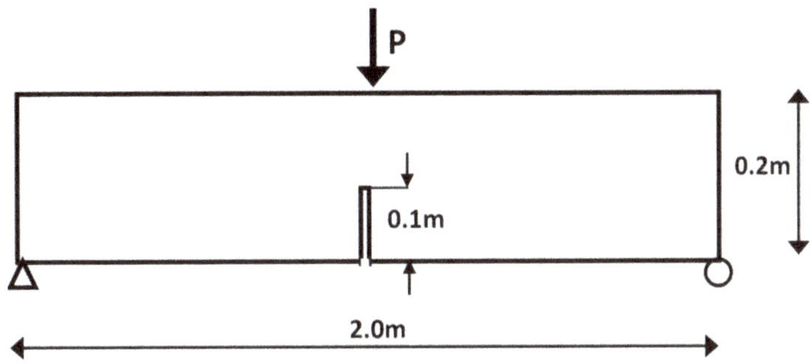

Figure 4: The notched beam subjected to the three-point bending test in [4].

Table 1: Mechanical properties according to the constitutive laws in [4].

σ-w curve	G_f (Nm^{-1})	f_t (MPa)	f_1 (MPa)	w_c (mm)	w_1 (mm)	E_c (MPa
One straight line (SL)	124	3.33	–	0.07450	–	30,000
Two straight lines (C)	124	3.33	1.11	0.13400	0.02979	30,000

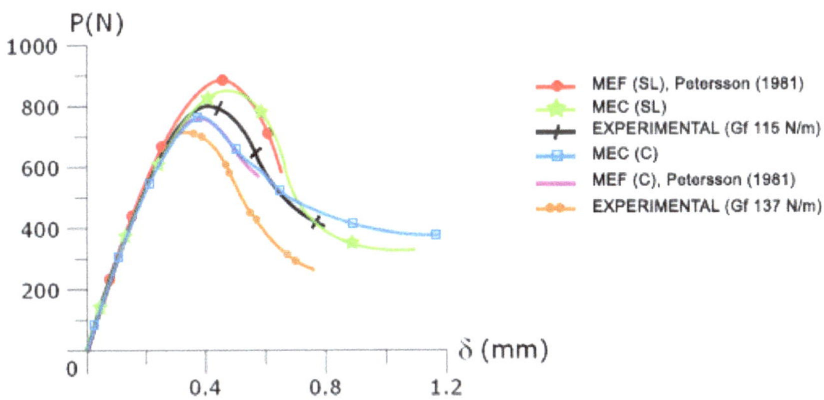

Figure 5: Comparison between the results obtained and those in [4] by Peterson.

of fracture energy (G_f): 115 Nm^{-1} and 137 Nm^{-1}. Experimental results were compared [4] with those obtained from a numerical analysis with the FEM using 124 Nm^{-1} for G_f, which was the median value of the fracture energy values of the experiments. Two constitutive laws were considered in the numerical analyses [4]: a) Straight line and b) Two straight lines.

The comparison between the experimental and numerical results presented in [4] with those obtained with the present formulation is shown in Fig. 5. There were included results obtained with the present formulation when one straight line was used because the code built for the constitutive law with two straight lines can be simplified to that with one straight line. The results obtained matched those obtained by Peterson [4] and the behavior of the present formulation was stable in the post-peak softening, as shown by the blue and green lines in Fig. 5.

The results obtained in analyses with FRC are compared with those presented in [31]. Fig. 6 presents the specimens used in [31], which the aspect ratio is different to that in Fig 4. The depth of the beam (d) is 125 mm, the width (b) is 80 mm, the ratio between notched depth and beam depth (a/d) is 0.3333, and the Poisson's ratio (ν) is 0.2. Table 2 presents the properties used in [31].

At the end of the crack propagation, there were employed 122 quadratic boundary elements (231 boundary nodes) with two boundary nodes placed at each corner and at the fictitious crack tip. The relative values of the peak load and the deflection were considered in Fig. 7 to show the difference between values obtained with SFRC specimens to that with plain concrete. The similar pre-peak behavior with respect to the plain concrete is shown in Fig. 7 whereas the comparison of the post-peak softening points out to the higher ductility for the fiber-reinforced specimens. The peak values for loads as well as the limit load after the softening are similar to those presented in [31].

5 CONCLUSIONS

The results obtained in the crack propagation analyses with the present DBEM formulation matched those in the literature. The meshes on the crack surfaces employed continuous quadratic boundary elements. The present code is different to that used in [26], [27], which employed linear boundary elements and considered different DBEM equations. Furthermore, the formulation using eqns (4) and (5) was stable when the constitutive law employed two straight lines with reference to that presented in [27] without reducing or changing benefits on the direct computation of the cohesive forces at each incremental loading step.

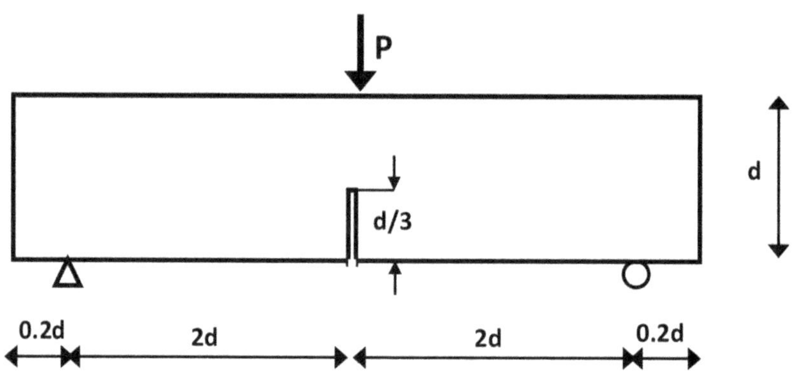

Figure 6: The notched beam subjected to the three-point bending test in [31].

Table 2: Mechanical properties used in [31].

σ-w curve	G_f (Nm^{-1})	f_t (MPa)	f_1 (MPa)	w_c (mm)	w_1 (mm)	E_c (MPa
Plain concrete	72	2.81	0.55	0.18	0.016	25,127
SFRC – 1	3,160	2.77	0.9	6.9	0.04	36,162
SFRC – 2	3,878	2.77	1.52	5.03	0.04	36,162
Micro – FRC	730	2.81	0.47	3.0	0.0175	25,127
Macro – FRC	22,520	2.81	0.9	50.0	0.014	25,127

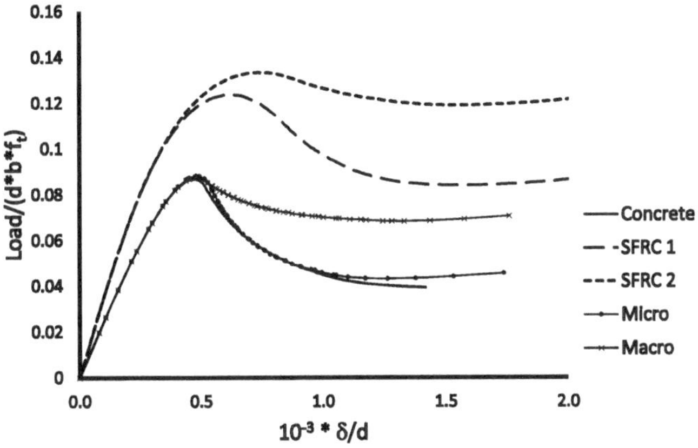

Figure 7: Results obtained using mechanical properties from Table 2 [31].

REFERENCES

[1] Hillerborg, A., Modeer, M. & Petersson, P.E., Analysis of crack formation and crack growth in concrete by means of fracture mechanics and finite elements. *Cement and Concrete Research*, **6**(6), pp. 773–781, 1976. DOI: 10.1016/0008-8846(76)90007-7.

[2] Hillerborg, A., Analysis of fracture by means of the fictitious crack model, particularly for fiber-reinforced concrete. *International Journal of Cement Composites*, **2**(4), pp. 177–184, 1980.

[3] Carpinteri, A., Post-peak and post-bifurcation analysis of cohesive crack propagation. *Engineering Fracture Mechanics*, **32**(2), pp. 265–278, 1989. DOI: 10.1016/0013-7944(89)90299-3.

[4] Peterson, P.E., *Crack Growth and Development of Fracture Zone in Plain Concrete and similar materials, Report N° TVBM-1006*, Division of Building Materials, Lund Institute of Technology: Lund Sweden, 1981.

[5] Olesen, J.F., Fictitious crack propagation in fiber-reinforced concrete beams. *Journal of Engineering Mechanics*, **127**(3), pp. 272–280, 2001. DOI: 10.1061/(asce)0733-9399(2001)127:3(272).

[6] Maalej, M. & Li, V.C., Flexural Strength of fiber cementitious composites. *Journal of Materials in Civil Engineering*, **6**(3), pp. 390–406, 1994.

[7] Li, V.C., Lin, Z. & Matsumoto, T., Influence of fiber bridging on structural size-effect. *International Journal of Solids and Structures*, **35**(31–32), pp. 4223–4238, 1998. DOI: 10.1016/s0020-7683(97)00311-9.

[8] Sousa, J.L.A.O. & Gettu, R., Determining the tensile stress–crack opening curve of concrete by inverse analysis. *Journal of Engineering Mechanics*, **132**(2), pp. 141–148, 2006. DOI: 10.1061/(asce)0733-9399(2006)132:2(141).

[9] Slowik, V., Villmann, B., Bretschneider, N. & Villmann, T., Computational aspects of inverse analyses for determining softening curves of concrete. *Computer Methods in Applied Mechanics and Engineering,* **195**(52), pp. 7223–7236, 2006. DOI: 10.1016/j.cma.2005.04.021.

[10] Brebbia, C.A., The birth of the boundary element method from conception to application. *Engineering Analysis with Boundary Elements*, **77**, pp. iii–x, 2017. DOI: 10.1016/j.enganabound.2016.12.001.

[11] Saleh, A.L. & Aliabadi, M.H., Crack growth analysis in concrete using boundary element method. *Engineering Fracture Mechanics*, **51**(4), pp. 533–545, 1995. DOI: 10.1016/0013-7944(94)00301-w.
[12] Saleh, A.L. & Aliabadi, M.H., Fracture mechanics analysis of crack on plain concrete using the boundary element methods. *Engineering Fracture Mechanics*, **69**(2), pp. 267–280, 2002. DOI: 10.1016/s0013-7944(01)00089-3.
[13] Chen, T., Wang, B., Cen, Z. & Wu, Z., A symmetric Galerkin multi-zone boundary elemento method for cohesive crack growth. *Engineering Fracture Mechanics*, **63**(5), pp. 591–609, 1999. DOI: 10.1016/s0013-7944(99)00036-3.
[14] Gospodinov, G., Boundary element modelling of cohesive cracks using displacement discontinuity method. *Advances in Boundary Element Techniques*, pp. 19–26, 2003.
[15] Karlis, G.F., Tsinopoulos, S.V., Polyzos, D. & Beskos, D.E., Boundary Element analysis of mode I and mixed mode (I and II) crack problems of 2-D gradient elasticity. *Computer Methods in Applied Mechanics and Engineering*, **196**(49–52), pp. 5092–5103, 2007. DOI: 10.1016/j.cma.2007.07.006.
[16] Leonel, E.D. & Venturini, W.S., Non-linear boundary element formulation with tangent operator to analyze crack propagation in quasi-brittle materials. *Engineering Analysis with Boundary Elements*, **34**(2), pp. 122–129, 2010. DOI: 10.1016/j.enganabound.2009.08.005.
[17] Távara, L., Mantic, V., Graciani, E. & París, F., BEM analysis of crack onset and propagation along fiber–matrix interface under transverse tension using a linear elastic–brittle interface model. *Engineering Analysis with Boundary Elements*, **35**(2), pp. 207–222, 2011. DOI: 10.1016/j.enganabound.2010.08.006.
[18] Trinh, Q.T., Mouhoubi, S., Chazallon, C. & Bonnet, M., Solving multizone and multicrack elastostatic problems: A fast multipole symmetric Galerkin boundary element method approach. *Engineering Analysis with Boundary Elements*, **50,** pp. 486–495, 2015. DOI: 10.1016/j.enganabound.2014.10.004.
[19] Benedetti, I. & Aliabad, M.H., Multiscale modeling of polycrystalline materials: A boundary element approach to material degradation and fracture. *Computer Methods in Applied Mechanics and Engineering*, **289**, pp. 429–453, 2015. DOI: 10.1016/j.cma.2015.02.018.
[20] Kupradze, V.D., *Three-Dimensional Problems of the Mathematical Theory of Elasticity and Thermoelasticity*, North Holland, 1979.
[21] Sladek, J. & Sladek, V., Three-dimensional curved crack in an elastic body. *International Journal of Solids and Structures*, **19**(5), pp. 425–436, 1983. DOI: 10.1016/0020-7683(83)90053-7.
[22] Bonnet, M., *Boundary Integral Equation Methods for Solids and Fluids*, John Wiley, 1999.
[23] Palermo, L. Jr., Almeida, L.P.C.P.F. & Gonçalves, P.C., *The Use of the Tangential Differential Operator in the Dual Boundary Element Equation, Structural Durability and Health Monitoring*, vol. 2, Tech Science Press, pp. 123–130, 2006.
[24] Palermo, L. Jr. & Almeida, L.P.C.P.F., On the use of the tangential differential operator in the traction boundary integral equation of the dual boundary element method for three dimensional problems. *ICCES*, **7**(2), pp. 83–87, 2008.
[25] Palermo, L. Jr., The tangential differential operator applied to a stress boundary integral equation for plate bending including the shear deformation effect. *Engineering Analysis with Boundary Elements*, **36**(8), pp. 1213–1225, 2012. DOI: 10.1016/j.enganabound.2012.02.010.

[26] Gonçalves, P.C., Palermo, L. Jr. & Proença, S.P.B., A boundary element formulation for crack analyses incorporating a cohesive zone model. *International Journal of Computational Methods and Experimental Measurements*, **5**(3), pp. 231–240, 2017. DOI: 10.2495/cmem-v5-n3-231-240.

[27] Gonçalves, P.C., Palermo, L. Jr. & Proença, S.P.B., Analysis of two cohesive zone models for crack propagation in notched beams using the BEM. *International Journal of Computational Methods and Experimental Measurements*, **6**(6), pp. 1057–1066, 2018. DOI: 10.2495/cmem-v6-n6-1057-1066.

[28] Palermo, L. Jr., Gonçalves, P.C. & Figueiredo, L.G., A simple implementation of the dual boundary element method using the tangential differential operator for plane problems. *XXXII Boundary Elements and other Reduction Methods*, WIT Press: Southampton, pp. 75–84, 2010.

[29] Aliabadi, M.H., *The Boundary Element Method, Applications in Solids and Structures,* vol. 2, John Wiley, 2002.

[30] Telles, J.C.F., A self-adaptive co-ordinate transformation for efficient numerical evaluation of general boundary element integrals. *International Journal for Numerical Methods in Engineering*, **24**(5), pp. 959–73, 1987. DOI: 10.1002/nme.1620240509.

[31] Bruggi, M. & Venini, P., A numerical investigation on the size effect of fiber-reinforced concrete specimens in crack propagation. *Computational Mechanics*, **50**(1), pp. 99–117, 2012. DOI: 10.1007/s00466-011-0677-3.

ITERATIVE BOUNDARY ELEMENT METHOD FOR CRACKS IN A THERMAL MEDIUM WITH EXACT CRACK FACE BOUNDARY CONDITIONS

QIAOYUN ZHANG[1,3], YAGUANG GUO[1] & MINGHAO ZHAO[1,2,3]
[1]School of Mechanics and Engineering Science, Zhengzhou University, China
[2]School of Mechanical Engineering, Zhengzhou University, China
[3]Henan Key Engineering Laboratory for Anti-Fatigue Manufacturing Technology, Zhengzhou University, China

ABSTRACT
An iterative boundary element method is proposed to analyze cracks with exact crack face boundary conditions in thermoelastic solids. Herein, the crack was opened under external loading, whereupon the opened cavity was considered a domain. The boundary element method for the crack-cavity domain and the sub-region boundary element method for the outer thermoelastic solid were used iteratively to obtain the real deformed crack faces. In this approach, the exact face boundary conditions on crack faces were used and the stress intensity factor and thermal flux density intensity factor near the crack tip were calculated.
Keywords: thermoelastic solid, boundary element method, crack, exact boundary conditions, stress intensity factor, thermal flux density intensity.

1 INTRODUCTION

Thermoelastic solids are sensitive to the presence of cracks, which cause a thermal flux concentration and affect the reliability and working life of materials and structures. Therefore, fracture problems in thermoelastic solids have attracted much attention [1]–[8].

Owing to the thermo-mechanical coupling property and the complex crack face boundary conditions in thermoelastic solids, it is difficult to find analytical solutions for some fracture problems, and thus numerical methods are adopted to study these problems [9]–[12]. Among these numerical methods, the boundary element method (BEM) [13], [14] reduces the solving dimensions and exhibits high accuracy, and is therefore very efficient to solve fracture problems in thermoelastic solids. Prasad et al. have studied several thermoelastic crack problems using dual BEM, wherein the stress intensity factors were obtained according to the independent *J*-integral [15]. Ekhlakov et al. have analyzed thermoelastic cracks in functionally-graded materials by BEM and have studied the influence that the material gradation, thermo-mechanical coupling, crack orientation and thermal shock loading have on the fracture behavior [16]. Pasternak has studied the inclined-crack model, edge crack model, the two-parallel-crack model and the kinked-crack model in two-dimensional anisotropic thermoelastic solids using BEM [17]. Finally, Zhao et al. have studied the extended intensity factors of an elliptical crack in isotropic three-dimensional thermoelastic solids [18].

Up to now, studies of thermal crack models have almost entirely been based on thermally-impermeable or -permeable crack face boundary conditions, but these two boundary conditions are too limiting for most cases. Zhong and Kang [19] and Zhong et al. [20] have therefore proposed a new partially-thermally-permeable crack model, which demonstrates that the crack face boundary conditions of the thermal crack model significantly affect the fracture behavior. In the present work, we consider the exact crack face boundary conditions for a thermal-medium crack model, and an iterative BEM to solve this nonlinear crack model is proposed. The paper is organized as follows: In Section 2, the basic equations for thermoelastic solids are given. In Section 3, the iterative BEM is proposed for a central crack

WIT Transactions on Engineering Sciences, Vol 122, © 2019 WIT Press
www.witpress.com, ISSN 1743-3533 (on-line)
doi:10.2495/BE410051

model in thermoelastic solids. In Section 4, numerical results are presented to demonstrate the relationships between the stress and thermal flux density intensity factors and the external mechanical and thermal loadings. Finally, in Section 5 the main conclusions are drawn.

2 BASIC EQUATIONS

In the absence of a body force or internal heat source, the governing equations for isotropic thermoelastic solids in the *Oxz* cartesian coordinate system are given by

$$\begin{aligned} \sigma_{ij,i} &= 0, \\ h_{i,i} &= 0, \end{aligned} \quad (i, j = 1, 2 \text{ or } x, z) \tag{1}$$

where σ_{ij} is the stress component and h_i is the thermal flux component. The constitutive equations are given by [1]

$$\begin{aligned} \sigma_x &= 2G\frac{\partial u}{\partial x} + \frac{Ev}{(1+v)(1-2v)}\left(\frac{\partial u}{\partial x} + \frac{\partial w}{\partial z}\right) - \frac{E}{1-2v}\alpha\theta, \\ \sigma_z &= 2G\frac{\partial w}{\partial z} + \frac{Ev}{(1+v)(1-2v)}\left(\frac{\partial u}{\partial x} + \frac{\partial w}{\partial z}\right) - \frac{E}{1-2v}\alpha\theta, \\ \tau_{xy} &= G\left(\frac{\partial u}{\partial z} + \frac{\partial w}{\partial x}\right), \\ h_x &= -\beta\frac{\partial \theta}{\partial x}, \quad h_z = -\beta\frac{\partial \theta}{\partial z}, \end{aligned} \tag{2}$$

where E is the elasticity modulus, G is the shear modulus, v is the Poisson ratio, θ is the temperature, α is the thermal expansion coefficient and β is the heat conductivity. Substituting eqn (2) into eqn (1), the governing equations can be rewritten as

$$\begin{aligned} &2G\frac{\partial^2 u}{\partial x^2} + \frac{Ev}{(1+v)(1-2\upsilon v)}\left(\frac{\partial^2 u}{\partial x^2} + \frac{\partial^2 w}{\partial x \partial z}\right) + \frac{E}{1-2v}\alpha\frac{\partial \theta}{\partial x} + G\left(\frac{\partial^2 u}{\partial z^2} + \frac{\partial^2 w}{\partial x \partial z}\right) = 0, \\ &2G\frac{\partial^2 u}{\partial z^2} + \frac{Ev}{(1+v)(1-2v)}\left(\frac{\partial^2 u}{\partial x \partial z} + \frac{\partial^2 w}{\partial z^2}\right) + \frac{E}{1-2v}\alpha\frac{\partial \theta}{\partial z} + G\left(\frac{\partial^2 u}{\partial x \partial z} + \frac{\partial^2 w}{\partial x^2}\right) = 0, \\ &-\beta\left(\frac{\partial^2 \theta}{\partial x^2} + \frac{\partial^2 \theta}{\partial z^2}\right) = 0. \end{aligned} \tag{3}$$

On the outer boundary of the solid, the displacement and surface traction boundary conditions are given by

$$\begin{aligned} u_i &= \overline{u}_i \ \text{ on } \Gamma_u, \\ t_i &= \sigma_{ij} n_j = \overline{t}_i \ \text{ on } \Gamma_t, \end{aligned} \tag{4}$$

where Γ_u and Γ_t represent the boundaries with prescribed displacements and prescribed tractions, respectively. For the temperature field, the boundary conditions are given by

$$\begin{aligned} \theta &= \overline{\theta} \ \text{ on } \Gamma_\theta, \\ q &= h_i n_i = \overline{q} \ \text{ on } \Gamma_q, \end{aligned} \tag{5}$$

where Γθ and Γq represent the boundary with prescribed temperature and prescribed thermal flux density, respectively; $\overline{u}_i, \overline{t}_i, \overline{\theta}$ and $\overline{q}$ are the prescribed displacement, traction, temperature and thermal flux density, respectively; and ni is the outward normal vector. Note that Γu + Γt = Γθ + Γq = Γ.

3 BOUNDARY ELEMENT METHOD FOR THERMOELASTIC SOLIDS WITH A CENTRAL CRACK

Consider a thermoelastic plate with a central crack, as shown in Fig. 1. The plate length is $2l$ and height is $2h$, and the crack length is $2c$. The symmetric mechanical loading p_0 and thermal loading q_0 are applied on the upper and lower surface of the plate, while the crack faces are load-free. Under the applied mechanical and thermal loadings, the crack opens and the crack cavity Ω_c is formed. The boundary of the crack cavity can be expressed as

$$\Gamma_c = \Gamma_{c1} + \Gamma_{c2}, \tag{6}$$

where Γ_{c1} and Γ_{c2} are the upper and lower faces of the crack cavity, respectively. Therefore, we adopted the sub-region BEM to divide the entire plate into the three domains Ω_1, Ω_2 and Ω_c, as shown in Fig. 2. The boundaries of Ω_1 and Ω_2 can be expressed as:

$$\begin{aligned} \Gamma_1 &= \Gamma_{12} + \Gamma_{1c} + B_1 + B_2 + B_3, \\ \Gamma_2 &= \Gamma_{21} + \Gamma_{2c} + S_1 + S_2 + S_3, \end{aligned} \tag{7}$$

where the divided regions are shown in Fig. 2. In the cavity of the opening crack, there exists only the temperature field, whose governing equation is yet the heat conduction equation, given as

$$-\beta\left(\frac{\partial^2 \theta^c}{\partial x^2} + \frac{\partial^2 \theta^c}{\partial z^2}\right) = 0, \tag{8}$$

where θ^c is the temperature field in the crack cavity.

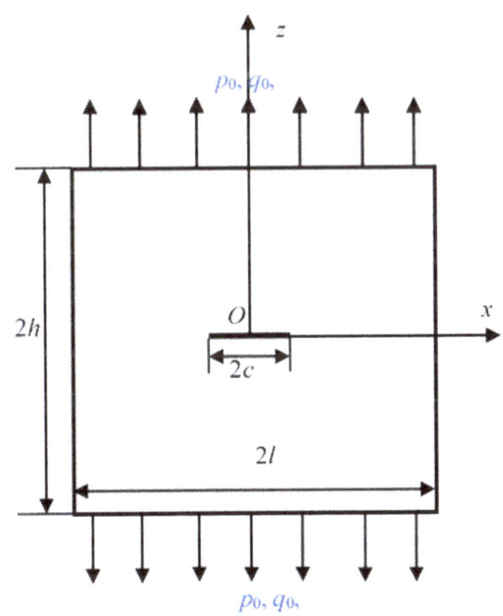

Figure 1: Schematic of a central crack in a thermoelastic rectangular plate under mechanical loading p_0 and thermal loading q_0.

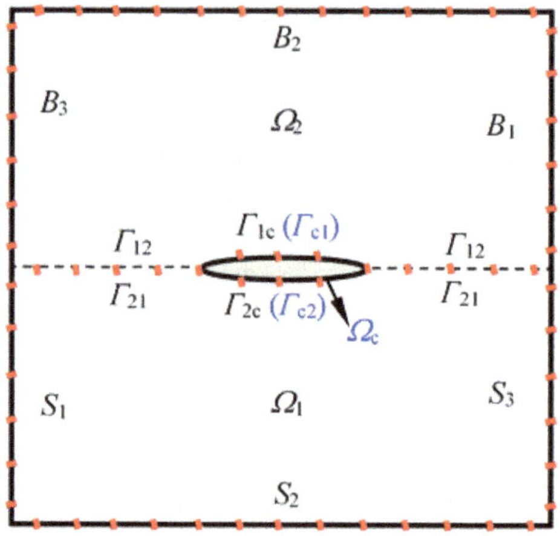

Figure 2: Schematic of the sub-regions of the central crack model and discretization of the involved boundaries.

The boundary conditions are given as follows:

(1) On the outer boundaries B_1, B_2, B_3, S_1, S_2 and S_3, the boundary conditions are

$$\begin{aligned} &t_1 = 0, \quad t_2 = p_0, \quad q = q_0, \quad (|x| \le l, z = h), \\ &t_1 = 0, \quad t_2 = -p_0, \quad q = -q_0, \quad (|x| \le l, z = -h), \\ &t_1 = 0, \quad t_2 = 0, \quad q = 0, \quad (x = \pm l, |z| \le h), \end{aligned} \tag{9}$$

where t_i ($i = x$, z) denotes the mechanical traction, p_0 is the external mechanical loading along the z-direction, and q_0 is the external thermal loading.

(2) On the common boundaries Γ_{12} and Γ_{21} between sub-regions Ω_1 and Ω_2, along the x-axis, the boundary conditions are

$$\begin{aligned} &u|_{\Gamma_{12}} = u|_{\Gamma_{21}}, \quad w|_{\Gamma_{12}} = w|_{\Gamma_{21}}, \\ &t_1|_{\Gamma_{12}} = -t_1|_{\Gamma_{21}}, \quad t_2|_{\Gamma_{12}} = -t_2|_{\Gamma_{21}}, \quad (c \le |x| \le l, z = 0). \\ &\theta|_{\Gamma_{12}} = \theta|_{\Gamma_{21}}, \quad q|_{\Gamma_{12}} = -q|_{\Gamma_{21}}, \end{aligned} \tag{10}$$

(3) On the common boundaries Γ_{1c} and Γ_{c1} (Γ_{2c} and Γ_{c2}) between sub-regions Ω_1 and Ω_c (Ω_2 and Ω_c), the boundary conditions are

$$\begin{aligned} &t_1 = 0, \quad t_2 = 0, \\ &\theta|_{\Gamma_{\alpha c}} = \theta|_{\Gamma_{c\alpha}}, \quad q|_{\Gamma_{\alpha c}} = -q|_{\Gamma_{c\alpha}}, \end{aligned} \quad (x, z) \in \Gamma_c, \quad \alpha = 1, 2. \tag{11}$$

We note that the thermal boundary conditions on the crack faces described in eqn (11) are the exact boundary conditions depending on the deformed crack faces.

3.1 Iterative approach

Owing to the existence of a temperature field in the crack cavity, the thermal flux density boundary condition on the upper and lower crack surfaces is affected by the deformed crack faces. This problem is a typical nonlinear problem that we propose to solve using an iterative BEM, whose iterative approach is described in detail in the following.

(1) The crack is initially treated as a slit that is considered to be thermally impermeable. Thus, the crack face boundary conditions are given by

$$t_1 = 0, \quad t_2 = 0, \quad q = 0. \tag{12}$$

In this case, only the two regions Ω_1 and Ω_2 remain in the thermoelastic plate. Thus, using the sub-region BEM and the boundary conditions in eqns (9), (10), and (12), the extended displacement fields ($u_i^{(1)}$, $\theta^{(1)}$) and a new crack cavity region $\Omega_c^{(1)}$ are obtained.

(2) Next, the crack cavity region $\Omega_c^{(1)}$ is treated as a single domain and the temperatures $\theta^{(1)}$ on the crack faces obtained in Step (1) are taken as the boundary conditions. Using the single-domain BEM, new values of the thermal flux density $q^{(1)}$ on the crack face are obtained.

(3) Next, the crack is again treated as a slit. Taking the extended stress fields $q^{(1)}$ on the crack faces obtained in Step (2) as the new boundary conditions and combining them with the boundary conditions in eqns (9) and (10), we obtain the extended displacement fields ($u_i^{(2)}$, $\theta^{(2)}$) and a new crack cavity region $\Omega_c^{(2)}$.

(4) Steps (2) and (3) are repeated until the preset precision for the thermal flux density on the crack face is satisfied at the kth step, such that

$$\left| \frac{q^{(k)} - q^{(k-1)}}{q^{(k-1)}} \right| \le \delta, \tag{13}$$

where δ is a preset error value. In the present paper, $\delta = 0.01\%$.

With this iterative method, we ultimately obtain the exact thermal boundary condition on the crack face and the size of the crack cavity. Simultaneously, the displacement field, temperature field, stress field and thermal flux density field of thermal-medium crack model under exact crack face boundary conditions are obtained.

3.2 Sub-region BEM for a crack in a thermoelastic solid

Employing the fundamental solutions and the Somigliana identity, the extended displacement boundary integral equations for the three sub-regions are established.

For thermoelastic solids, when the temperature in the solid is changed, an initial strain occurs owing to the thermal expansion. This initial strain is given by

$$\varepsilon_{jk}^{0} = \alpha \Delta T \delta_{jk}, \tag{14}$$

where α denotes the thermal expansion coefficient, ΔT denotes the variations in temperature and δ denotes the Kronecker delta function. Eqn (14) indicates that the temperature induces a purely volume expansion. Further, the initial stress caused by the temperature field in the solids is given by

$$\sigma_{jk}^{0} = 2G \frac{1+\upsilon}{1-2\upsilon} \alpha \Delta T \delta_{jk}. \tag{15}$$

For the sub-regions Ω_1 and Ω_2, the fundamental solutions are identical. Regarding the initial temperature stress as the body force, the extended displacement boundary integral equation for an arbitrary point Y in a thermal solid with a body force f_j within domain Ω_1 or Ω_2 can be written as

$$u_i(Y)=\int_\Gamma\left[U_{ij}^*(x,Y)t_j(x)-T_{ij}^*(x,Y)u_j(x)\right]d\Gamma(x)+\int_\Omega U_{ij}^*(x,Y)f_j d\Omega, \quad (16)$$

where x is the field point; j=1,2,3; $u_3=\theta$; $t_3=q$; $f_1=f_2=0$ and $f_3=\sigma_{ij}^0$; $\Gamma=\Gamma_1$ or Γ_2; and $\Omega=\Omega_1$ or Ω_2. Finally, U_{ij}^* and T_{ij}^* are the extended displacement and traction matrices of the fundamental solutions, which can be expressed as in [21]

$$U_{ij}^*(x,Y)=-\frac{1}{8\pi\mu(1-\nu)}\left[(3-4\nu)\ln(r)\delta_{ij}-r_{,i}r_{,j}\right],$$
$$T_{ij}^*(x,Y)=\frac{1}{4\pi(1-\nu)}\frac{1}{r}\left\{(1-2\nu)(r_{,i}n_j-r_{,j}n_i)-r_{,n}\left[(1-2\nu)\delta_{ij}+2r_{,i}r_{,j}\right]\right\}, \quad (17)$$

where xi and Yi denote the coordinates of the field point and source point, respectively; and the radius vector $\vec{r}$ between the source point and field point is defined as

$$r_{,i}=\frac{\partial r}{\partial x_i}=\frac{r_i}{r},\quad r_{,n}=\frac{\partial r}{\partial n}=r_{,i}n_i,$$
$$r_i=x_i-Y_i,\quad r=\sqrt{r_i r_i}. \quad (18)$$

When Y approaches the boundary Γ, eqn (16) becomes

$$C_{ij}(Y)u_i(Y)=\int_\Gamma\left[U_{ij}^*(x,Y)t_j(x)-T_{ij}^*(x,Y)u_j(x)\right]d\Gamma(x)+\int_\Omega U_{ij}^*(x,Y)f_j d\Omega, \quad (19)$$

where C_{ij} is the boundary coefficient matrix wherein $C_{ij}=\delta_{ij}/2$ when the boundary is smooth. Employing the Gauss divergence theorem, the domain integral in eqn (19) can be transformed into the boundary integral

$$\int_\Omega U_{ij}^*(x,Y)f_j d\Omega=\int_\Gamma\left[(R_i(x,Y)\theta(x)-Q_i(x,y)\frac{\partial\theta(x)}{\partial n}\right]d\Gamma, \quad (20)$$

where $\partial\theta(x)/\partial n$ denotes the normal temperature gradients of the temperature $\theta(x)$ on the boundary Γ. Further, $R_i(x,Y)$ and $Q_i(x,Y)$ are given by [21]

$$R_i(x,Y)=\frac{\alpha(1+\nu)}{4\pi(1-\nu)}\left[(\ln\frac{1}{r}-\frac{1}{2})n_i-r_{,i}r_{,n}\right],$$
$$Q_i(x,Y)=\frac{\alpha(1+\nu)}{4\pi(1-\nu)}(\ln\frac{1}{r}-\frac{1}{2})r_i, \quad (21)$$

whereupon the extended displacement boundary integral equation can be given by

$$C_{ij}(Y)u_i(Y)=\int_\Gamma\begin{bmatrix}U_{ij}^*(x,Y)t_j(x)-T_{ij}^*(x,Y)u_j(x)\\ +R_i(x,Y)\theta(x)-Q_i(x,Y)\dfrac{\partial\theta(x)}{\partial n}\end{bmatrix}d\Gamma(x). \quad (22)$$

For the sub-regions Ω_c, eqn (8) can be treated as a potential problem, whereby the integral equation for the temperature of any arbitrary point Y within the crack cavity Ω_c is given by

$$\theta(Y)=\int_{\Gamma}[\Phi^{*}(x,Y)\frac{\partial\theta(x)}{\partial n}-\frac{\partial\Phi^{*}(x,Y)}{\partial n}\theta(x)]d\Gamma(x), \tag{23}$$

where

$$\Phi^{*}\left(x,Y\right)=\frac{1}{2\pi}\ln\frac{1}{r},\quad \frac{\partial\Phi^{*}(x,Y)}{\partial n}=-\frac{1}{2\pi r}r_{,i}n_{i}. \tag{24}$$

When Y approaches the boundary Γ, eqn (23) becomes

$$C(Y)\theta(Y)=\int_{\Gamma}[\Phi^{*}(x,Y)\frac{\partial\theta(x)}{\partial n}-\frac{\partial\Phi^{*}(x,Y)}{\partial n}\theta(x)]d\Gamma(x). \tag{25}$$

When the crack is treated as a slit, only two sub-regions exist; i.e., Ω_1 and Ω_2. Using linear elements to discretize the boundaries of Ω_1 and Ω_2, the total linear system of algebraic equations for all nodes on the boundaries Γ_1 and Γ_2 can be written in the matrix form

$$\begin{bmatrix}\mathbf{A}^{(1)} & 0\\ 0 & \mathbf{A}^{(2)}\end{bmatrix}\begin{bmatrix}\mathbf{u}^{(1)}\\ \mathbf{u}^{(2)}\end{bmatrix}=\begin{bmatrix}\mathbf{B}^{(1)} & 0\\ 0 & \mathbf{B}^{(2)}\end{bmatrix}\begin{bmatrix}\mathbf{t}^{(1)}\\ \mathbf{t}^{(2)}\end{bmatrix}, \tag{26}$$

where $\mathbf{u}^{(1)}$ and $\mathbf{u}^{(2)}$ denote the extended displacement fields (u_i, θ) of Ω_1 and Ω_2, respectively; and $\mathbf{t}^{(1)}$ and $\mathbf{t}^{(2)}$ denote the extended stress fields (σ_{ij}, h_i) of Ω_1 and Ω_2, respectively. The compatibility conditions are given as

$$\mathbf{u}_{cb}^{(1)}=\mathbf{u}_{cb}^{(2)},\quad \mathbf{t}_{cb}^{(1)}=-\mathbf{t}_{cb}^{(2)}, \tag{27}$$

where $\mathbf{u}_{cb}^{(1)}$ and $\mathbf{u}_{cb}^{(2)}$ denote the displacement fields (u_i, θ) on the common boundaries of Ω_1 and Ω_2, respectively; and $\mathbf{t}_{cb}^{(1)}$ and $\mathbf{t}_{cb}^{(2)}$ denote the extended stress fields (σ_{ij}, h_i) of the common boundaries of Ω_1 and Ω_2, respectively. Substituting the boundary conditions given in eqns (9)–(11) and the compatibility conditions given in eqn (27) into eqn (26), the unknown extended displacements (u_i, θ) and the extended stress fields (σ_{ij}, h_i) on the boundaries Γ_1, Γ_2 and common boundaries Γ_{12} (Γ_{21}) can be obtained.

Similarly, in the crack cavity Ω_c, the total linear system of algebraic equations for all nodes on the crack face can be written in the matrix form

$$\mathbf{H}\boldsymbol{\theta}^{c}=\mathbf{G}\mathbf{h}^{c}, \tag{28}$$

where $\boldsymbol{\theta}^c$ and $\mathbf{h}^c$ respectively denote the temperature fields and the thermal flux density fields in the crack cavity Ω_c. Substituting the boundary conditions given in eqn (11) into eqn (28), the unknown temperature field θ and the thermal flux density h in the crack cavity Ω_c are obtained.

4 NUMERICAL ANALYSIS: A CENTRAL CRACK IN A RECTANGULAR THERMOELASTIC PLATE

The material of the plate was ZrO_2 ceramic, with the material constants

$$\begin{aligned}&E=200\times10^{9}\text{N/m}^{2},\quad \nu=0.29,\\ &\alpha=10.6\times10^{-6}\text{m/K},\quad \beta\text{=4.2W/(m.K)}.\end{aligned} \tag{29}$$

The iterative sub-region BEM proposed in Section 3 was used to analyze the central crack model.

4.1 Validation of the iterative BEM

The plate dimensions used were a height $2h$=100 mm and width $2l$=100 mm, and a central crack length $2c$=10 mm. Further, the extended external loadings were

$$p_0 = 100\text{MPa}, \quad q_0 = 0.1\text{W/m}^2. \tag{30}$$

In this case, we assumed that the central crack in the thermal plate was thermally impermeable, the iteration times satisfying the accuracy requirement in eqn (13) are ten. The calculated crack opening displacement (COD) and the temperature across the crack face were in good agreement with those obtained by the finite element method (FEM), as shown in Figs 3 and 4. This comparison of the results demonstrates that the proposed method is effective for fracture analysis of thermoelastic solids with a central crack.

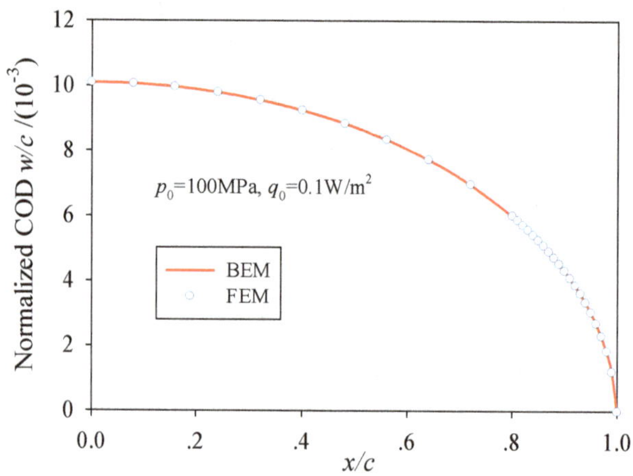

Figure 3: Normalized crack opening displacement (COD) w/c across the crack face under a mechanical loading of p_0=100 MPa and thermal loading of q_0=0.1 W/m^2.

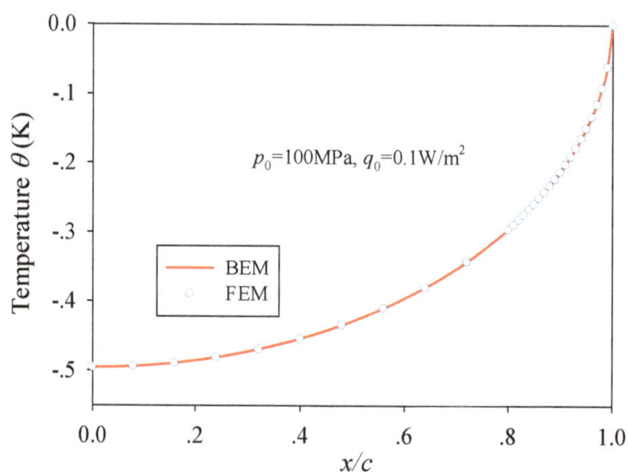

Figure 4: Temperature θ across the crack face under a mechanical loading of p_0=100 MPa and thermal loading of q_0=0.1 W/m^2.

4.2 Effect of crack length on the extended displacements of a central crack in a thermoelastic solid

As an example, we studied the effect of the crack size on the COD and the temperature across the crack face. In this case, the ratio of the plate height to crack length h/c was fixed as 10, the crack face possessed impermeable boundary conditions, and the external mechanical loading p_0 and thermal loading q_0 were the same as given in eqn (30).

The COD and temperature across the crack face are plotted in Figs 5 and 6, respectively. It can be seen that, with increasing crack length, the COD increased while the temperature across the crack face decreased.

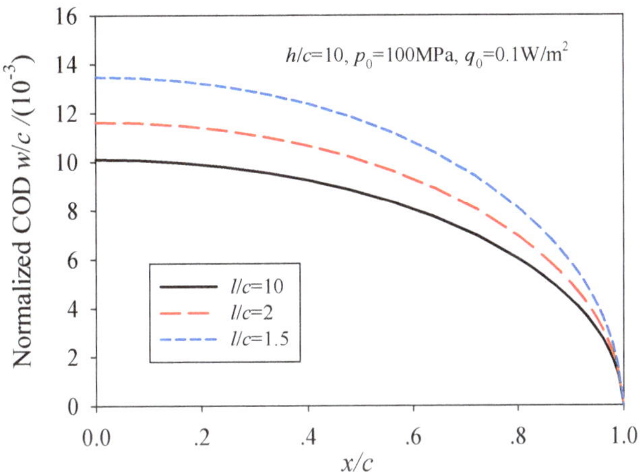

Figure 5: Normalized crack opening displacement (COD) w/c across the crack face with various l/c and fixed h/c=10 under a mechanical loading of p_0=100 MPa and thermal loading of q_0=0.1 W/m^2.

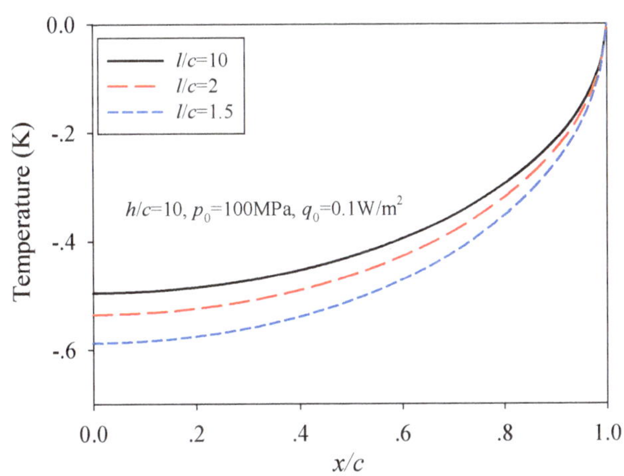

Figure 6: Temperature θ across the crack face with various l/c and fixed h/c=10 under a mechanical loading of p_0=100 MPa and thermal loading of q_0=0.1 W/m^2.

4.3 Effects of extended loads on the extended stress intensity factors of a central crack in a thermoelastic solid

Using the proposed iterative BEM, the crack opening displacements and temperatures near the crack tip were obtained. Then, the stress intensity factor K_{I} and thermal flux density intensity factor K_{T} near the crack tip were calculated using [19]

$$K_{\mathrm{I}} = \frac{E}{8(1-\nu^2)} \lim_{\rho \to 0} \sqrt{\frac{2\pi}{\rho}} \|w\|, \quad K_{\mathrm{T}} = -\frac{\beta}{4} \lim_{\rho \to 0} \sqrt{\frac{2\pi}{\rho}} \|\theta\|, \quad (\|w\| = w^+ - w^-, \|\theta\| = \theta^+ - \theta^-), \tag{31}$$

where ρ is the distance between a point on the crack face and the crack tip; $\|w\|$ is the displacement discontinuity across the crack face; and $\|\theta\|$ is the temperature discontinuity across the crack face; whose subscripts "+" and "–" denote the upper and lower crack faces, respectively. In our numerical calculation, $\|w\|$ and $\|\theta\|$ were obtained by fitting the corresponding values for two elements close to the crack tip, and the normalized stress intensity factor and heat flux density intensity factor are given by

$$F_{\mathrm{I}} = K_{\mathrm{I}} / (p_0 \sqrt{\pi c}), \quad F_{\mathrm{T}} = K_{\mathrm{T}} / (q_0 \sqrt{\pi c}). \tag{32}$$

In the following calculation, the geometric parameters h/c and l/c were fixed as 10. The crack face boundary conditions are the exact boundary condition as in eqn (11) and the external mechanical loading p_0 and thermal loading q_0 are the same as that in eqn. (30).

It can be observed that the normalized stress intensity factors F_{I} and thermal flux intensity factor F_{T} remain the same with the increase of the normalized mechanical loading p/p_0, as shown in Fig. 7.

It is found that the normalized thermal flux intensity factor F_{T} remains constant with increasing normalized thermal loading q/q_0. The normalized stress intensity factor F_{I}, however, increases with increasing normalized thermal loading q/q_0, as shown in Fig. 8.

We can see in Figs 7 and 8 that a one-way coupling property exists in the thermoelastic solids. Specifically, for the thermoelastic solids, the mechanical characteristic does not affect the thermal characteristic, while the thermal characteristic does affect the mechanical characteristic.

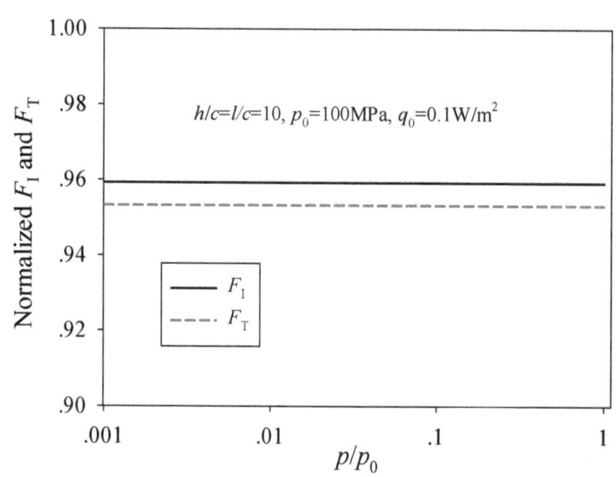

Figure 7: Normalized intensity factors F_{I} and F_{T} versus normalized p/p_0 with $h=l=10c$, under a mechanical loading of $p_0=100$ MPa and thermal loading of $q_0=0.1$ W/m^2.

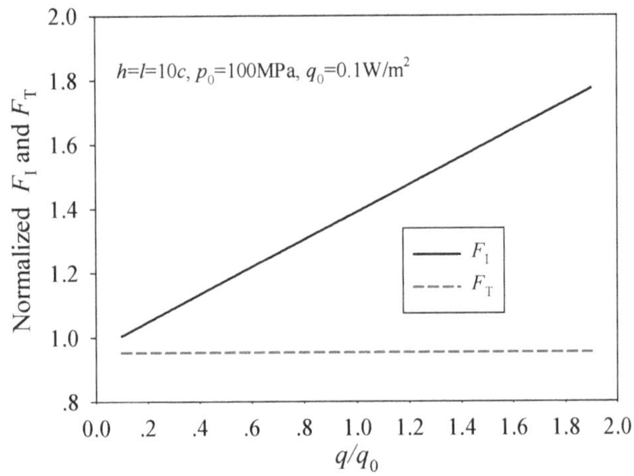

Figure 8: Normalized intensity factors F_I and F_T versus normalized q/q_0 with $h=l=10c$, under a mechanical loading of p_0=100 MPa and thermal loading of q_0=0.1 W/m^2.

5 CONCLUSIONS

An iterative BEM was proposed to analyze the cracking in thermoelastic solids using exact crack boundary conditions. The proposed method was validated by comparing the COD and temperature across the crack face with the results obtained by FEM. Numerical results demonstrated that the crack size greatly influences the COD and temperature across the crack face and that a one-way mechanical–thermal coupling property exists in thermoelastic solids. The proposed iterative BEM is an efficient approach for solving the fracture problem of thermoelastic solids.

ACKNOWLEDGEMENT

The work was supported by the National Natural Science Foundation of China (Nos. 11572289, 11702251).

REFERENCES

[1] Nowacki, W., *Thermoelasticity*, Pergamon Press: New York, 1962.

[2] Khaund, A.K., Krstic, V.D. & Nicholson, P.S., Influence of elastic and thermal mismatch on the local crack-driving force in brittle composites. *Journal of Materials Science*, **12**(11), pp. 2269–2273, 1977. DOI: 10.1007/bf00552248.

[3] Noda, N. & Jin, Z.H., Thermal stress intensity factors for a crack in a strip of a functionally gradient material. *International Journal of Solids and Structures*, **30**(8), pp. 1039–1056, 1993. DOI: 10.1016/0020-7683(93)90002-o.

[4] Erdogan, F. & Wu, B.H., Crack problems in FGM layers under thermal stresses. *Journal of Thermal Stresses*, **19**(3), pp. 237–265, 1996. DOI: 10.1080/01495739608946172.

[5] Chen, J., Soh, A.K., Liu, J. & Liu, Z., Thermal fracture analysis of a functionally graded orthotropic strip with a crack. *International Journal of Mechanics and Materials in Design*, **1**(2), pp. 131–141, 2004.

[6] Chen, Z.T. & Hu, K.Q., Thermo-elastic analysis of a cracked half-plane under a thermal shock impact using the hyperbolic heat conduction theory. *Journal of Thermal Stresses*, **35**(4), pp. 342–362, 2012. DOI: 10.1080/01495739.2012.663685.

[7] Dang, H.Y., Zhao, M.H., Fan, C.Y. & Chen, Z.T., Analysis of an arbitrarily shaped interface crack in a three-dimensional isotropic thermal elastic bi-material. Part 2: Numerical method. *International Journal of Solids and Structures*, **99**, pp. 48–56, 2016. DOI: 10.1016/j.ijsolstr.2016.08.007.

[8] Wang, P., Wang, K.F., Wang, B.L. & Cui, Y.J., Effective thermoelectric conversion properties of thermoelectric composites containing a crack/hole. *Composite Structures*, **191**, pp. 180–189, 2018. DOI: 10.1016/j.compstruct.2018.02.049.

[9] Yu, X.M., Shah, K. & Mukherjee, S., Thermoelastic fracture mechanics with regularized hypersingular boundary integral equations. *Engineering Analysis with Boundary Elements*, **23**(1), pp. 89–96, 1999. DOI: 10.1016/s0955-7997(98)00064-2.

[10] Shiah, Y.C. & Tan, C.L., Exact boundary integral transformation of the thermoelastic domain integral in BEM for general 2D anisotropic elasticity. *Computational Mechanics*, **23**(1), pp. 87–96, 1999. DOI: 10.1007/s004660050387.

[11] Duflot, M., The extended finite element method in thermoelastic fracture mechanics. *International Journal for Numerical Methods in Engineering*, **74**(5), pp. 827–847, 2008. DOI: 10.1002/nme.2197.

[12] Sladek, J., Sladek, V., Solek, P. & Zhang, C.Z., Two- and three-dimensional transient thermoelastic analysis by the MLPG method. *Computer Modeling in Engineering and Sciences*, **47**(1), pp. 61–95, 2009.

[13] Brebbia, C.A., *The Boundary Element Method for Engineers*, Halsted Press: New York, 1978.

[14] Portela, A., Aliabadi, M.H. & Rooke, D.P., The dual boundary element method: Effective implementation for crack problems. *International Journal for Numerical Methods in Engineering*, **33**(6), pp. 1269–1287, 2010.
DOI: 10.1002/nme.1620330611.

[15] Prasad, N.N.V., Aliabadi, M.H. & Rooke, D.P., The dual boundary element method for thermoelastic crack problems. *International Journal of Fracture*, **33**(3), pp. 255–272, 1994. DOI: 10.1007/bf00042588.

[16] Ekhlakov, A., Khay, O., Zhang, C., Sladek, J., Sladek, V. & Gao, X.W., Thermoelastic crack analysis in functionally graded materials and structures by a BEM. *Fatigue and Fracture of Engineering Materials and Structures*, **35**(8), pp. 742–766, 2012.
DOI: 10.1111/j.1460-2695.2011.01657.x.

[17] Pasternak, I., Boundary integral equations and the boundary element method for fracture mechanics analysis in 2D anisotropic thermoelasticity. *Engineering Analysis with Boundary Elements*, **36**(12), pp. 1931–1941, 2012.
DOI: 10.1016/j.enganabound.2012.07.007.

[18] Zhao, M.H., Dang, H.Y., Li, Y., Fan, C.Y. & Xu, G.T., Displacement and temperature discontinuity boundary integral equation and boundary element method for analysis of cracks in three-dimensional isotropic thermoelastic media. *International Journal of Solids and Structures*, **81**, pp. 179–187, 2016. DOI: 10.1016/j.ijsolstr.2015.11.024.

[19] Zhong, X.C. & Kang, Y.L., A thermal-medium crack model. *Mechanics of Materials*, **51**, pp. 110–117, 2012. DOI: 10.1016/j.mechmat.2012.04.013.

[20] Zhong, X.C., Long, X.Y. & Zhang, L.H., An extended thermal-medium crack model. *Applied Mathematical Modelling*, **56**, pp. 202–216, 2018.
DOI: 10.1016/j.apm.2017.11.016.

[21] Gao, X.W., Boundary element analysis in thermoelasticity with and without internal cells. *International Journal for Numerical Methods in Engineering*, **57**(7), pp. 975–990, 2003. DOI: 10.1002/nme.715.

BEM ANALYSIS OF PLANE WAVE COUPLING TO THREE-PHASE POWER LINE

DRAGAN POLJAK & DAMIR CAVKA
University of Split, Croatia

ABSTRACT

The paper deals with BEM analysis of plane wave coupling to three-phase power line. The formulation is based on the set of Pocklington equations for arbitrarily shaped wires. The presence of a lossy ground is taken into account via rigorous Sommerfeld integral formulation. The set of Pocklington equations is solved by the Galerkin–Bubnov Indirect Boundary Element Method (GB-IBEM) using isoparametric elements. Some computational examples for the currents induced along phase conductors and shield wire of the 3-phase power line are presented. The numerical results obtained via GB-IBEM are compared to the results calculated via NEC (Numerical Electromagnetic Code). The results obtained via different approaches agree satisfactorily.
Keywords: BEM analysis, antenna theory, set of Pocklington equations, power lines, current distribution.

1 INTRODUCTION

The knowledge of an induced current distribution along considered curved wire is a prerequisite to understand the behaviour of wire configurations such as antenna arrays, power lines, communication cables, etc. The simplest approach is to assume the waveform of the current distribution along the given wire structure [1]. Furthermore, Transmission Line (TL) approach, which neglects radiation effects, is an efficient approximation in engineering practice for a wide range of problems [1]–[5]. However, some applications still require a more accurate approach related to the solution of a corresponding integral expression arising from scattering theory (full wave models) [6].

Thus, the use of full wave model represents the most rigorous approach to analyze electromagnetic field coupling to arbitrary wire configurations in the presence of a lossy ground.

The present paper deals with the assessment of electromagnetic interference (EMI) induced along three-phase power line illuminated by the plane wave incident electric field by using the via full wave model.

The frequency domain formulation is based on the corresponding set of Pocklington integral equations for arbitrarily shape overhead wires above a lossy ground, thus providing one to account for the conductor sag. The effects of a lossy half-space are taken into account via the rigorous Sommerfeld integral approach.

Numerical solution of corresponding integro-differential equations is carried out by means of the Galerkin–Bubnov Indirect Boundary Element Method (GB-IBEM) featuring linear and quadratic isoparametric elements.

The use of integral approach and isoparametric elements for an efficient treatment of curved wires is an important advantage of the full wave model, which is not readily the case with a TL approximation.

Some illustrative numerical results pertaining to some practical scenarios involving realistic power line configurations are given in the paper.

WIT Transactions on Engineering Sciences, Vol 122, © 2019 WIT Press
www.witpress.com, ISSN 1743-3533 (on-line)
doi:10.2495/BE410061

2 FORMULATION VIA SET OF INTEGRO-DIFFERENTIAL EQUATIONS

The currents $I_n(s')$ induced along the N_w curved wires located above a lossy ground (Fig. 1) due to a plane wave electric field E^{exc} (Fig. 2) are governed by the set of Pocklington integro-differential equations [6].

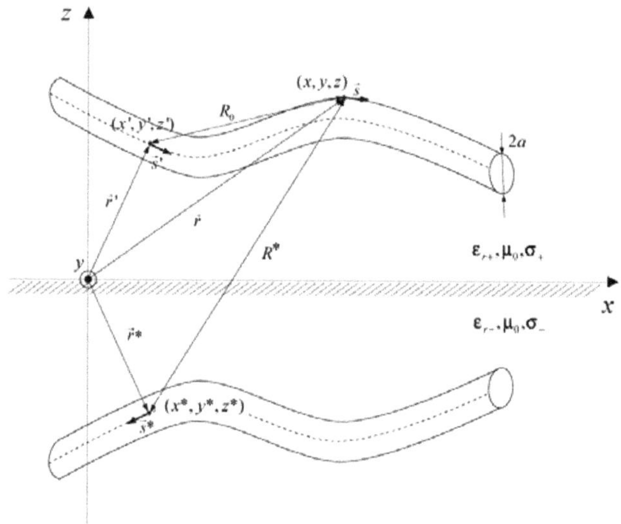

Figure 1: Buried wire of arbitrary shape and its image.

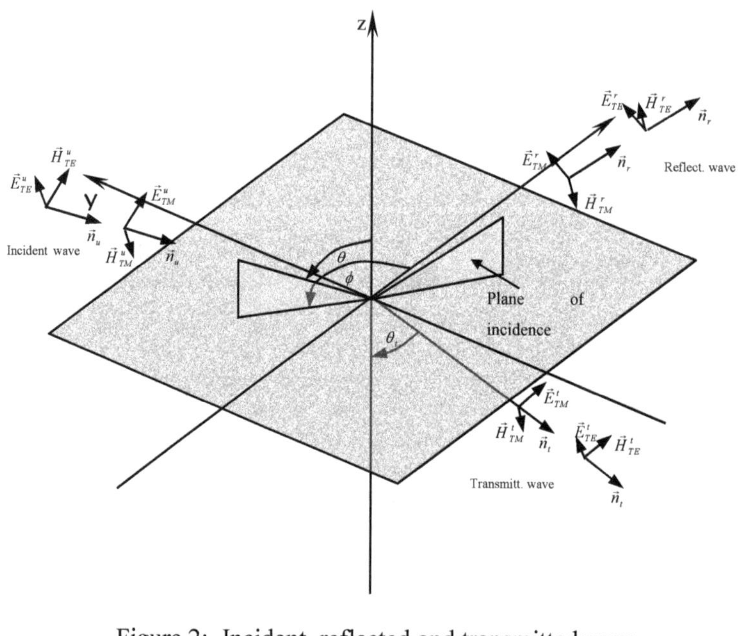

Figure 2: Incident, reflected and transmitted wave.

WIT Transactions on Engineering Sciences, Vol 122, © 2019 WIT Press
www.witpress.com, ISSN 1743-3533 (on-line)

The set of Pocklington equations are given by [6]

$$E_{sm}^{exc}(s) = -\frac{1}{j4\pi\omega\varepsilon_0}\sum_{n=1}^{N_W}\left[\begin{array}{l}\int\limits_{C_n'} I_n(s')\cdot\vec{s}\cdot\vec{s}'\cdot\left[k_0^2+\nabla\nabla\right]g_{0n}(s_m,s_n')ds'+ \\ \frac{k_g^2-k_0^2}{k_g^2+k_0^2}\int\limits_{C_n'} I_n(s_n')\cdot\vec{s}\cdot\vec{s}^*\cdot\left[k_0^2+\nabla\nabla\right]g_{in}(s_m,s_n^*)ds'+ \\ \int\limits_{C_n'} I_n(s')\cdot\vec{s}\cdot\vec{G}_s(s_m,s_n')ds'\end{array}\right] + Z_S\cdot I_m(s), \quad (1)$$

where the corresponding Green functions are given by:

$$g_{0mn}(s_m,s_n') = \frac{e^{-jk R_{1mn}}}{R_{1mn}}, \quad (2)$$

$$g_{imn}(s_m,s_n') = \frac{e^{-jk R_{2mn}}}{R_{2mn}}, \quad (3)$$

where R_{1mn} and R_{2mn} are distances from the source point (from real and image wire axis, respectively) to the observation point (at the real wire surface)

$$\vec{G}_s(s,s') = (\vec{e}_x\cdot\vec{s}')\cdot\left(G_\rho^H\cdot\vec{e}_\rho + G_\phi^H\cdot\vec{e}_\phi + G_z^H\cdot\vec{e}_z\right) + (\vec{e}_z\cdot\vec{s}')\cdot\left(G_\rho^V\cdot\vec{e}_\rho + G_z^V\cdot\vec{e}_z\right) \quad (4)$$

while the related wave numbers are:

$$k_0^2 = \omega^2\mu_0\varepsilon_0, \quad (5)$$

$$k_g^2 = \omega^2\mu_0\varepsilon_{eff} = \omega^2\mu_0\left(\varepsilon_0\varepsilon_{rg} - j\frac{\sigma_g}{\omega}\right), \quad (6)$$

where ε_{rg} and σ_g are relative permittivity and conductivity of the ground, respectively, and ω is the frequency of interest. The kernel terms are, as follows:

$$G_\rho^V = \frac{\partial^2}{\partial\rho\partial z}k_g^2 V^R, \quad (7)$$

$$G_z^V = \left(\frac{\partial^2}{\partial z^2} + k_0^2\right)k_g^2 V^R, \quad (8)$$

$$G_\rho^H = \cos\phi\left(\frac{\partial^2}{\partial\rho^2}k_0^2 V^R + k_0^2 U^R\right), \quad (9)$$

$$G_\phi^H = -\sin\phi\left(\frac{1}{\rho}\frac{\partial}{\partial\rho}k_0^2 V^R + k_0^2 U^R\right), \quad (10)$$

$$G_z^H = -j4\pi\omega\varepsilon_0\cos\phi G_\rho^V. \quad (11)$$

And the Sommerfeld integral terms are given by:

$$U^R = \int_0^\infty D_1(\lambda)\, e^{-\gamma_0|z+z'|}\, J_0(\lambda\rho)\lambda\, d\lambda, \tag{12}$$

$$V^R = \int_0^\infty D_2(\lambda)\, e^{-\gamma_0|z+z'|}\, J_0(\lambda\rho)\lambda\, d\lambda, \tag{13}$$

$$D_1(\lambda) = \frac{2}{\gamma_0 + \gamma_g} - \frac{2k_0^2}{\gamma_0\left(k_0^2 + k_g^2\right)}, \tag{14}$$

$$D_2(\lambda) = \frac{2}{k_g^2\gamma_0 + k_0^2\gamma_g} - \frac{2}{\gamma_0\left(k_0^2 + k_g^2\right)}, \tag{15}$$

$$\gamma_0 = \sqrt{\lambda^2 - k_0^2}; \qquad \gamma_g = \sqrt{\lambda^2 - k_g^2}, \tag{16}$$

where J_0 denotes the Bessel function. The set of Pocklington IEs is handled via the GB-IBEM with isoparametric elements.

3 NUMERICAL SOLUTION

The set of Pocklington IEs is handled via the GB-IBEM with isoparametric elements. The unknown current along the n-th wire segment is expressed by a sum of independent basis functions with unknown complex coefficients:

$$I_n^e(s') = \sum_{i=1}^{n} I_{ni} f_{ni}(s') = \{f\}_n^T \{I\}_n. \tag{17}$$

The use of isoparametric elements yields:

$$I_n^e(\zeta) = \sum_{i=1}^{n} I_{ni} f_{ni}(\zeta) = \{f\}_n^T \{I\}_n. \tag{18}$$

Having performed BEM discretization procedure the set of coupled Pocklington equations is transformed into the following matrix equation

$$\sum_{n=1}^{N_w}\sum_{\mathrm{i}=1}^{N_n} [Z]_{\mathrm{ji}}^{\mathrm{e}} \{I\}_{\mathrm{i}}^{\mathrm{e}} = \{\mathrm{V}\}_{\mathrm{j}}^{\mathrm{e}},\ \mathrm{m} = 1,2,\dots,N_w; \qquad \mathrm{j} = 1,2,\dots,N_m, \tag{19}$$

where the local matrix and right-side vector are given by:

$$\begin{aligned}
[Z]_{ij}^e = &-\int_{-1}^{1}\int_{-1}^{1} \{D\}_j \{D'\}_i^T\, g_{0nm}(s_n, s_m^{'}) \frac{ds_m^{'}}{d\xi'} d\xi' \frac{ds_n}{d\xi} d\xi \\
&+ k_0^2 \vec{s}_n \cdot \vec{s}_m^{\,'} \int_{-1}^{1}\int_{-1}^{1} \{f\}_j \{f'\}_i^T\, g_{0nm}(s_n, s_m^{'}) \frac{ds_m^{'}}{d\xi'} d\xi' \frac{ds_n}{d\xi} d\xi - \\
&- \frac{k_g^2 - k_0^2}{k_g^2 + k_0^2} \int_{-1}^{1}\int_{-1}^{1} \{D\}_j \{D'\}_i^T\, g_{inm}(s_n, s_m^{*}) \frac{ds_m^{'}}{d\xi'} d\xi' \frac{ds_n}{d\xi} d\xi \\
&+ \frac{k_g^2 - k_0^2}{k_g^2 + k_0^2} k_0^2 \vec{s}_n \cdot \vec{s}_m^{\,'} \int_{-1}^{1}\int_{-1}^{1} \{f\}_j \{f'\}_i^T\, g_{inm}(s_n, s_m^{*}) \frac{ds_m^{'}}{d\xi'} d\xi' \frac{ds_n}{d\xi} d\xi + \\
&+ \vec{s}_m^{\,'} \int_{-1}^{1}\int_{-1}^{1} \{f\}_j \{f'\}_i^T\, \vec{G}_{snm}(s_n, s_m^{*}) \frac{ds_m^{'}}{d\xi'} d\xi' \frac{ds_n}{d\xi} d\xi + \frac{j}{4\pi\omega\varepsilon_0} \int_{-1}^{1} Z_T^{'} \{f\}_j \{f'\}_j^T \frac{ds_n}{d\xi},
\end{aligned} \tag{20}$$

$$\{V\}_j^n = -j4\pi\omega\varepsilon_0 \int_{-1}^{1} E_{s_n}^{exc}(s_n)\{f\}_j \frac{ds_n}{d\xi} d\xi_n. \tag{21}$$

Note that the physical meaning of (20) is the mutual impedance, while (21) represents the voltage.

4 COMPUTATIONAL EXAMPLE

The computational example is related to a plane wave (f=5MHz) coupling to 3-phase power line, shown in Fig. 3, consisting of 4 wires (phase conductors+shield wire) above a lossy ground (ε_r=10, σ=0.001S/m). The separation between towers iz 300m. The radius of perfectly conducting (PEC) conductors is a=5mm.

More details about this particular power line configuration can be found elsewhere, e.g. in [7].

4.1 Case No 1: Straight conductors

The power line system consisting of straight conductors illuminated by the plane wave incidence (α=0°, θ=60°and φ=30°) with amplitude E_0=1V/m, and frequency f=5MHz is considered. Fig. 4(a)–(d) show the real and imaginary parts of the current distribution along the wires calculated via GB-IBEM and compared to the results computed via NEC software package [8].

An excellent agreement between the results obtained via GB-IBEM and NEC is considered.

4.2 Case No 2: Curved conductors

In present example the wire sag (23m for phase conductors, 13.5m for shield wire) is taken into account. The power line system is illuminated by the plane wave incidence (α=0°, θ=0°and φ=0) with amplitude E_0=1V/m, and frequency f=5MHz. Fig. 5 shows the curresnt distribution along the wires calculated via GB-IBEM (linear and quadratic elements) and compared to the results obtained via NEC software package.

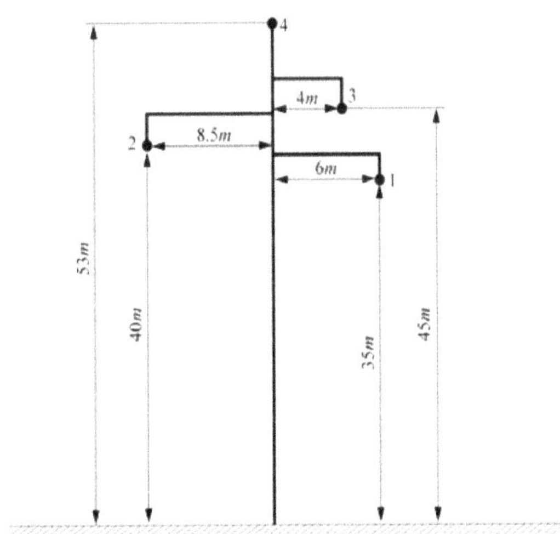

Figure 3: Three-phase power line system.

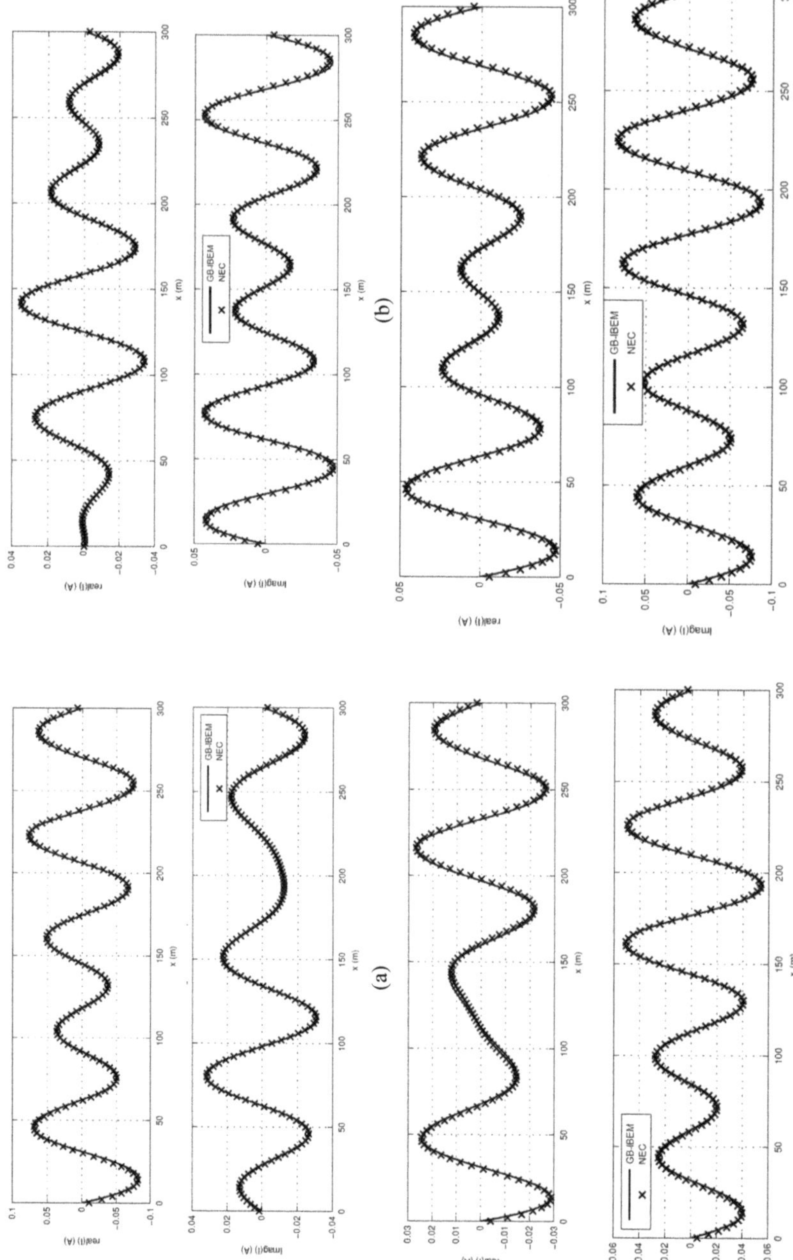

Figure 4: (a) Real and imag. part of the current distribution along the conductor No 1; (b) Real and imag. part of the current distribution along the conductor No 2; (c) Real and imag. part of the current distribution along the conductor No 3; (d) Real and imag. part of the current distribution along the shielded wire.

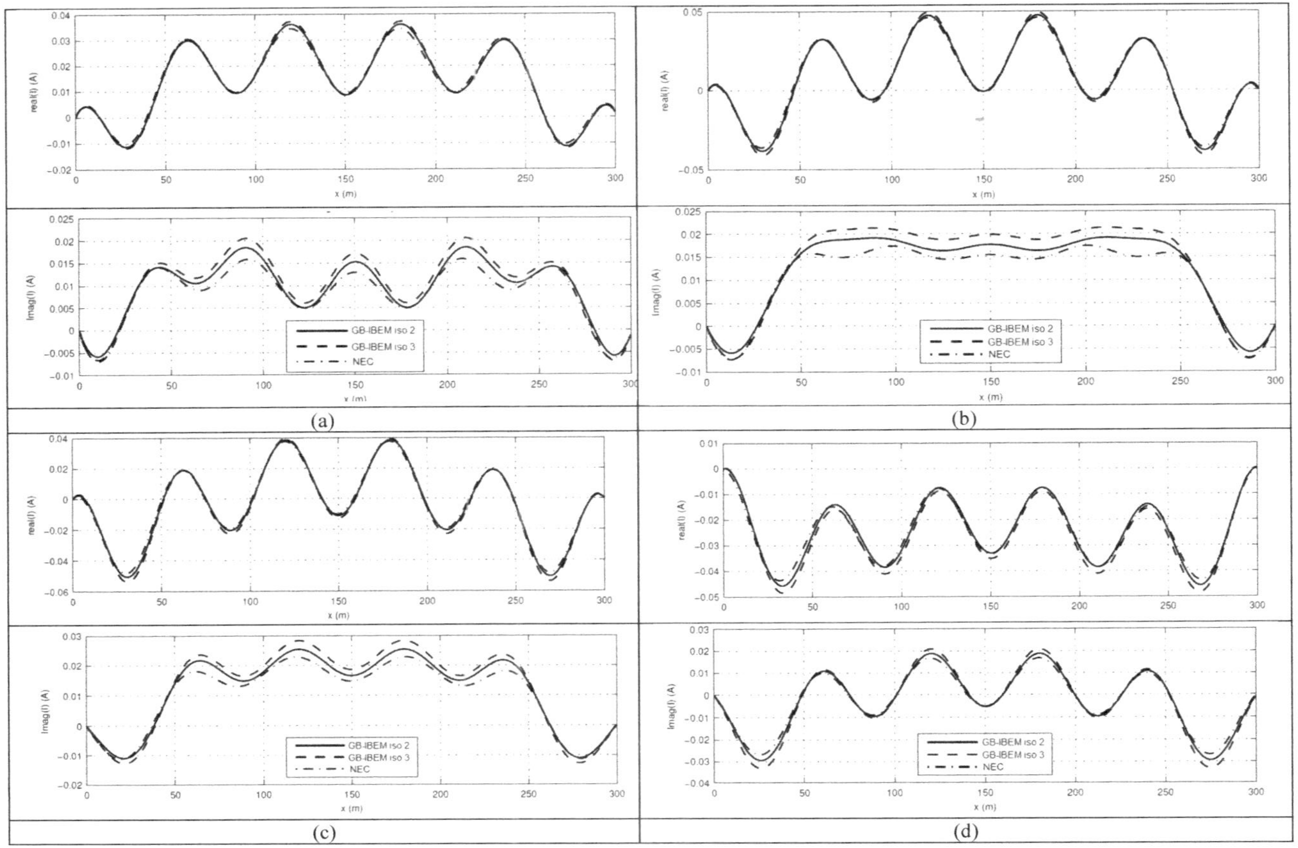

Figure 5: (a) Real and imag. part of the current distribution along the conductor No 1; (b) Real and imag. part of the current distribution along the conductor No 2; (c) Real and imag. part of the current distribution along the conductor No 3; (d) Real and imag. part of the current distribution along the shielded wire.

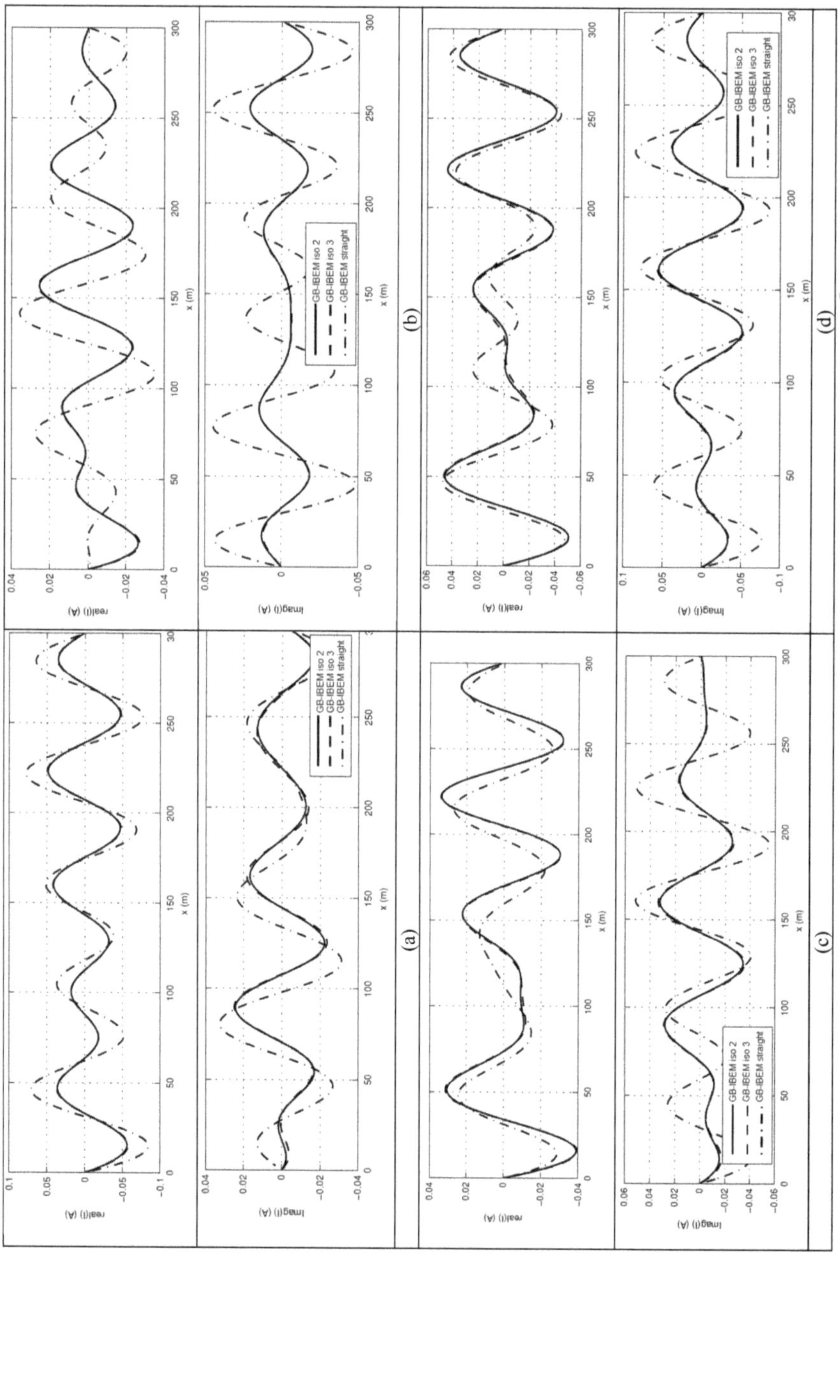

Figure 6: (a) Real and imag. part of the current distribution along the conductor No 1; (b) Real and imag. part of the current distribution along the conductor No 2; (c) Real and imag. part of the current distribution along the conductor No 3; (d) Real and imag. part of the current distribution along the shielded wire.

Some discrepancies are noticeable in imaginary part of current distribution along the conductors.

4.3 Case No 3: straight and curved conductors

Fig. 6 shows the comparison of currents induced on the power line system for the case of straight and curved conductors respectively. The excitation is unit amplitdue plane wave (α=0°, θ=60° and φ=30°) of frequency 5MHz. The current distribution along the straight conductors is calculated by using linear interpolation only, while the current distribution along the curved wires is calculated by using linear and quadratic approximation, respectively.

The influence of the conductor sag is clearly demonstrated in Fig. 6(a)–(d).

5 CONCLUDING REMARKS

BEM modeling of electromagnetic field coupling to 3-phase power line is carried out in this work. The formulation is based on the set of Pocklington integro-differential equations for curved wires. The presence of a lossy half-space is taken into account via rigorous Sommerfeld integrals.

The set of Pocklington integro-differential equations is handled via the frequency domain variant of the GB-IBEM featuring the use of isoparametric elements.

Some illustrative numerical results for the currents induced along phase conductors and shield wire of the 3-phase power line are given in the paper.

The numerical results obtained via GB-IBEM are compared to the results calculated via commercial software package NEC (Numerical Electromagnetic Code) for the case of straight conductors. The results obtained via different approaches agree satisfactorily. The influence of the conductor sag becomes clearly visible when the results obtained for straight and curved conductors are compared.

REFERENCES

[1] Tesche, F.M., Ianoz, M. & Karlsson, T., *EMC Analysis Methods and Computational Models*, Wiley Interscience: New York, 1997.

[2] Agrawal, A.K., Price, H.J. & Gurbaxani, S.H., Transient response of multiconductor transmission lines excited by a nonuniform electromagnetic field. *IEEE Transactions on Electromagnetic Compatibility*, **22**(2), pp. 119–129, 1980. DOI: 10.1109/temc.1980.303824.

[3] Poljak, D., *Advanced Modeling in Computational Electromagnetic Compatibility*, Wiley: New York, 2007.

[4] Poljak, D. & Khamlichi Drissi, K.El., Electromagnetic field coupling to overhead wire configurations: antenna model versus transmission line approach. *International Journal of Antennas and Propagation*, **2012**, pp. 1–18, 2012. DOI: 10.1155/2012/730145.

[5] Bridges G.E. & Shafai, L., Plane wave coupling to multiple conductor transmission lines above a lossy earth. *IEEE Transactions on Electromagnetic Compatibility*, **31**(1), pp. 21–33, 1989. DOI: 10.1109/15.19904.

[6] Poljak, D. & Cavka, D., Electromagnetic compatibility aspects of wind turbine analysis and design. *Properties and Characterization of Modern Materials,* eds. A. Ochsner & H. Altenbach, Springer, 2017.

[7] Kraljevic, S., Poljak, D. & Doric, V., A simplified method for the assessment of ELF magnetic fields from three-phase power lines, *Proc. Boundary Elements XXVII, Orlando, USA,* pp. 551–562, 2005.
[8] Burke, G.J., Poggio, A.J., Logan, I.C. & Rockway, J.W., Numerical electromagnetics code: A program for antenna system analysis. *Proc. EMC 1979 Symp.*, Rotterdam, The Netherlands, pp. 89–94.

MODELLING OF GROUNDING SYSTEM PLACED INTO VERTICALLY LAYERED SOIL

ADNAN MUJEZINOVIĆ, NEDIS DAUTBAŠIĆ & IRFAN TURKOVIĆ
Faculty of Electrical Engineering, University of Sarajevo, Bosnia and Herzegovina

ABSTRACT

The main purpose of the substations grounding systems is to ensure integrity of substations equipment and safety of personnel in and outside of substation at the maximum fault currents. To meet safety requirements, grounding system should have a low as possible resistance. In order to achieve low resistance, grounding systems are designed in a way to achieve as large as possible contact surface between the grounding system conductors and the surrounding soil. On the other hand, cost – efficiency of the proposed solution must be taken into account. Therefore, to meet the technical criteria on the one hand and economic criteria on the other hand, grounding systems are composed from large number of horizontal, vertical and inclined galvanic connected unisolated conductors that in most practical cases form complex geometries. Additionally, soil in which grounding systems are placed is almost always composed of a number of layers with different electric conductivity. In this paper, numerical model based on the indirect boundary element method is presented for calculation of grounding system parameters placed into the vertically layered soil.

Keywords: grounding system, vertically layered soil, Fredholm's integral equation of first kind, Green's functions, Touch voltage, Step voltage.

1 INTRODUCTION

Adequate design of substation grounding systems of any voltage level is of crucial importance for the safety of personnel and integrity of the equipment [1]. Safety conditions must be met both at normal operation of the system and during the flow of the highest fault currents. In order to meet the safety requirements, the grounding system with the surrounding soil must provide as low as possible resistance [2]. Low resistance grounding system ensures that the fault current is dissipated to the surrounding soil [3]. Also, geometry of the grounding system must be designed in such a way that during fault current dissipation into the soil, potential difference between the two close points (at a distance of 1 meter [4]) on the earh surface does not exceed the permissible values [5].

In order to achieve the lowest possible grounding resistance, grounding systems are designed in such a way to achieve as large as possible contact surface between the grounding system conductors and the surrounding soil. Therefore, grounding systems are formed from the horizontal, vertical and inclined galvanic connected unisolated conductors that in most cases form very complex geometries. Length and position of the grounding system conductors depends on the available space, substations gear disposition and surrounding soil resistivity [1].

Grounding system parameters that need to be calculated are grounding resistance, touch voltage, step voltage, mesh voltage and electric potential distribution on the soil surface [6]. In order to obtain accurate values of grounding system parameters, soil parameters need to be introduced in the calculation method. In this paper calculation method based on the indirect boundary element method for the grounding system placed into the vertically layered soil is presented.

WIT Transactions on Engineering Sciences, Vol 122, © 2019 WIT Press
www.witpress.com, ISSN 1743-3533 (on-line)
doi:10.2495/BE410071

2 MATHEMATICAL MODEL

Calculation methodology for obtaining the relevant grounding system parameters is based on the calculation of the electric potential distribution in the soil around analysed grounding system when fault current flows. Electric potential distribution in the soil can be calculated by using following set of equations [7]:

$$-\nabla(\gamma \cdot \nabla \varphi) = 0 \text{ in } \Omega, \tag{1}$$

$$\varphi = \varphi_\Gamma \text{ on } \Gamma, \tag{2}$$

$$\varphi \to 0 \text{ for } d \to \infty, \tag{3}$$

where φ is electric potential, γ is electric conductivity of soil, Ω is soil, Γ is surface of grounding conductor-surrounding soil interface, d is distance from grounding grid and symbol ∇ is Nabla operator.

Eqn (1) is a well-known Laplace partial differential equation for static current field, while other two eqns (2) and (3) are imposed boundary conditions. Previous given boundary conditions treat grounding system to soil interface and infinite boundary. For modelling of the surrounding soil additional boundary conditions must be applied. In most practical situation soil is inhomogeneous and can be treated as multi-layered [8]. It can be assumed that multi-layer soil is composed of two or more soil layers of different soil conductivity. At boundary between soil layers, following boundary condition can be applied [9]:

$$\vec{n} \cdot \left(\vec{J}_1 - \vec{J}_2\right) = 0, \tag{4}$$

where $\vec{n}$ is a normal unit vector, $\vec{J}_1$ and $\vec{J}_2$ is a vector of current density of soil layer 1 and 2, respectively.

Same boundary condition can be applied at the boundary soil – air, but since air can be treated as insulator, boundary condition on this interface takes a following form:

$$\vec{n} \cdot \vec{J}_1 = 0, \tag{5}$$

where $\vec{J}_1$ is vector of current density in soil layer *1* which borders with air.

3 INDIRECT BOUNDARY ELEMENT METHOD

By applying the well-known Green's identity over the eqn (1), Fredholm's integral equation of the first kind can be obtained, in the following form:

$$\varphi(q) = \frac{1}{\gamma} \iint_\Gamma J(p) \cdot G(p,q) d\Gamma, \tag{6}$$

where p is source point, q is observation point, J is normal component of the surface current density and $G(p,q)$ is Green's function (integral kernel).

To solve integral equation by indirect boundary element method double numerical integration must be applied. This can be reduced to one numerical integration by applying thin wire approximation on the grounding system conductors. Thin wire approximation is suitable for grounding system analysis because diameter to length ratio of grounding system conductors is small [7]. After application of the thin wire approximation Fredholm's integral eqn (6) takes the following form:

$$\varphi(q)=\frac{1}{\gamma}\int_{l}\sigma(p)\cdot G(p,q)dl\,, \tag{7}$$

where σ is unknown line current density and l is length of grounding system conductors.

3.1 Selection of appropriate Green's function

The choice of an adequate soil model is of crucial importance for the calculation of all relevant parameters of the grounding system. In most practical situation, soil is composed of more than one soil layer. Boundary between soil layers of different electric conductivity can be discretized by boundary elements. This further leads to the large number of boundary elements and consequently large calculation time. The discretization of the boundary between two soil layers can be avoided by proper selection of the Green's function.

When homogeneous soil model is adopted than boundary between soil and air must be taken into account [8]. Appropriate Green's function for homogeneous soil model is given by following relation:

$$G(p,q)=\frac{1}{4\pi}\cdot\left(\frac{1}{\|p-q\|}+\frac{1}{\|p_h-q\|}\right), \tag{8}$$

where p_h denotes horizontal image of the source point p.

For two vertical layer soil model, two situations must be taken into account. Therefore, two Green's function must be considered based on the position of the source point p and observation point q. For case when both source point p and observation point q are in same layer of soil, based on image theory, Green's function takes the following form:

$$G(p,q)=\frac{1}{4\pi}\cdot\left(\frac{1}{\|p-q\|}+\frac{\beta}{\|p_v-q\|}+\frac{1}{\|p_h-q\|}+\frac{\beta}{\|p_{hv}-q\|}\right), \tag{9}$$

where p_v denotes vertical image of source point p, p_{hv} denotes horizontal image of the p_v.

For case when source point p and observation point q are in different layers of vertically layered soil, Green's function takes the following form [6]:

$$G(p,q)=\frac{1}{4\pi}\cdot(1-\beta)\cdot\left(\frac{1}{\|p-q\|}+\frac{1}{\|p_h-q\|}\right), \tag{10}$$

where reflection coefficient β is defined by following equation:

$$\beta = \frac{\gamma_1 - \gamma_2}{\gamma_1 + \gamma_2}. \tag{11}$$

Graphical representation of the source point and its images in the vertical two layer soil model is given on the Fig. 1.

3.2 Solution technique

After selection of appropriate Green's function, Fredholm's integral eqn (7) can be solved by using boundary element technique. First step is to discretized grounding system conductors on the 1D boundary elements. In this step it is necessary to take care that 1D boundary elements are not too short, since thin wire approximation was used [10]. After discretization, Fredholm's integral equation needs to be solved for every boundary element.

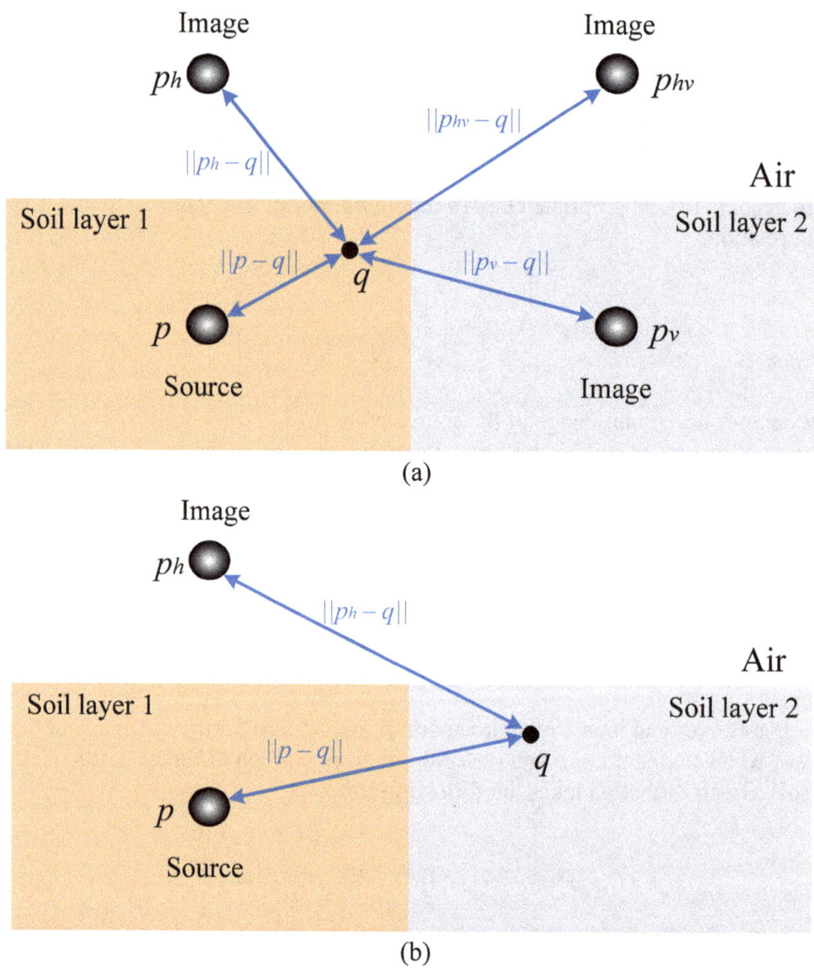

Figure 1: Graphical representation of source point and its images in the vertical two-layer soil model. (a) Source point and observation point in same soil layers; (b) Source point and observation point in different soil layers.

After discretization of the grounding system conductor its geometry and unknown line current density is approximated with shape functions. In this paper second order polynomial function was used:

$$f^e(\xi) = \sum_{i=1}^{n_n} f_i^e \cdot \psi_i(\xi), \tag{12}$$

where $f_e(\xi)$ is value of function (geometry of line current density) at any point of boundary element, $\psi_i(\xi)$ is the shape function, f^e_i is value of function at i-th point of boundary element and n_n is number of collocation points on the boundary element.

In order to solve integral equations, it is appropriate to introduce coordinate transformation. After application of the coordinate transformation, Fredholm's integral eqn (7) takes the following form [11]:

$$\varphi(q) = \frac{1}{\gamma} \sum_{e=1}^{n_e} \int_{-1}^{1} \sum_{i=1}^{n_n} \left(\sigma_i^e \cdot \psi_i(\xi)\right) G^e(\xi, q) \det J(\xi) d\xi \ , \tag{13}$$

where: n_e is number of boundary elements, σ_i^e is line current density of the e-th boundary element and det $J(\xi)$ is Jacobian determinant.

After application of the Gauss-Legendre's numerical integration on the eqn (13), following matrix equation is obtained [12]:

$$[G] \cdot \{\sigma\} = \{\varphi\}, \tag{14}$$

where $[G]$ is coefficient matrix, $\{\sigma\}$ is matrix vector of unknown line current densities and $\{\varphi\}$ is matrix of unknown electric potentials.

Since in matrix eqn (14) two matrix vectors are unknown, equation cannot be solved in this form. In ordered to solve calculate line current densities and electric potential of grounding system, additional equation must be introduced. The fault current that flows into the grounding system leaks into the surrounding soil. Therefore, mathematical link between the fault current and line current density that leaks from the grounding system to the sounding soil can be given by following equation [1], [11], [12]:

$$I_F = \sum_{e=1}^{n_e} \int_{l^e} \sigma^e dl^e = \sum_{e=1}^{n_e} \sum_{i=1}^{n_n} \sigma_i^e \cdot l_i^e \ , \tag{15}$$

where: l_i^e is calculated length in the i-th collocation point of e-th boundary element and I_F is the fault current which flows into the grounding system

Substituting eqn (15) in eqn (14), following matrix equation can be written [1], [11]–[13]:

$$\left[\begin{array}{cccc|c} G_{11} & G_{12} & \cdots & G_{1n_{cp}} & -1 \\ G_{21} & G_{22} & \cdots & G_{2n_{cp}} & -1 \\ \vdots & \vdots & \ddots & \vdots & \vdots \\ G_{n_{cp}1} & G_{n_{cp}2} & \cdots & G_{n_{cp}n_{cp}} & -1 \\ \hline l_1 & l_2 & \cdots & l_{n_{cp}} & 0 \end{array}\right] \cdot \left\{\begin{array}{c} \sigma_1 \\ \sigma_2 \\ \vdots \\ \sigma_{n_{cp}} \\ \hline \varphi_G \end{array}\right\} = \left\{\begin{array}{c} 0 \\ 0 \\ 0 \\ 0 \\ \hline I_F \end{array}\right\}, \tag{16}$$

where: φ_G is electric potential of the grounding system.

3.3 Calculation of touch and step voltage

When designing grounding systems, the two most important parameters needs to be calculated are the touch voltage and the step voltage. Based on these two parameters, it can be assessed whether the safety conditions are met within and around the substation.

To calculate touch and step voltage, electric potential on the earth surface must be calculated. Electric potential on the earth surface can be calculated by using Fredholm's integral eqn (7) with observation points q are placed on the earth surface. Since in this stage, current densities of all segments are known, electric potential of all points on the earth surface can be calculated by using following matrix eqn [11]:

$$[G'] \cdot \{\sigma\} = \{\varphi_e\}, \tag{17}$$

where $[G']$ is coefficient matrix for case when observation points q are placed on the earth surface, $\{\sigma\}$ is matrix vector of line current densities and $\{\varphi_e\}$ is matrix of unknown electric potentials on earth surface.

When electric potential of the grounding system and on the earth surface is known, touch and step voltages can be calculated. For calculation of the touch and step voltages square grid with dimensions of 1 m on the earth surface is used. Observation points q are placed on the edges of every square. Then the touched and the step voltage can be calculated by using following equations [11], [14]:

$$U_t = \max\left[\left|\varphi_G - \varphi_e(x_0 \pm 1(m), y_0)\right|, \left|\varphi_G - \varphi_e(x_0, y_0 \pm 1(m))\right|\right], \tag{18}$$

$$U_s = \max\left[\left|\varphi_e(x_0, y_0) - \varphi_e(x_0 \pm 1(m), y_0)\right|, \left|\varphi_e(x_0, y_0) - \varphi_e(x_0, y_0 \pm 1(m))\right|\right], \tag{19}$$

where U_t is calculated touch voltage, U_s is calculated step voltage, x_0 and y_0 are coordinates of the point on the earth surface wherein touch and step voltage is determined.

In order to determine whether the grounding system meets the safety requirements, the calculated values of the touch and the steps voltage at each point on the earth surface needs to be compared with the allowed values of the touch and step voltage. Allowed values of the touch and step voltage are defined by IEEE Std. 80-2013 [4].

4 CASE STUDY

Previously presented mathematical model was used for calculation of the parameters of L-shape grounding system with vertical grounding electrodes, taken from literature [4]. Analysed grounding system is placed in one layer of two vertically layered soil. Geometry of analysed grounding system is given on the Fig. 1.

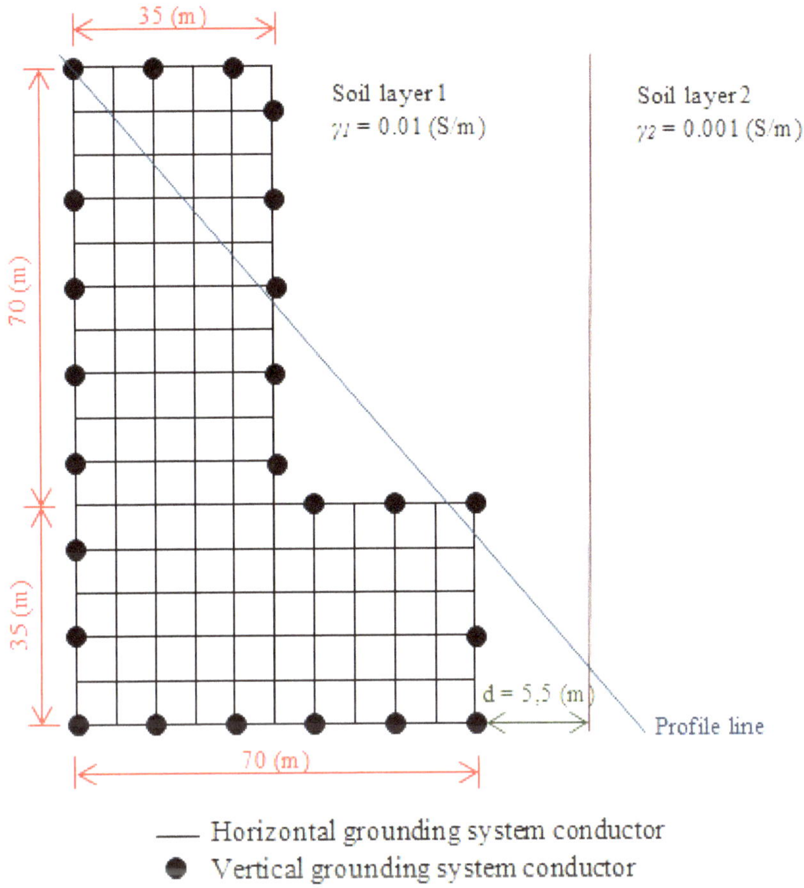

Figure 2: Geometry of analysed grounding system.

Other parameters relevant for the calculation are given in the Table 1.

Table 1: Input parameters for the calculation.

Parameter	Value
Fault current I_F:	1908 (A)
First layer soil conductivity γ_1:	0.01 (S/m)
Second layer soil conductivity γ_2:	0.001 (S/m)
The depth of the grounding system h:	1 (m)
Length of vertical electrodes l_e:	7,5 (m)
Duration of the fault current t:	0.35 (s)
Conductivity of the crush rock on the surface γ_c:	3.3 10^{-4} (S/m)
Thickness of the crush rock on the surface h_c:	0.1 (m)
Human body mass m:	70 kg

Results of the electric potential distribution on the earth surface, distribution of the touch and the step voltage are given on the Figs 3, 4 and 5, respectively.

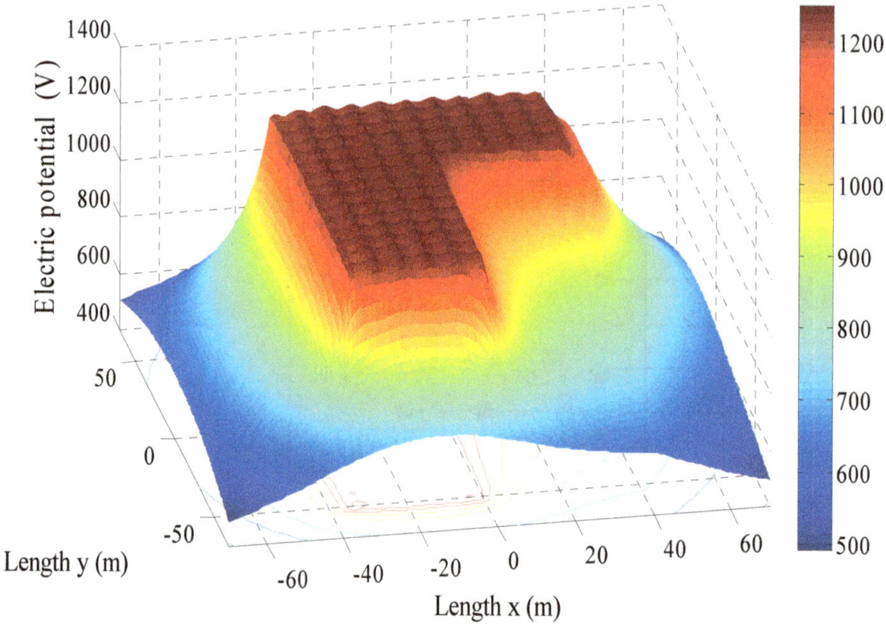

Figure 3: Electric potential distribution on the earth surface.

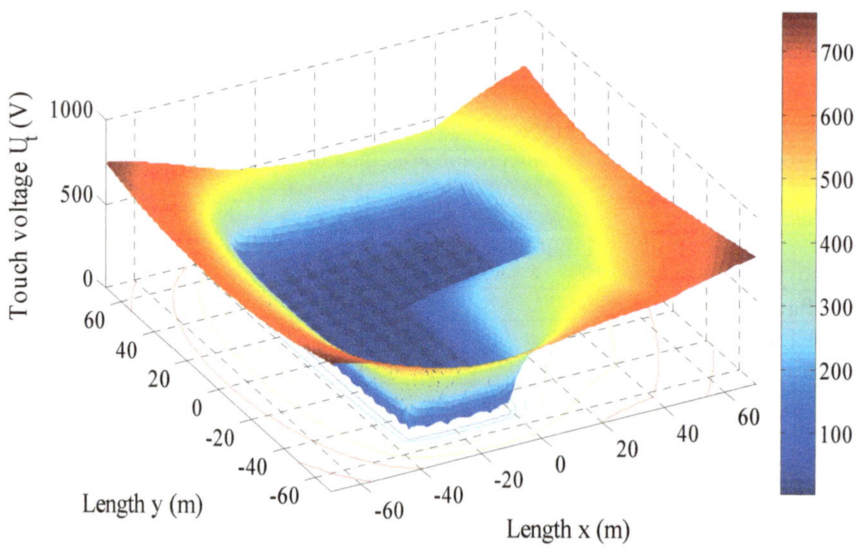

Figure 4: Distribution of the touch voltage.

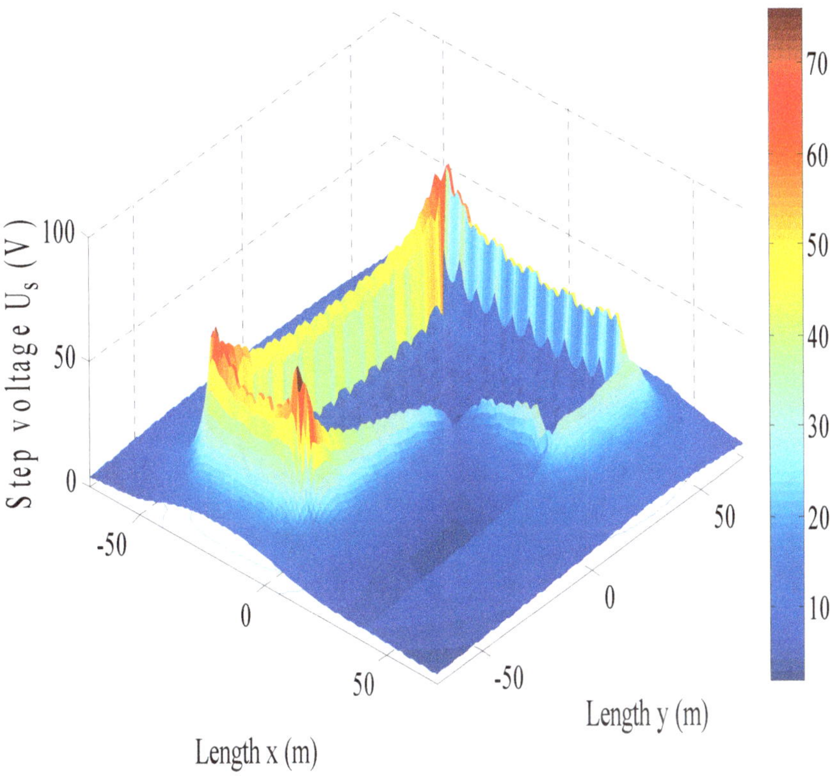

Figure 5: Distribution of the step voltage.

As it can be noted from results given on the Fig. 3 highest values of the electric potential on the earth surface is directly above the grounding system. Maximum value of the electric potential is 1250.3 V. Highest electric potential gradient is on the edges and outside of the grounding system, while lowest electric potential gradient is directly above grounding system. Also, it can be noted that electric potential gradient is slightly higher in the soil layer with lower electric conductivity.

From Fig. 4 it can be noted that values of the touch voltages are lowest on area that is directly above the grounding system and that it increases with increasing distance from the grounding system.

Highest values of the step voltage are on the earth surface directly above the edges of the grounding system and on the boundary between two soil layers. It can be noted that value of the step voltage on the boundary between two adjacent soil layers is increased on areas that are closer to the grounding system.

On the Figs 6 and 7 changes of the touch and step voltages over the profile line outlined on the Fig. 2 are given. Also maximal allowable values of the touch and step voltages for the body weight of 70 kg and fault clearing time of 0.35 sec.

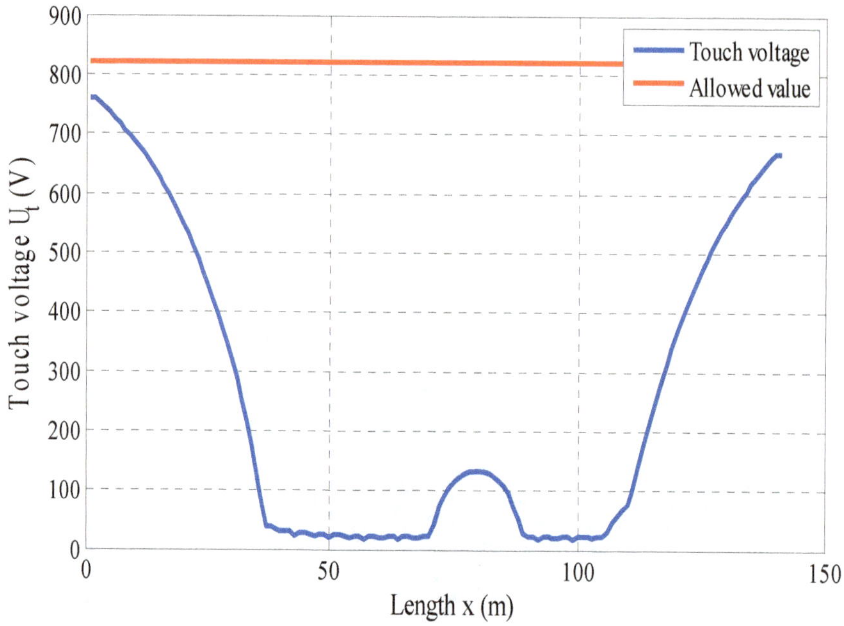

Figure 6: Change of the touch voltage over the profile line.

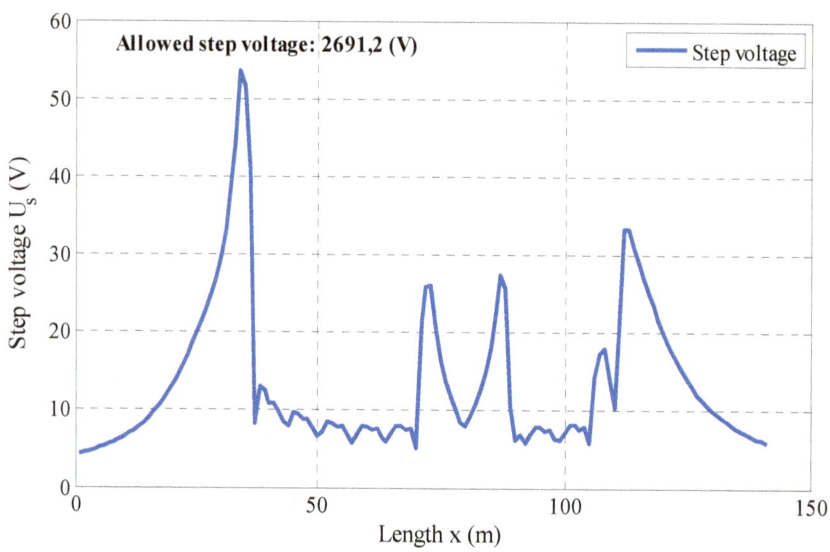

Figure 7: Change of the step voltage over the profile line.

From the results given on the Figs 6 and 7 it is noticeable that touch and step voltages are below maximum allowed limits.

5 CONCLUSION

Grounding system has a key role in ensuring safety of personnel and equipment in the substation and therefore calculation of its parameters requires application of very accurate method. One of the main parameter that can affect computation results is characteristic of the soil in which grounding system is placed. This paper presents mathematical model based on the indirect boundary element method for the calculation of parameters of the grounding system placed into vertically layered soil. In presented mathematical model impact of the soil structure was taken into account with applying adequate Green's functions.

Presented mathematical model was used for the calculation of the illustrative grounding system placed in vertically layered soil.

ACKNOWLEDGEMENT

This research has been supported by the Ministry of Education and Science of the Federation of Bosnia and Herzegovina.

REFERENCES

[1] Mujezinović, A., Software tool for grounding system design. eds. M. Hadžikadić & S. Avdaković, *Advanced Technologies, Systems, and Applications. Lecture Notes in Networks and Systems*, Springer: Cham, **3**, 2017.

[2] Mujezinović, A., Analysis of discontinuity of electrical conductivity of soil on grounding resistance, MSc. thesis, Faculty of Electrical Engineering, University of Sarajevo: Bosnian, 2011.

[3] Colominas, I., Navarrina, F. & Casteleiro, M., Analysis of transferred earth potentials in grounding systems: A BEM numerical approach. *IEEE Transactions on Power Delivery*, **20**(1), pp. 339–345, 2005. DOI: 10.1109/tpwrd.2004.835035.

[4] IEEE Std. 80—2013, IEEE Guide for Safety in AC Substation Grounding, Dec. 2013.

[5] Colominas, I., Navarrina, F. & Casteleiro, M., Numerical simulation of transferred potentials in earthing grids considering layered soil models. *IEEE Transaction on Power Delivery*, **22**(3), pp. 1514–1522, 2007. DOI: 10.1109/tpwrd.2007.899282.

[6] He, J., Zeng, R. & Zhang, B., *Methodology and Technology for Power System Grounding*, John Wiley & Sons: Singapore, 2013.

[7] Colominas, I., Navarrina, F. & Casteleiro, M., A numerical formulation for grounding analysis in stratified soils. *IEEE Transactions on Power Delivery*, **17**(2), pp. 587–595, 2002. DOI: 10.1109/61.997943.

[8] Barić, T., Boras, V. & Nikolovski, S., Analysing the effect of discontinuity of electrical conductivity of soil on ground resistance. *Journal of Energy*, **55**(4), pp. 454–473, 2006.

[9] Mujezinović, A., Muharemović, A., Muharemović, A., Turković, I. & Bajramović, Z., Application of finite element method in calculation of large and complex grounding systems. *International Conference and Exposition on Electrical and Power Engineering (EPE)*, pp. 688–692, Romania, Iasi, Oct. 2012.

[10] Navarrina, F., Colominas, I. & Casteleiro, M., Why do computer methods for grounding analysis produce anomalous results? *IEEE Transaction on Power Delivery*, **18**(4), pp. 1192–1202, 2003. DOI: 10.1109/tpwrd.2003.817499.

[11] Mujezinović, A., Mulaosmanović, A., Muharemović, A., Turković, I. & Bajramović, Z., A combined boundary element and an analytical approach to grounding mesh modeling in a multi-layer soil. *Elektrotehniški Vestnik* **84**(1–2), pp. 39–47, 2017.

[12] Zildzo, H., Muharemović, A., Turković, I. & Matoruga, H., Numerical Calculation of Floating Potentials for Large Earthing Systems. *XXII International Symposium on Information,* Communication and Automation Technologies: Sarajevo, Oct. 2009.

[13] Freschi, F., Mitolo, M. & Tartaglia, M., Interferences phenomena between separate grounding systems, industry applications society annual meeting. *IEEE*, 2013.
[14] Poljak, D., *The Theory of Electromagnetic Fields with Applications in Engineering*, Školska knjiga: Zagreb, 2014.

BOUNDARY INTEGRAL METHODS IN BIOELECTROMAGNETICS AND BIOMEDICAL APPLICATIONS OF ELECTROMAGNETIC FIELDS

DRAGAN POLJAK[1], MARIO CVETKOVIĆ[1], DAMIR CAVKA[1], ANDRES PERATTA[2], CRISTINA PERATTA[2], HRVOJE DODIG[3] & AKIMASA HIRATA[4]
[1]University of Split, Split, Croatia
[2]Wessex Institute of Technology, Southampton, UK
[3]Marine Center for Electronics, Split, Croatia
[4]Nagoya Institute of Tech., Japan

ABSTRACT

The paper reviews some integral methods used in static, low frequency (LF), and high frequency (HF) dosimetry, respectively, and biomedical applications of electromagnetic fields. Laplace equation/Boundary element method (BEM) is applied to certain electrostatic and LF exposure scenarios. Surface Integral Equation (SIE)/Method of Moments (MoM) and tensor volume integral equation/MoM and Stratton Chu integral formulation coupled with vector Helmholtz equations/hybrid Boundary element method (BEM)/Finite element method (FEM) scheme are used to analyze the exposure to HF radiation. In addition to human exposure to undesired electromagnetic radiation sources, this work also deals with biomedical applications of electromagnetic fields, by using SIE/MoM approach. Some illustrative numerical results for induced current densities, electric field and SAR, are also given in the paper.
Keywords: human exposure to electromagnetic fields, integral formulation, numerical methods, Static, low and high frequency dosimetry, biomedical applications of electromagnetic fields.

1 INTRODUCTION

Human exposure to non-ionising radiation has caused public concern regarding possible adverse health effects [1]. The assessment of low frequency (LF) exposure involves the calculation of internal current density [2] and internal electric field [3], while high frequency (HF) exposure requires the calculation of specific absorption rate (SAR) and resultant temperature increase in a tissue. Measurement of internal fields is generally not possible and human exposure assessment is undertaken by means of sophisticated anatomically based computational models [4]–[7] mostly related to the use of the Finite Difference Time Domain (FDTD) methods. The Finite Element Method (FEM), Boundary Element Method (BEM), Method of Moments (MoM) and some other methods are used to a somewhat lesser extent.

Integral equation approaches, using the Green integral representation, are based on the fundamental solution of the leading operator for the governing equation thus being competitive with other well-established methods, such as Finite Difference Methods (FDM), FDTD (suffering from so-called staircasing approximation) or FEM in terms of accuracy and efficiency [5]. Effects of numerical artifacts in low frequency (LF) dosimetry are studied by working group 2 (WG 2) of IEEE/ICES TC95 SC6 EMF Dosimetry Modeling [8].

Applying integral approaches, such as BEM, one avoids; the staircasing error (by use of isoparametric elements), implementation of absorbing boundary conditions and volume meshes for large-scale problems.

WIT Transactions on Engineering Sciences, Vol 122, © 2019 WIT Press
www.witpress.com, ISSN 1743-3533 (on-line)
doi:10.2495/BE410081

Some disadvantages of integral methods, such as BEM are; more complex fomulation (particularly for non-homogeneous domains) and related numerical implementation (dense matrices, Green function singularities/quasisingularities).

This paper is an extension of recent review papers [5] and [9] and within the framework of WG2 SC6 TC95 tasks gives an overview to the use of some integral methods in electromagnetic dosimetry and biomedical applications of electromagnetic fields. Thus, Laplace equation formulation /BEM scheme for the exposure of human head to electrostatic fields and whole body to LF electric fields is presented. Furthermore, surface integral equation/Moment Method (MoM) approach for the brain exposure is described. This is followed by the tensor volume integral equation/MoM approach applied to the exposure of the whole body to HF radiation. The eye exposure to plane wave is analyzed by using the hybrid BEM/ FEM approach.

2 INTEGRAL METHODS IN LF DOSIMETRY

The quasistatic formulation for LF exposures used in this work is based on the Laplace (quasi-static) version of the continuity equation for scalar potential φ giveny by [5]:

$$\nabla \cdot \left[(\sigma + j\omega\varepsilon) \nabla \varphi \right] = 0 , \tag{1}$$

where ω is the frequency of the incident field, σ and ε_r are the conductivity and permittivity of the material.

For the case of human exposure to electrostatic fields (1) simplifies into standard Laplace equation:

$$\nabla^2 \varphi = 0 . \tag{2}$$

Knowing the scalar potential along the body, the induced current density is computed from the differential form of Ohm's Law:

$$\vec{J} = (\sigma + j\omega\varepsilon) \nabla \phi . \tag{3}$$

The electric field E can be readily obtained as follows:

$$\vec{E} = -\nabla \phi . \tag{4}$$

First computational example is related to the human head exposed to video display unit (VDU) at φs=15kV at the distance ls=0.4m.

The boundary integral representation of (2) is given, as follows [5]:

$$c_i \phi_i = \int_\Gamma \Psi \frac{\partial \phi}{\partial n} d\Gamma - \int_\Gamma \phi \frac{\partial \Psi}{\partial n} d\Gamma , \tag{5}$$

where ψ is the fundamental solution Laplace equation:

$$\psi = (4\pi R)^{-1} . \tag{6}$$

with R being the distance between the source point and observation point in Γ, and:

$$c_i = \begin{cases} 1, & i \in \Omega \\ \frac{1}{2}, & i \in \Gamma \ (smooth\, boundary) \end{cases} . \tag{7}$$

Fig. 1 shows the electrostatic field induced at the surface of the female face calculated via BEM and FEM, respectively. The details of the head model can be found in [5] and [10]. Comparison of the BEM and FEM results for specific points on the head are shown in Fig. 2.

Comparison of the BEM and FEM results for specific points on the head are shown in Fig. 2.

The details of the FEM solution can be found elsewhere, e.g. in [10]. Differences on the certain points, (nose tip in particular) between the BEM and the FEM solution arise due to difficulties to accomplish sufficiently refined mesh around complex geometries, such as the human face.

Next example is the pregnant woman/foetus exposure to power line electric field E=10 kV/m. The amniotic fluid (AF) has the highest conductivity which varies depending on the period of gestation. Kidney, muscle bone cortical, bladder, spleen, skin have conductivity very close to 0.1 S/m, while the ovary and cartilage conductivity is around 0.2 S/m. More details on electrical properties of the pregnant woman tissue can be found elsewhere, e.g. in [11].

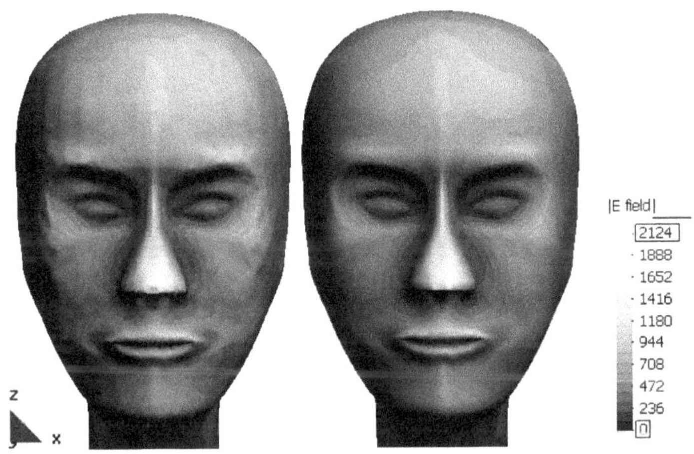

Figure 1: Electrostatic field on the female face. (a) FEM solution; and (b) BEM solution.

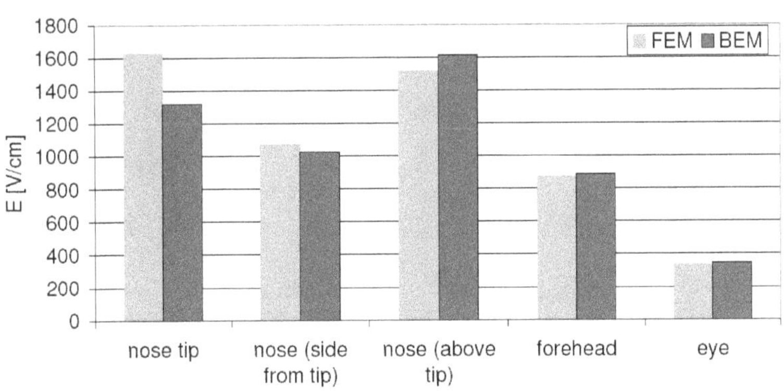

Figure 2: Electrostatic field at the specific locations of the female face.

Fig. 3 shows the sliced model of the pregnant woman at 26th gestational week with the electric field/scalar potential lines for the foetus in cephalic and breach presentation, respectively.

It is visible that the uterus, due to the higher conductivity of the amniotic fluid compared to the maternal tissue, tends to concentrate the field lines.

The maximal value of current density in the foetus occurs during the 8th week, while the maximum current density induced in the foetus is 7.4 mA/m^2 for an external field exposure of 10kV/m. Since restriction recommended for public exposure by ICNRP [5] is 2 mA/m2. Then through the modelling it is possible to set restrictions on external field exposure, which translate to a maximum external field of 2.7 kV/m in order to limit the current density exposure in the earlier months of development.

3 INTEGRAL METHODS IN HF DOSIMETRY AND BIOMEDICAL APPLICATIONS

The main task of HF dosimetry is to quantify thermal effects, by determining the level and distribution of the electromagnetic energy absorbed by the body which is expressed in terms of the specific absorption rate (*SAR*). Standard definition of *SAR* is given by the rate of energy *W* absorbed by the unit body mass *m*:

$$SAR = \frac{dP}{dm} = \frac{\sigma}{2\rho}|E|^2, \tag{8}$$

where P is the dissipated power, E is the peak value of the electric field, respectively, ρ is the tissue density and σ is the tissue conductivity.

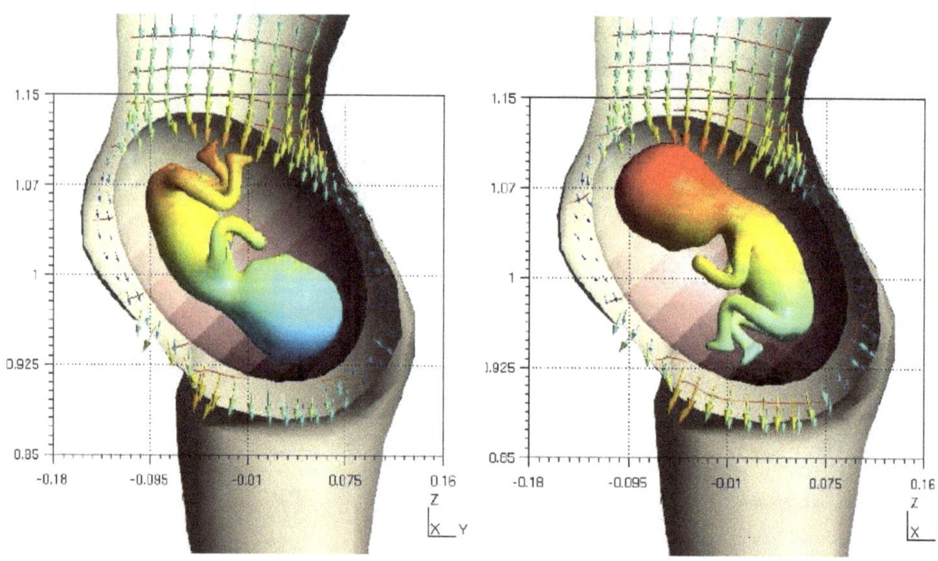

Figure 3: Lateral view of the pregnant woman at 26th gestational week. (a) Foetus in the cephalic presentation; and (b) Foetus in the breach presentation.

3.1 Set of coupled surface integral equations (SIE) – solution via Method of Moments (MoM)

The brain exposed to HF radiation is analysed by solving the coupled surface integral equations (SIE) [12]:

$$
\begin{aligned}
& j\omega\mu_n \iint_S \vec{J}(\vec{r}\,')G_n(\vec{r},\vec{r}\,')\,\mathrm{d}S' - \\
& - \frac{j}{\omega\varepsilon_n}\iint_S \nabla'_S \cdot \vec{J}(\vec{r}\,')\nabla G_n(\vec{r},\vec{r}\,')\,\mathrm{d}S' + \\
& + \iint_S \vec{M}(\vec{r}\,') \times \nabla' G_n(\vec{r},\vec{r}\,')\,\mathrm{d}S' = \begin{cases} \vec{E}^{inc}, & n = 1 \\ 0, & n = 2, \end{cases}
\end{aligned}
\tag{9}
$$

where $\vec{J}$ and $\vec{M}$ is quivalent electric and magnetic current density, respectively, k_n is the wave number of a medium n, while G_n is the interior/exterior Green function for the homogeneous medium [12]:

$$
G_n(\vec{r},\vec{r}\,') = \frac{e^{-jk_n R}}{4\pi R}; \quad R = |\vec{r} - \vec{r}\,'|, \tag{10}
$$

and R is the distance from the source to observation point, respectively while kn denotes the wave number of a medium n, (n = 1; 2).

Set of integral eqn (9) is solved via the Method of Moments (MoM) procedure presented in [12].

The equivalent electric and magnetic current densities $\vec{J}$ and $\vec{M}$ in (9) are expressed in terms of a linear combination of basis functions $\vec{f}_n$ and $\vec{g}_n$:

$$
\vec{J}(\vec{r}) = \sum_{n=1}^{N} J_n \vec{f}_n(\vec{r}), \tag{11}
$$

$$
\vec{M}(\vec{r}) = \sum_{n=1}^{N} M_n \vec{g}_n(\vec{r}), \tag{12}
$$

where J_n and M_n are unknown coefficients, while N is the total number of triangular elements.

Utilizing the weighted residual approach, i.e. multiplying (9) by the set of a weight functions $\vec{f}_m$ and integrating over the surface S, after certain mathematical manipulations, it follows:

$$
\begin{aligned}
& j\omega\mu_i \sum_{n=1}^{N} J_n \iint_S \vec{f}_m(\vec{r}) \cdot \iint_{S'} \vec{f}_n(\vec{r}\,')G_i \, dS' \, dS + \\
& + \frac{j}{\omega\varepsilon_i} \sum_{n=1}^{N} J_n \iint_S \nabla_S \cdot \vec{f}_m(\vec{r}) \iint_{S'} \nabla'_S \cdot \vec{f}_n(\vec{r}\,')G_i \, dS' \, dS + \\
& \pm \sum_{n=1}^{N} M_n \iint_S \vec{f}_m(\vec{r}) \cdot [\hat{n} \times \vec{g}_n(\vec{r}\,')] \, dS + \\
& + \sum_{n=1}^{N} M_n \iint_S \vec{f}_m(\vec{r}) \cdot \iint_{S'} \vec{g}_n(\vec{r}\,') \times \nabla' G_i \, dS' \, dS = \\
& \qquad = \begin{cases} \iint_S \vec{f}_m(\vec{r}) \cdot \vec{E}^{inc} \, dS & , i = 1 \\ 0 & , i = 2, \end{cases}
\end{aligned}
\tag{13}
$$

where subscript i is the index of the medium. The details of the procedure could be found elsewhere, e.g. in [12].

Fig. 4 shows the SAR distribution in the brain at f=900MHz due to the vertically polarized incident plane wave with power density P = 5mW/cm^2. The brain electrical parameters are: ε_r=46, σ=0.8S/m [12].

The obtained numerical results for peak and average SAR values are: SAR_{max}=0.866W/kg and SAR_{avg}=0.158W/kg.

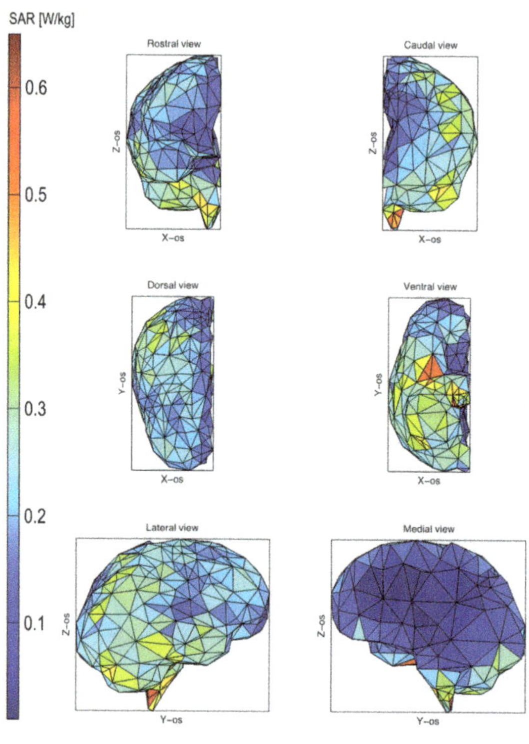

Figure 4: SAR distribution at f=900MHz.

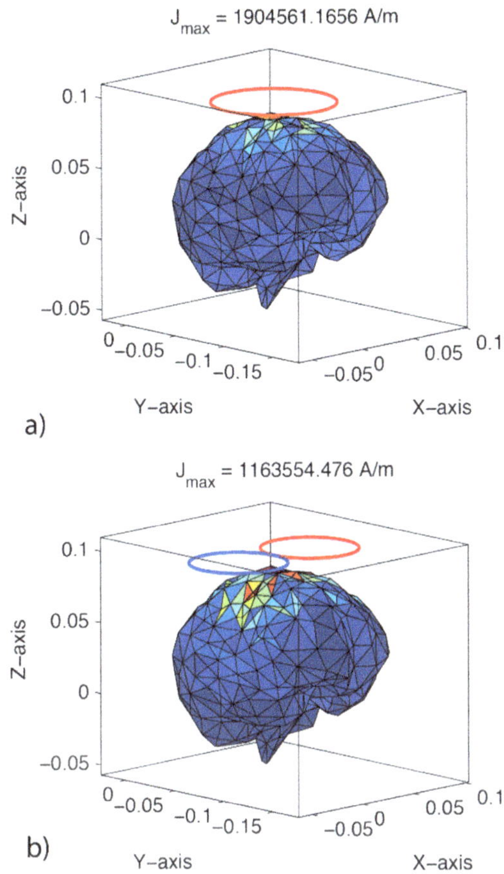

Figure 5: Equivalent electric current density. (a) Circular coil; and =(b) No 8 coil.

Additional example deals with the transcranial magnetic stimulation (TMS) treatment for the case of exposure to circular coil and No 8 coil. The numerical results for the electric current density induced in the brain are presented in Fig. 5.

More details can be found elsewhere, e.g. in [13].

3.2 Tensor volume integral equation/MoM approach

The whole-body exposure to HF radiation can be analysed via tensor integral eqn (14). The internal field due to plane wave incident field E^i is governed by the tensor volume integral eqn (14):

$$\left[1+\frac{\tau(\vec{r})}{3j\varpi\varepsilon_0}\right]\vec{E}(\vec{r}) - PV\int_v \tau(\vec{r})\vec{E}(\vec{r})\cdot\underline{G}(\vec{r},\vec{r}')dv' = \vec{E}^i(\vec{r}), \tag{14}$$

where PV is the principal value and $\tau(\vec{r}) = \sigma + j\varpi\left[\varepsilon(\vec{r}) - \varepsilon_0\right]$.

Integral eqn (14) is solved via (MoM) by discretising the body into N subvolumes V_m with E and τ assumed to be constant within each cell. If $\vec{r}_m$ is the center of the mth cell, it follows:

$$\left[1+\frac{\tau(\vec{r})}{3j\varpi\varepsilon_0}\right]E_{x_p}(\mathbf{r}_m)-\sum_{q=1}^{3}\left[\sum_{q=1}^{3}\tau(\vec{r})PV\int_{V_m}G_{x_px_q}(\vec{r}_m,\vec{r}\,')dv'\right]E_{x_q}(\vec{r}_n)=E_{x_p}(\vec{r}_m), \qquad (15)$$

where $G_{x_px_q}$ is the tensor Green function given by:

$$G_{x_px_q}(\vec{r},\vec{r}\,')=-j\varpi\mu_0\left[\delta_{pq}+\frac{1}{k_0^2}\frac{\partial^2}{\partial x_p\partial x_q}\right]G_0(\vec{r},\vec{r}\,'),\ \text{p, q}=1,2,3, \qquad (16)$$

where $m, n= 1, 2,..., N,\ p, q= 1, 2, 3$.

Numerical solution details can be found elsewhere, e.g. in [14]. The computational example is related to the whole-body exposure to UMTS base station antenna system at f=2.14 GHz mounted on roof at the height h=15m. The human being is located approximately 24 m from the base station. The measured value of the electric field illuminating the body is 14.87V/m. The human body parameters are: ε_r=53.11 and σ=1.54 S/m, i.e. it follows: τ = 1.54 + j 6.1978 [9]. The maximum value of SAR induced in the body is 0.31mW/kg, in the region of the waist. The computed value of the whole-body average SAR is 82.4701μW/kg which stay far below the ICNIRP exposure limit.

3.3 Straton Chu formulation/hybrid BEM/FEM solution

Plane wave incident on the corneal part of the eye, treated as an exterior unbounded scattering problem is formulated via the Stratton-Chu formulation, i.e. the time-harmonic electric field at the exterior domain is expressed by the following boundary integral eqn (5):

$$\alpha\vec{E}=\vec{E}_i+\oint_{\partial V'}\vec{n}'\times\left(\nabla'\times\vec{E}\right)GdS'+\oint_{\partial V'}\left[\left(\vec{n}'\times\vec{E}\right)\times\nabla'G+\left(\vec{n}'\cdot\vec{E}\right)G\right]dS', \qquad (17)$$

where E_i is the incident electric field.

The interior domain is governed by the vector Helmholtz equation [5]:

$$\nabla\times\left(\frac{1}{k_B}\nabla\times\vec{E}\right)-k_A\vec{E}=0, \qquad (18)$$

where subscripts A and B denote the exterior and interior region, respectively.

This coupled formulation (17), (18) is treated via hybrid BEM/FEM technique. The eye parameters and mathematical details are available elsewhere, e.g. in [5].

Fig. 6 shows the *SAR* induced in the exposed to plane wave with power density of 10 W/m^2.

The whole eye averaged *SAR* at f=1 GHz is 0.3352 W/kg being below ICNIRP limits [2].

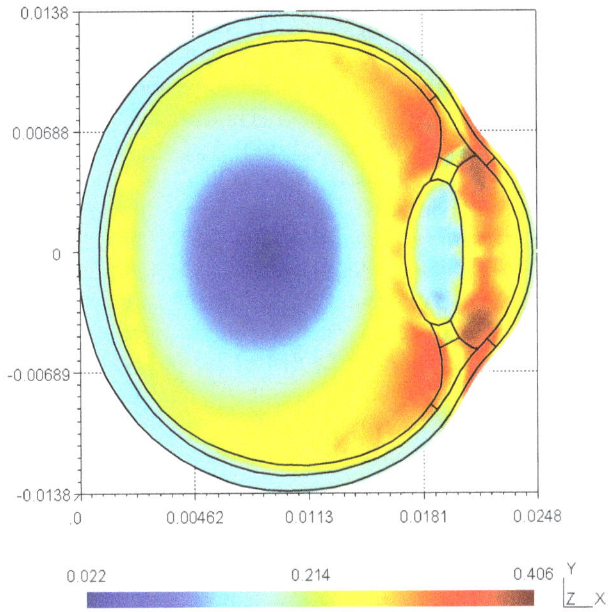

Figure 6: SAR in the eye due to plane wave exposure 10 W/m^2 at f=1 GHz.

4 CONCLUDING REMARKS

An overview of the use of certain boundary integral methods of solution in bioelectromagnetics and biomedical applications of electromagnetic fields is given in the paper. A trade-off between integral boundary and differential domain methods, respectively, is given, as well.

Thus, static and LF dosimetry techniques based on the Laplace equation /BEM solution are analysed. Furthermore, HF dosimetry methods are analysed; Surface Integral Equation/Metod of Moments (MoM), tensor volume integral equation/MoM and Stratton-Chu integral formulation/Helmholtz equation treated with hybrid boundary element method/finite element method (BEM/FEM) scheme. Some illustrative computational examples are also presented.

REFERENCES

[1] Poljak, D., Electromagnetic fields: Environmental exposure. *Encyclopedia of Environmental Health*, ed. J.O. Nriagu, Elsevier: Burlington, **2**, pp. 259–268, 2011.

[2] International Commission on Non-Ionizing Radiation Protection, "Guidelines for limiting exposure to time-varying electric, magnetic and electromagnetic fields (up to 300 GHZ)". *Health Physics*, **74**(4), pp. 494–522, Apr. 1998.

[3] International Commission on Non-Ionizing Radiation Protection, "Guidelines for limiting exposure to time-varying electric and magnetic fields (1 HZ—100 kHZ)". *Health Physics*, **99**(6), pp. 818–836; Dec. 2010.

[4] Hand, J.W., Modeling the interaction of electromagnetic fields (10 MHz–10 GHz) with the human body: methods and applications. *Physics in Medicine and Biology,* **53**(16), pp. 243–286, 2008. DOI: 10.1088/0031-9155/53/16/r01.

[5] Poljak, D., Cavka, D., Dodig, H., Peratta, C. & Peratta, A., On the use of boundary element analysis in bioelectromagnetics. *Engineering Analysis with Boundary Elements,* **49**, pp. 2–14, 2014. DOI: 10.1016/j.enganabound.2014.02.008.
[6] Singh, K.D., Longan, N.S. & Gilmartin, B., Three dimensional modeling of the human eye based on magnetic resonance imaging. *Investigative Opthamology & Visual Science*, **47**(6), pp. 2272–2279, 2006. DOI: 10.1167/iovs.05-0856.
[7] Hirata, A., Temperature increase in human eyes due to near-field and far-field exposures at 900 MHz, 1.5 GHz, and 1.9 GHz. *IEEE Transactions on Electromagnetic Compatibility,* **47**(1), pp. 68–76, 2005. DOI: 10.1109/temc.2004.842113.
[8] Reilly, J.P. & Hirata, A., Low-frequency electrical dosimetry: Research agenda of the IEEE international committee on electromagnetic safety. *Physics in Medicine and Biology,* **61**(12), R138–R149, 2016. DOI: 10.1088/0031-9155/61/12/r138.
[9] Poljak, D., Cvetković, M., Peratta, A., Peratta, C., Dodig, H. & Hirata, A.*, On Some Integral Approaches in Electromagnetic Dosimetry, The Joint Annual Meeting of The Bioelectromagnetics Society and the European BioElectromagnetics Association.* BioEM: Gent, 289–295, 2016.
[10] Čavka, D., Poljak, D. & Peratta, A., Comparison between finite and boundary element methods for analysis of electrostatic field around human head generated by video display units. *Journal of Communications Software and Systems,* **7**(1), pp. 22–28, 2011. DOI: 10.24138/jcomss.v7i1.184.
[11] Poljak, D., On the assessment of pregnant woman/foetus exposure to electromagnetic fields by using numerical methods. *Paediatria Croatica*, **57**(Supl 1), pp. 1–9, 2013.
[12] Cvetković, M. & Poljak, D., An efficient integral equation based dosimetry model of the human brain. *International Symposium on Electromagnetic Compatibility (EMC Europe 2014)*, Gothenburg, Sweden, pp. 375–380, Sep. 1–4, 2014.
[13] Poljak, D. et al., Integral equation models in some biomedical applications of electromagnetic fields: Transcranial magnetic stimulation (TMS), nerve fiber stimulation. *International Multidisciplinary Conference on Computer and Energy Science (SpliTech),* pp. 1–6, 2016. DOI: 10.1109/splitech.2016.7555936.
[14] Sadiku, M.N.O., *Numerical Techniques in Electromagnetics*, 2nd ed., CRC Press: New York, 2001.

ON THE COMPUTATION OF SINGULAR INTEGRALS OVER TRIANGULAR SURFACES IN $\mathbb{R}^3$

HRVOJE DODIG[1], MARIO CVETKOVIĆ[2] & DRAGAN POLJAK[2]
[1]University of Split, Faculty of Maritime Studies, Croatia
[2]University of Split, Faculty of Electrical Engineering, Croatia

ABSTRACT

Various integral equation formulations and the related numerical solutions either via Boundary Element Method (BEM) or Method of Moments (MoM) require tedious calculation of double surface integrals arising from the use of vector triangular basis functions. This paper presents an accurate technique for computation of these integrals by first converting the surface integrals to contour integrals facilitating the decomposition of boundary integral to the sum of line integrals over triangle edges. It was shown that application of this technique to a Laplace type of equations yields expressions having analytical solutions. Moreover, although the same was not possible to achieve in case of integrals involving Helmholtz kernels, nonetheless, the technique enabled the computation of surface integrals to a machine accuracy by employing the adaptive quadrature rules. This approach could be found useful in the high frequency computational dosimetry.

Keywords: boundary element method, Helmholtz equation, Laplace equation, adaptive quadrature, contour integrals.

1 INTRODUCTION

Boundary element method (BEM) is a domain reduction method for solving partial differential equations (PDE's). In contrast to finite element method (FEM), BEM requires surface-only discretization. The versatility of BEM is well known, and nowadays BEM found applications in elasticity problems [1], [2], acoustics [3], [4], electromagnetism [5], [6], seismology [7], [8] and in significant number of other engineering applications [9]–[12].

There is a number of three-dimensional (3D) BEM formulations where the domain can be reduced using Green's theorems and the field $\phi(\vec{r})$ at observation point $\vec{r}$ can be expressed as the surface integral [13]–[15]:

$$\alpha(\vec{r})\phi(\vec{r}) = \oint_{\partial V} dS' \left[G(\vec{r},\vec{r}\,')\frac{\partial\phi(\vec{r}\,')}{\partial\vec{n}\,'} - \phi(\vec{r}\,')\frac{\partial G(\vec{r},\vec{r}\,')}{\partial\vec{n}\,'} \right], \tag{1}$$

where ∂V is bounding surface of computational domain V, $\vec{n}$ is an outward unit normal vector and $\vec{r}\,'$ is source point. Depending on the observation point position, $\alpha(\vec{r})$ is either 0, 1/2 or 1 [16].

The Green's function $G(\vec{r},\vec{r}\,')$ represents the fundamental solution to the $\mathcal{L}\left[G(\vec{r},\vec{r}\,')\right] = -\delta\left(\vec{r},\vec{r}\,'\right)$. In case of BEM applied to Laplace or Helmholtz problem, the fundamental solution is a function of $R = |\vec{r} - \vec{r}\,'|$, R being the distance between the observation and the source point, respectively [17], [18].

After having discretized the boundary ∂V followed by interpolation of $\phi(\vec{r})$ and it's normal derivative using shape functions N_i, the linear system of equations is obtained, containing the following two types of integrals:

$$I_1(\vec{r}) = \int_{S_\Delta} N_i(\vec{r}\,')G(R)dS', \tag{2}$$

$$I_2(\vec{r}) = \int_{S_\Delta} N_i(\vec{r}\,')\frac{\partial G(R)}{\partial\vec{n}\,'}dS', \tag{3}$$

WIT Transactions on Engineering Sciences, Vol 122, © 2019 WIT Press
www.witpress.com, ISSN 1743-3533 (on-line)
doi:10.2495/BE410091

where S_Δ is the surface of element into which the surface ∂V is discretized.

In case of regular integrals, where $\vec{r} \neq \vec{r}\,'$, the numerical integration of (2) and (3) is usually sufficient. For nearly singular integrals, often times it is sufficient to increase the number of integration points or the use of finer discretization is necessary. However, in case of singular integrals, where $\vec{r} = \vec{r}\,'$, and depending on the type of singularity, it is required to perform certain additional steps [19], such as singularity extraction [20], [21] or the so called singularity cancellation technique [22], [23], which includes transformation to polar coordinates or domain transformation, resulting in an extra computational burden if aiming to achieve accuracy.

Hence, an accurate technique for computation of these integrals to a desirable precision could be found useful. As showed in [24], accurate computation of (2) and (3) is particularly important when BEM or MoM, respectively, are utilized in the computational dosimetry of human exposure to electromagnetic field (EMF) since the resulting numerical artifacts will significantly affect the reliability of the numerical technique for this purpose. To overcome this, the singular integrals arising in these type of problems can be computed by first converting the surface integrals to contour integrals facilitating the decomposition of boundary integral to the sum of easily and quickly computed line integrals over triangle edges.

This paper is organized as follows: following the introductory part, the shape functions over the triangular elements are given in terms of the position vector and the kernel modification is given in the second section. Some details on the procedure for converting surface integrals to contour integrals are given in the following sections. Moreover, the application to a Laplace type of equations and integrals involving Helmholtz kernels is also given as well as some numerical results for the contour integration on several test examples. The final part is related to the concluding remarks and some consideration for the future work.

2 MATHEMATICAL FORMULATION

2.1 Shape functions over triangular elements

When considering BEM for the solution to the boundary value problem in 3D, the boundary surface is usually decomposed into triangular or quadrilateral elements. In case of first order triangular element, as shown in Fig. 1, the shape functions N_i can be expressed in terms of area coordinates such as [25]:

$$N_i = \frac{A_{jk}}{A}, \quad i = 1,2,3;\ j = 2,3,1;\ k = 3,1,2, \tag{4}$$

where A and A_{jk} are the area of first order triangular element and areas of sub-triangles, respectively, as denoted in Fig. 1.

In order to express the shape functions N_i using the position vector $\vec{r}\,'$, the sub-areas A_{jk} are expressed as:

$$A_{ij} = \frac{1}{2}\vec{n}\,' \cdot (\vec{r}_j - \vec{r}\,') \times (\vec{r}_i - \vec{r}\,'), \tag{5}$$

whereas the area A of triangle $\triangle 123$ can be written as:

$$A = \frac{1}{2}\vec{n}\,' \cdot (\vec{r}_2 - \vec{r}_1) \times (\vec{r}_3 - \vec{r}_1). \tag{6}$$

Inserting (5) and (6) into (4) the shape functions can be written as:

$$N_i(\vec{r}\,') = \frac{\vec{n}\,' \cdot (\vec{r}_j - \vec{r}\,') \times (\vec{r}_k - \vec{r}\,')}{2A}, \quad i = 1,2,3;\ j = 2,3,1;\ k = 3,1,2. \tag{7}$$

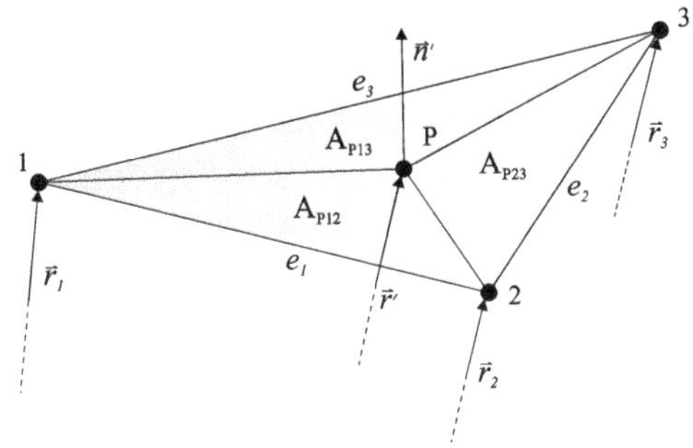

Figure 1: First order triangular element. Vectors $\vec{r}_i$, $i = 1, 2, 3$, are the position vectors of vertices 1,2 and 3 whereas vector $\vec{r}\,'$ is the position vector of point P. Vector $\vec{n}\,'$ is the constant unit normal vector to triangle $\triangle 123$. Triangle edges are e_1, e_2 and e_3.

After expanding the cross product in (7), and using $\vec{r}\,' \times \vec{r}\,' = 0$, (7) can be rewritten as:

$$N_i(\vec{r}\,') = \frac{\vec{n}\,' \cdot [\vec{r}_j \times \vec{r}_k + \vec{r}\,' \times (\vec{r}_j - \vec{r}_k)]}{2A}, \quad i = 1, 2, 3;\ j = 2, 3, 1;\ k = 3, 1, 2. \tag{8}$$

After adding and subtracting $\vec{r} \times (\vec{r}_j - \vec{r}_k)$ to the numerator of (8), where $\vec{r}$ is the observation point Q (shown in Fig. 2), yields the following:

$$N_i(\vec{r}\,', \vec{r}) = \vec{n}\,' \cdot (\vec{r}\,' - \vec{r}) \times \vec{a}_i + b_i(\vec{r}), \tag{9}$$

where $\vec{a}_i$ is a constant vector given by:

$$\vec{a}_i = \frac{\vec{r}_j - \vec{r}_k}{2A}, \tag{10}$$

while coefficients $b_i(\vec{r})$ are given as:

$$b_i(\vec{r}) = \frac{\vec{n}\,' \cdot \vec{r}_j \times \vec{r}_k}{2A} + \vec{n}\,' \cdot \vec{r} \times \vec{a}_i, \tag{11}$$

where $i = 1, 2, 3;\ j = 2, 3, 1;\ k = 3, 1, 2.$

2.2 Modification of the kernel

In order to convert the surface integrals (2) and (3) to contour integrals, the integrand first has to be appropriately transformed so the Gauss' theorem for the surface could be applied. This is done by rewritting the function $G(R)$ in the following way:

$$G(R) = \vec{n} \cdot \nabla' \times [u(R)\vec{q}\,], \tag{12}$$

where $R = |\vec{r} - \vec{r}\,'|$, ∇' is del operator acting only on primed coordinates and function $u(R)$ is given by the following integral:

$$u(R) = \int RG(R)dR. \tag{13}$$

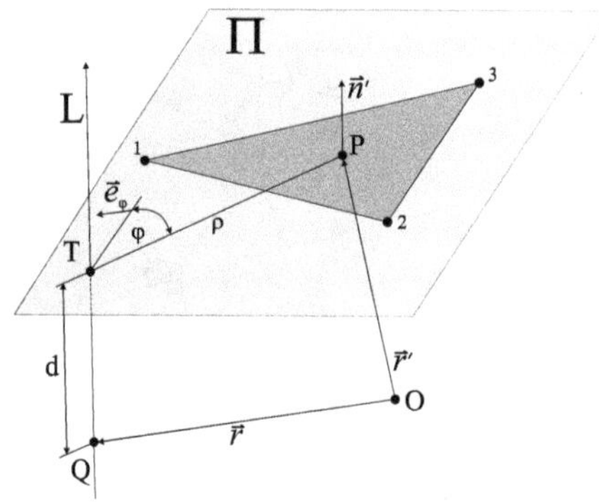

Figure 2: Triangle located in plane Π defined by unit normal vector $\vec{n}'$. The source and observation points are denoted by P and Q, respectively. Point T is the projection of Q on the triangle plane, while d is the distance between points Q and T.

Furthermore, vector $\vec{q}$ can be written as:

$$\vec{q} = \frac{1}{\rho^2}\vec{n}' \times (\vec{r}' - \vec{r}), \tag{14}$$

where ρ is the distance from point T to the source point P, as shown in Fig. 2.

Applying the vector identity $\nabla \times \left(f\vec{A}\right) = \nabla f \times \vec{A} + f\nabla \times \vec{A}$, the right-hand side of (12) can be rewritten as:

$$\vec{n}' \cdot \nabla' \times [u(R)\vec{q}] = \vec{n}' \cdot \nabla' u(R) \times \vec{q} + u(R)\vec{n}' \cdot \nabla' \times \vec{q}. \tag{15}$$

The gradient $\nabla' u(R)$ can be evaluated using the chain rule as:

$$\nabla' u(R) = \sum_{i=1}^{3} \frac{\partial u(R)}{\partial R}\frac{\partial R}{\partial x_i}\vec{e}_i = \frac{\partial u(R)}{\partial R}\nabla' R, \tag{16}$$

where x_i and $\vec{e}_i$ are the coordinates and the unit ort vectors in the Cartesian coordinate system, respectively.

As the partial derivative of $u(R)$ is equal to $RG(R)$, and $\nabla' R = (\vec{r}' - \vec{r})/R$, eqn (16) thus becomes:

$$\nabla' u(R) = (\vec{r}' - \vec{r})G(R). \tag{17}$$

Inserting (14) and (17) in (15), the first term on the right hand-side term can now be rewritten as:

$$\vec{n}' \cdot \nabla' u(R) \times \vec{q} = \vec{n}' \cdot (\vec{r}' - \vec{r}) \times [\vec{n}' \times (\vec{r}' - \vec{r})] \frac{G(R)}{\rho^2}. \tag{18}$$

It is easily showed that the triple product from (18) is equal to:

$$\begin{aligned} &\vec{n}' \cdot (\vec{r}' - \vec{r}) \times [\vec{n}' \times (\vec{r}' - \vec{r})] = \\ &= \vec{n}' \cdot \vec{n}' [(\vec{r}' - \vec{r}) \cdot (\vec{r}' - \vec{r})] - [\vec{n}' \cdot (\vec{r}' - \vec{r})]^2 \\ &= R^2 - d^2 = \rho^2, \end{aligned} \tag{19}$$

where d is the distance between points Q and T, as shown on Fig. 2, thus canceling the denominator in (18):

$$\vec{n}' \cdot \nabla' u(R) \times \vec{q} = G(R). \tag{20}$$

The second term on the right-hand side of (15) is equal to zero, as vector $\vec{q}$ can be expressed as a gradient of the cylindrical angle φ, shown on Fig. 2. As curl of any gradient vector field is equal to zero at any point except at point $\rho = 0$, this leads to $\nabla' \times \vec{q} = 0$, and thus corroborates the expression (12). However, when $\rho = 0$, the point can be excluded from the surface integrals I_1 and I_2, as it will be shown later.

3 CONVERTING SURFACE INTEGRAL $\int N_i G dS$ TO CONTOUR INTEGRAL

To convert surface integral (2) to contour integral the shape function formulation (9) is used to write the integrand $N_i G$ as:

$$N_i(\vec{r}', \vec{r})G(R) = \vec{n}' \cdot G(R)\,(\vec{r}' - \vec{r}) \times \vec{a}_i + b_i(\vec{r})G(R). \tag{21}$$

Inserting (12) and (17) in (21), followed by some mathematical manipulation, the following is obtained:

$$N_i(\vec{r}', \vec{r})G(R) = \vec{n}' \cdot \nabla' \times \left[u(R)\,(\vec{a}_i + b_i(\vec{r})\vec{q})\right]. \tag{22}$$

After inserting (22) in (2) followed by application of Stokes' theorem yields:

$$I_1(\vec{r}) = \int_{S_\Delta} dS'\vec{n}' \cdot \nabla' \times \left[u(R)\,(\vec{a}_i + b_i(\vec{r})\vec{q})\right] = \oint_{\partial S_\Delta} u(R)\left[\vec{a}_i + b_i(\vec{r})\vec{q}\right] \cdot d\vec{l}', \tag{23}$$

where $d\vec{l}'$ is the line element along contour ∂S_Δ.

The integral (23) can be broken into sum of line integrals over triangle edges:

$$I_1(\vec{r}) = \sum_{i=1}^{3} \int_{e_i} u(R)\left[\vec{a}_i + b_i(\vec{r})\vec{q}\right] \cdot d\vec{l} = \sum_{i=1}^{3} I_1^i(\vec{r}), \tag{24}$$

where e_i are edges of the triangle shown in Fig. 1 and I_1^i is the line integral along edge e_i. Line integral I_1^i can be evaluated using:

$$I_1^i(\vec{r}) = \int_{x_1}^{x_2} u(R)\left[\vec{a} \cdot \vec{e}_x' - b_i(\vec{r})\frac{y'}{x'^2 + y'^2}\right] dx', \tag{25}$$

where integration bounds x_1 and x_2 can be determined using $x_i = (\vec{r}_i - \vec{r}) \cdot \vec{e}_x'$, $i = 1, 2$. The derivation of (25) is omitted for brevity.

In case when point T is inside the triangle (shown on Fig. 2), we have to exclude the small area surrounding the point T and solve for the residual using some limiting process, i.e.:

$$c_i = -\lim_{\epsilon \to 0} \int_0^{\alpha} u(R)\left[\epsilon \vec{a}_i \cdot \vec{e}_\varphi + b_i(\vec{r})\right] d\varphi, \tag{26}$$

where the angle α depends on the location of point T, being $\alpha = 2\pi$ (T inside the triangle), $\alpha = \pi$ (T on the edge), or a part of the circle (when T is in the vertex).

4 CONVERTING SURFACE INTEGRAL $\int N_i \frac{\partial G}{\partial \vec{n}} dS$ TO CONTOUR INTEGRAL

Integral (3) can be dealt with in a similar manner. The first step, though, is with a little effort, working out the partial derivative by taking the del operator outside the integral, i.e. (3) can be written as:

$$I_2(\vec{r}) = \int_{S_\Delta} N_i(\vec{r}\,') \frac{\partial G(R)}{\partial \vec{n}\,'} dS' = -\vec{n}\,' \cdot \nabla \int_{S_\Delta} N_i(\vec{r}\,') G(R) dS', \tag{27}$$

where $\nabla G(R) = -\nabla' G(R)$ have been used.

Inserting (23) in (27), leads to:

$$I_2(\vec{r}) = -\vec{n}\,' \cdot \nabla \oint_{\partial S_\Delta} u(R) \left[\vec{a}_i + b_i(\vec{r})\vec{q}\right] \cdot d\vec{l}\,'. \tag{28}$$

As nabla does not act on primed coordinates, we can write $-\vec{n}\,' \cdot \nabla$ under the integral and then use expressions $\nabla u(R) = -\nabla' u(R)$ and $\nabla(\vec{q} \cdot d\vec{l}\,') = -\nabla'(\vec{q} \cdot d\vec{l}\,')$ to obtain:

$$\begin{aligned} I_2(\vec{r}) \;=\; & \oint_{\partial S_\Delta} \left[\vec{n}\,' \cdot \nabla' u(R)\right] \left[\vec{a}_i + b_i(\vec{r})\vec{q}\right] \cdot d\vec{l}\,' - \oint_{\partial S_\Delta} b_i(\vec{r}) u(R) \vec{n}\,' \cdot \nabla(\vec{q} \cdot d\vec{l}\,') - \\ & \oint_{\partial S_\Delta} \left[\vec{n}\,' \cdot \nabla b_i(\vec{r})\right] \vec{q} \cdot d\vec{l}\,'. \end{aligned} \tag{29}$$

It is easily showed that the last term from eqn (29) is equal to zero, since $\vec{n}\,' \cdot \nabla b_i(\vec{r}) = \vec{n}\,' \cdot \vec{a} \times \vec{n}\,' = 0$.

The same could be also showed for the second integral from eqn (29), although with somewhat more elaborate work. For the sake of brevity, it is suffice to say that the trick is to use the fact that vectors $\vec{q}$ and $\vec{n}\,'$ are perpendicular. So, with some clever mathematical manipulations, integrand could be rearranged so that the following could be used:

$$\vec{n}\,' \cdot \vec{q} = \vec{n}\,' \cdot \frac{1}{\rho^2} \vec{n}\,' \times (\vec{r}\,' - \vec{r}) = 0. \tag{30}$$

This means that (29) is reduced to:

$$I_2(\vec{r}) = \oint_{\partial S_\Delta} \frac{\partial u(R)}{\partial \vec{n}\,'} \left[\vec{a}_i + b_i(\vec{r})\vec{q}\right] \cdot d\vec{l}\,', \tag{31}$$

where we have used $\frac{\partial u(R)}{\partial \vec{n}\,'} = \vec{n}\,' \cdot \nabla' u(R)$.

Integral (31) can be broken down to the sum of the easily solved line integrals over triangle edges, while in case when point T falls within the triangle, we have to perform additional computation on the residual term:

$$d_i = -\lim_{\epsilon \to 0} \int_0^\alpha \frac{\partial u(R)}{\partial \vec{n}\,'} \left[\epsilon \vec{a}_i \cdot \vec{e}\varphi + b_i(\vec{r})\right] d\varphi, \tag{32}$$

that, when $\rho = 0$ reduces to:

$$d_i = -b_i(\vec{r}) \frac{\partial u(|z'|)}{\partial z'} \alpha. \tag{33}$$

5 APPLICATION TO INTEGRALS INVOLVING LAPLACE KERNELS

Integrals (24) and (31) have analytical solutions for Laplace and Poisson kernels while the fundamental solution of the 3D Poisson problem is given by $G(R) = 1/4\pi R$.

If the function $u(R)$ is expressed as $u(R) = R/4\pi$, the integrand of (24) can be rewritten in the appropriate form facilitating the conversion into such contour integral having the analytical solution given as:

$$\int_{x_1}^{x_2} u(R)\vec{a}_i \cdot d\vec{l} = (\vec{a}_i \cdot \vec{e}_x') \frac{x'R + (y'^2 + z'^2)\ln(x' + R)}{8\pi}\Bigg|_{x_1}^{x_2}, \tag{34}$$

$$\int_{x_1}^{x_2} u(R)b_i(\vec{r})\vec{q} \cdot d\vec{l} = -\, b_i(\vec{r}) \frac{z'\arctan\left(\frac{x'z'}{y'R}\right) + y'\ln(x' + R)}{4\pi}\Bigg|_{x_1}^{x_2}, \tag{35}$$

where $\vec{a}_i$ and $b_i(\vec{r})$ are given in (10) and (11), respectively.

The integral (31), involving normal derivative of $u(R)$, can be similarly rewritten, i.e. leading to the analytical solution in the form of:

$$\int_{x_1}^{x_2} \frac{\partial u(R)}{\partial \vec{n}'}\vec{a}_i \cdot d\vec{l} = (\vec{a}_i \cdot \vec{e}_x') \frac{z'\ln(x + R)}{4\pi}\Bigg|_{x_1}^{x_2}, \tag{36}$$

$$\int_{x_1}^{x_2} \frac{\partial u(R)}{\partial \vec{n}'} b_i(\vec{r})\vec{q} \cdot d\vec{l} = -\, b_i(\vec{r}) \frac{\arctan\left(\frac{x'z'}{y'R}\right)}{4\pi}\Bigg|_{x_1}^{x_2}, \tag{37}$$

where $\vec{a}_i$ and $b_i(\vec{r})$ are given in (10) and (11), respectively.

6 COMPUTATION OF INTEGRALS INVOLVING HELMHOLTZ KERNELS

When the differential operator is Helmholtz operator, i.e. when function $G(R)$ satisfies:

$$\nabla^2 G(R) + k^2 G(R) = -\delta(\vec{r}, \vec{r}') \tag{38}$$

the fundamental solution is of the form:

$$G(R) = \frac{e^{-ikR}}{4\pi R}. \tag{39}$$

From (13) the function $u(R)$ can easily be computed as:

$$u(R) = \int R \frac{e^{-ikR}}{4\pi R} dR = i\frac{e^{-ikR}}{4\pi k}. \tag{40}$$

The contour integrals (24) and (31) when $u(R)$ is given by (40) do not have analytical solutions, thus, the standard Gaussian quadrature is necessary to compute the contour integrals for geometric configurations as shown in Fig. 3. However, in case when y' becomes very small, the number of Gaussian points have to be increased.

The numerical results of the contour integration for several examples as shown in Fig. 3 are given in Table 1. The results obtained using the proposed technique are compared to the numerical results obtained using the Wolfram Mathematica software.

In order to handle the singularity problem when $y' \to 0$, Gauss-Patterson adaptive quadrature [26] has been used to compute the contour integrals (24) and (31) to machine

Table 1: Comparison of contour integral results with numerical results from Wolfram Mathematica software. Number of Gauss points per triangle's edge $N = 20$.

Test case	Integral	Mathematica	Contour Integral
(a)	$\int N_1 GdS$	$0.00528364 - j0.000198499$	$0.00528364 - j0.000198499$
	$\int N_2 GdS$	$0.00528364 - j0.000198499$	$0.00528364 - j0.000198499$
	$\int N_3 GdS$	$0.00529556 - j0.0000995714$	$0.00529556 - j0.0000995714$
(b)	$\int N_1 GdS$	$0.00172243 - j0.0284663$	$0.00172243 - j0.0284663$
	$\int N_2 GdS$	$-0.00432864 + j0.0129573$	$-0.00432864 + j0.0129573$
	$\int N_3 GdS$	$0.0051389 + j0.00317481$	$0.0051389 + j0.00317481$
(c)	$\int N_1 GdS$	$0.0276113 - j0.137714$	$0.0276113 - j0.137714$
	$\int N_2 GdS$	$-0.0324167 + j0.00762088$	$-0.0324167 + j0.00762088$
	$\int N_3 GdS$	$-0.0744389 + j0.00563727$	$-0.0744389 + j0.00563727$
(d)	$\int N_1 GdS$	$0.00971219 - j0.00284063$	$0.00971219 - j0.00284063$
	$\int N_2 GdS$	$0.00797566 - j0.00559024$	$0.00797566 - j0.00559024$
	$\int N_3 GdS$	$0.00854506 - j0.00499818$	$0.00854506 - j0.00499818$
(e)	$\int N_1 GdS$	$0.00370118 - j0.00108283$	$0.00370118 - j0.00108283$
	$\int N_2 GdS$	$0.00334767 - j0.00171923$	$0.00334767 - j0.00171923$
	$\int N_3 GdS$	$0.00367605 - j0.00116374$	$0.00367605 - j0.00116374$
(f)	$\int N_1 GdS$	$-0.0138075 - j0.0103983$	$-0.0138075 - j0.0103983$
	$\int N_2 GdS$	$-0.0117517 + j0.00130501$	$-0.0117517 + j0.00130501$
	$\int N_3 GdS$	$-0.0148588 - j0.00791919$	$-0.0148588 - j0.00791919$

Table 2: The number of bisections and the number of evaluation points (EP) for adaptive Gauss-Patterson numerical integration method (AGP) as function of y'. The integral computed is $\int N_i GdS$ for Helmholtz equation using $u(R)$ given by (40).

Method	y'	Bisections	EP
AGP	1×10^{-1}	0	511
AGP	1×10^{-2}	6	604
AGP	1×10^{-4}	20	821
AGP	1×10^{-8}	52	1061
AGP	1×10^{-16}	120	2035
AGP	1×10^{-32}	260	7389

accuracy. As shown in Table 2, the drawback to this approach is the increased number of bisections when y' tends to zero. Also, in some cases one can note the decreasing numbers

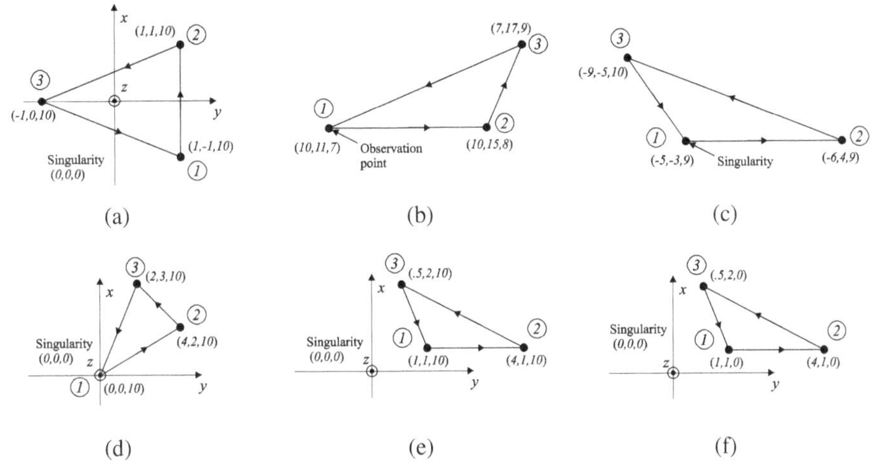

Figure 3: Several test examples for integration of $\int N_i G$ using contour integration. The integration path is denoted with arrows. Propagation constant is chosen to be $k = \frac{2\pi}{\lambda} = \frac{2\pi}{5}$. In (b) and (c) observation point is located at vertex 1.

of bisections. However, the numerical accuracy of the integration procedure converged to machine accuracy.

7 CONCLUSIONS

In this paper the technique for converting surface integrals containing Green's function and its derivative to contour integrals is shown. This conversion to contour integral is possible when Green's function is $G = G(R)$ and when indefinite integral $\int RG(R)dR$ can be computed analytically, such as in the case of Helmholtz, Laplace and Poisson type problems. It was shown that for Poisson and Laplace type of problems this technique yields analytical solutions for integrals of this type. The advantage of computation of contour integrals to surface integrals is the possibility to precisely control the accuracy of the numerical computation of these integrals when the analytical solution is not available. It has been found out that in the relatively rare cases the number of evaluation points have to be increased if the adaptive Gauss-Patterson quadrature is used. This issue is expected to be addressed in the future work within the new scheme for adaptive computation of such integrals.

REFERENCES

[1] Simpson, R.N., Bordas, S.P.A., Trevelyan, J. & Rabczuk, T., A two-dimensional isogeometric boundary element method for elastostatic analysis. *Computer Methods in Applied Mechanics and Engineering*, **209–212**, pp. 87–100, 2012.

[2] Elleithy, W.M., Al-Gathani, H.J. & El-Gebeily, M., Iterative coupling of BE and FE methods in elastostatics. *Engineering Analysis with Boundary Elements*, **25**, pp. 685–695, 2001.

[3] Chen, J., Chen, K. & Chen, C.T., Adaptive boundary element method of time-harmonic exterior acoustics in two dimensions. *Computer Methods in Applied Mechanics and Engineering*, **191**, pp. 3331–3345, 2002.

[4] Fischer, M., Gauger, U. & Gaul, L., A multipole Galerkin boundary element method for acoustics. *Engineering Analysis with Boundary Elements*, **28**, pp. 155–162, 2004.

[5] Dodig, H., Lalléchère, S., Bonnet, P., Poljak, D. & El Khamlichi Drissi, K., Stochastic sensitivity of the electromagnetic distributions inside a human eye modeled with a 3d hybrid BEM/FEM edge element method. *Engineering Analysis with Boundary Elements*, **49**, pp. 48–62, 2014.

[6] Dodig, H., Poljak, D. & Peratta, A., Hybrid BEM/FEM edge element computation of the thermal rise in the 3D model of the human eye induced by high frequency EM waves. *International Conference on Telecommunications and Computer Networks (SoftCOM)*, IEEE, pp. 1–5, 2012.

[7] Kampitsis, A., Sapountzakis, E., Giannakos, S. & Gerolymos, N., Seismic soil-pile-structure kinematic and inertial interaction: A new beam approach. *Soil Dynamics and Earthquake Engineering*, **55**, pp. 211–224, 2013.

[8] António, J. & Tadeu, A., 3D seismic response of a limited valley via BEM using 2.5D analytical Green's functions for an infinite free-rigid layer. *Soil Dynamics and Earthquake Engineering*, **22**, pp. 659–673, 2002.

[9] Benedetti, I., Aliabadi, M.H. & Milazzo, A., A fast BEM for the analysis of damaged structures with bonded piezoelectric sensors. *Computer Methods in Applied Mechanics and Engineering*, **199**, pp. 490–501, 2010.

[10] Chen, X. & Liu, Y., An advanced 3D boundary element method for characterizations of composite materials. *Engineering Analysis with Boundary Elements*, **29**, pp. 513–523, 2005.

[11] Sutradhar, A. & Paulino, G.H., The simple boundary element method for transient heat conduction in functionally graded materials. *Computer Methods in Applied Mechanics and Engineering*, **193**, pp. 4511–4539, 2004.

[12] Bush, The interaction between a crack and a particle cluster. *International Journal of Fracture*, **88**, pp. 215–232, 1997.

[13] Brebbia, C.A. & Dominguez, J., Boundary element methods for potential problems. *Applied Mathematical Modelling*, **1**, pp. 372–378, 1977.

[14] Cheng, A.H.-D. & Cheng, D.T., Heritage and early history of the boundary element method. *Engineering Analysis with Boundary Elements*, **29**, pp. 368–302, 2005.

[15] Liu, Y.J., On the simple-solution method and non-singular nature of the BIE/BEM a review and some new results. *Engineering Analysis with Boundary Elements*, **29**, pp. 789–795, 2000.

[16] Liu, Y.J. & Nishimura, N., The fast multipole boundary element method for potential problems: A tutorial. *Engineering Analysis with Boundary Elements*, **30**, pp. 371–381, 2006.

[17] Jaswon, M.A., Integral equation methods in potential theory. I. *Proceedings of the Royal Society of London A*, **275**, pp. 23–32, 1963.

[18] Wu, H., Liu, Y. & Jiang, W., Analytical integration of the moments in the diagonal form fast multipole boundary element method for 3-D acoustic wave problems. *Engineering Analysis with Boundary Elements*, **36**, pp. 248–254, 2012.

[19] Cvetković, M., Poljak, D. & Drissi, K.E.K., Some computational aspects of calculation of integrals arising within the framework of method of moments-application to bioelectromagnetics. *Applied Electromagnetics and Communications (ICECOM), 2016 22nd International Conference on*, IEEE, pp. 1–6, 2016.

[20] Khayat, M. & Wilton, D., Numerical evaluation of singular and near-singular potential integrals. *IEEE Transactions on Antennas and Propagation*, **53**(10), pp. 3180–3190, 2005.

[21] Jarvenpaa, S., Taskinen, M. & Yla-Oijala, P., Singularity subtraction technique for high-order polynomial vector basis functions on planar triangles. *IEEE Transactions*

on Antennas and Propagation, **54**(1), pp. 42–49, 2006.
[22] Graglia, R., On the numerical integration of the linear shape functions times the 3-D Green's function or its gradient on a plane triangle. *IEEE Transactions on Antennas and Propagation*, **41**(10), pp. 1448–1455, 1993.
[23] Li, L. & Eibert, T.F., Radial-angular singularity cancellation transformations derived by variable separation. *IEEE Transactions on Antennas and Propagation*, **64**(1), pp. 189–200, 2016.
[24] Poljak, D., Cvetković, M., Bottauscio, O., Hirata, A., Laakso, I., Neufeld, E., Reboux, S., Warren, C., Giannopoulos, A. & Costen, F., On the use of conformal models and methods in dosimetry for nonuniform field exposure. *IEEE Transactions on Electromagnetic Compatibility*, **60**(2), pp. 328–337, 2018.
[25] Volakis, J.L., Chatterjee, A. & Cambell, L.C., *Finite Element Method for Electromagnetics*. The IEE/OUP series on electromagnetic wave theory, IEEE Press: New York, USA, 1998.
[26] Patterson, T.N.L., An algorithm for generating interpolatory quadrature rules of the highest degree of precision with preassigned nodes for general weight functions. *ACM Transactions on Mathematical Software*, **15**, pp. 123–136, 1989.

HIGH PERFORMANCE OF LOCAL MESHFREE METHOD WITH REDUCED INTEGRATION

WILBER VELEZ, THIAGO ARAUJO & ARTUR PORTELA
Departamento de Engenharia Civil, Universidade de Brasília, Brazil

ABSTRACT

The Meshfree Method with reduced integration (ILMF) is derived through the work theorem of structures theory. In the formulation of the ILMF, the kinematically-admissible strain field is an arbitrary rigid-body displacement; as a consequence, the domain term is canceled out and the work theorem is reduced to regular local boundary terms only. The moving least squares (MLS) approximation of the elastic field is used to construct the trial function in this local meshfree formulation. ILMF has a high performance in problems with irregular nodal arrangement leading to accurate numerical results. This paper presents the size effect of the irregularity nodal arrangement parameter (c_n) on three different nodal discretization to solve the Timoshenko cantilever beam using values fixed for the local support domain (α_s) and the local quadrature domain (α_q). Results obtained are optimal for 2D plane stress problems when compared with the exact solution.
Keywords: work theorem, reduced integration, local meshless method.

1 INTRODUCTION

Numerical methods based in grid, like Finite Element Method (FEM), are widely used for scientific research. Grid-based methods required high quality meshes to solve fracture mechanics problems with material discontinuity, large deformation where excessive mesh distortion takes place and other situations. The meshless methods were generated with the expectation of providing more adaptive, accurate and stable numerical solutions that can deal with problems where conventional methods are not suitable [1]. Generally, their formulation is based in the weighted-residual method [2].

Different meshless methods have been developed during the last 20 years [3]. Some methods based on a weighted-residual weak-form formulation were applied in solid mechanics such as the Diffuse Element Method (DEM) [4], the Reproducing Kernel Particle Method (RKPM) [5], and the Element-free Galerkin (EFG) [6]. Other methods emerged based on local weighted-residual weak forms, such as the Meshless Local Petrov–Galerkin Method (MLPG) [7], [8], the Meshless Local Boundary Integral Equation (MLBIE) [9], the Local Point Interpolation Method (LPIM) [10], Local Radial Point Interpolation Method (LRPIM) [11], the Meshless Finite Volume Method (FVM) [12], the Rigid-Body Displacement Meshfree (RBDMF) and the Generalized-Strain Meshfree (GSMF) [13].

The Meshfree Method with reduce integration (ILMF) formulation presented by [14], the kinematically-admissible strain field is chosen as the one corresponding to an arbitrary rigid-body displacement; as a consequence, the domain term is canceled out and the work theorem is reduced to regular local boundary terms only. The Moving Least Squares (MLS) approximation of the elastic field is used to construct the trial function in this local meshfree formulation. ILMF and the popular MLPG using the MLS approximation, this fact allowing having more precise conclusions when comparing the two methods.

The Meshfree Methods were developed to solve large displacements issues that the Finite Element Methods could not handle [15]. This Meshfree approach can be implemented to solve any real engineering problems. The ILMF and MLPG presented are suitable to work with inhomogeneous materials, since it has no elements with continuous physical properties,

www.witpress.com, ISSN 1743-3533 (on-line)
doi:10.2495/BE410101

and it has intrinsic ability to adapt to irregular geometries, once the nodes can be automatically generated anywhere inside and in the edges of the problem domain.

This paper is focused on the size effect of the irregularity nodal arrangement parameter (c_n) when is used a fixed value for the local support domain (α_s) and the local quadrature domain (α_q). It presents a comparison of the energy and displacement relative error for three different irregular nodal distribution to solve the Timoshenko cantilever beam. The results are compared with the exact solution and the popular Meshless Local Petrov–Galerkin (MLPG). Optimal results have been obtained.

2 MODELLING STRATEGY

Let Ω be the domain of a body and Γ its boundary subdivided in Γ_u and Γ_t that is $\Gamma = \Gamma_t \cup \Gamma_u$, as represented in Fig. 1. The general fundamental boundary value problem of linear elastostatics aims to determine the distribution of stresses $\boldsymbol{\sigma}$, strains $\boldsymbol{\varepsilon}$ and displacements $\boldsymbol{u}$, throughout the body, when it has constrained displacements $\bar{\boldsymbol{u}}$, on Γ_u and is loaded by an external system of distributed surface and body forces with densities denoted, respectively by $\bar{\boldsymbol{t}}$, on Γ_t and $\boldsymbol{b}$, in Ω.

In the domain of the body, loaded by a system of external distributed surface and body forces with densities denoted, respectively by $\bar{\boldsymbol{t}}$, on the boundary Γ_t and $\boldsymbol{b}$, in the domain Ω, consider a statically admissible stress field $\boldsymbol{\sigma}$, that is any stress field that satisfies equilibrium with the system of applied external forces which therefore satisfies

$$\boldsymbol{L}^T\boldsymbol{\sigma} + \boldsymbol{b} = \boldsymbol{0}, \tag{1}$$

in the domain Ω, with boundary conditions

$$\boldsymbol{t} = \boldsymbol{n}\boldsymbol{\sigma} = \bar{\boldsymbol{t}}, \tag{2}$$

on the static boundary Γ_t, in which $\boldsymbol{L}$ is a matrix differential operator; $\boldsymbol{t}$ is the vector of the traction components; $\bar{\boldsymbol{t}}$ is the vector of the prescribed tractions and $\boldsymbol{n}$ is the matrix of the components of the unit outward normal to the boundary.

In the domain Ω, with boundary $\Gamma = \Gamma_u \cup \Gamma_t$, consider an arbitrary local domain Ω_Q, assigned to a reference point $Q \in \Omega_Q$, with local boundary $\Gamma_Q = \Gamma_{Qi} \cup \Gamma_{Qt} \cup \Gamma_{Qu}$, in which Γ_{Qi} is the interior local boundary, while Γ_{Qt} and Γ_{Qu} are local boundaries that share the global boundaries, respectively the static boundary Γ_t and the kinematic boundary Γ_u, as represented

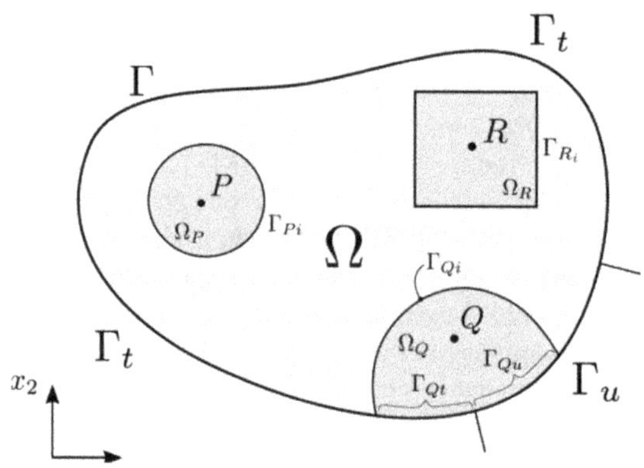

Figure 1: Schematic representation of the domain **Ω** of the body.

in Fig. 1. The work theorem is derived as a local form that is valid in an arbitrary domain Ω_Q, associated to the reference point Q.

The general work theorem of the theory of structures establishes an energy relationship, valid in an arbitrary local domain $\Omega_Q \cup \Gamma_Q \in \Omega \cup \Gamma$, between two independent elastic fields that can be defined in the body which are, respectively a statically admissible stress field that satisfies equilibrium with a system of external distributed surface and body forces, and a kinematically admissible strain field that satisfies compatibility with a set of constrained displacements.

Expressed as an integral local form, defined in the local domain $\Omega_Q \cup \Gamma_Q$, the work theorem can be written in a compact form, simply as

$$\int_{\Gamma_Q} \boldsymbol{t}^T \boldsymbol{u}^* d\Gamma + \int_{\Omega_Q} \boldsymbol{b}^T \boldsymbol{u}^* d\Omega = \int_{\Omega_Q} \boldsymbol{\sigma}^T \varepsilon^* d\Omega, \tag{3}$$

in which the stress field $\boldsymbol{\sigma}$ and the strain field $\boldsymbol{\varepsilon}^*$ are not linked by any constitutive relationship and therefore, they are independent of each other.

Kinematic formulations consider, in the work theorem, a particular and convenient specification of the kinematically admissible strain field, leading thus to an equation of mechanical equilibrium that is used in numerical models, to generate the respective stiffness matrix of each model. A simple case of local equilibrium equations, based on a kinematically admissible strain field generated by a rigid-body displacement, is presented here.

Bearing in mind the key feature of the work theorem, which is the complete independence of the statically admissible stress field $\boldsymbol{\sigma}$ and the kinematically admissible strain field $\boldsymbol{\varepsilon}^*$, the strain field can be conveniently defined by a rigid-body displacement that is

$$\boldsymbol{u}^*(\boldsymbol{x}) = \boldsymbol{c}, \tag{4}$$

where $\boldsymbol{c}$ is a constant vector that conveniently leads to null strains that is

$$\boldsymbol{\varepsilon}^*(\boldsymbol{x}) = \boldsymbol{0}. \tag{5}$$

When the kinematically admissible strain field generated by the arbitrary rigid-body displacement (4) is considered, the local form of the work theorem, eqn (3), simply leads to the equation

$$\int_{\Gamma_Q - \Gamma_{Qt}} \boldsymbol{t} d\Gamma + \int_{\Gamma_{Qt}} \bar{\boldsymbol{t}} d\Gamma + \int_{\Omega_Q} \boldsymbol{b} d\Omega = \boldsymbol{0}, \tag{6}$$

which states an integral form of mechanical equilibrium, of tractions and body forces, in the local domain $\Omega_Q \cup \Gamma_Q$.

The modelling strategy adopted in this paper, is based in the application of the work theorem, in the set of kinematically admissible strain fields, to solve the actual elastic problem.

Consider the local form of the work theorem, eqns (3), (6). To derive the equilibrium equations of the numerical model, the kinematic formulation of the local form is carried out through the specification of an appropriate kinematically admissible strain field $\boldsymbol{\varepsilon}^*$.This paper considers the arbitrary rigid-body displacement formulation that leads to the local form of equilibrium (eqn (6)). which are then used to generate the stiffness matrix of the meshfree numerical model.

The statically admissible stress field $\boldsymbol{\sigma}$, which is required to satisfy equilibrium with a system of external forces, is assumed as the stress field that settles in the body, when it is loaded by the actual system of external distributed surface and body forces, with the actual displacement constraints.

Recall that the elastic field that settles in the body is the only totally admissible elastic field that satisfies the given problem. Therefore, besides satisfying static admissibility, through eqns (1), (2), that is the same as satisfying equilibrium through eqn (6), generated by the weak form (3) of the work theorem, this unique totally admissible elastic field also satisfies kinematic admissibility defined as

$$\boldsymbol{\varepsilon} = \boldsymbol{L}\boldsymbol{u}, \tag{7}$$

in the domain Ω, with boundary conditions

$$\boldsymbol{u} = \overline{\boldsymbol{u}}, \tag{8}$$

on the kinematic boundary Γ_u, in which the displacement $\boldsymbol{u}$ is assumed continuous with small derivatives, to allow for geometrical linearity of the strain field $\boldsymbol{\varepsilon}$. Hence, eqn (8), which specifies the constraints of the actual unique solution of the elastic problem must be fulfilled.

For the sake of simplicity, this paper considers the formulation of the meshfree numerical methods in the absence of body forces. Consequently, the equations of equilibrium are always defined only on the boundary of the local domain.

The essential feature of meshfree numerical methods is that they perform the discretization of the problem domain and boundaries with a set of scattered field nodes that do not require any mesh for the approximation of the field variables. The meshfree method ILMF, presented in this paper, is based on the moving least-squares (MLS) approximation.

Each node of the meshfree discretization is associated with its local domain. In general, this local domain is a circular or rectangular region, centered at the respective node, where the rigid-body displacement formulation of the work theorem is defined as a local form of mechanical equilibrium.

The local character of the MLS approximation is a consequence of the compact support of each node, where the respective shape functions are defined. The size of the compact support sets out, in a neighborhood of a sampling point, the respective domain of MLS approximation at this point. The domain of definition contains all the nodes whose MLS shape functions do not vanish at this sampling point. Therefore, the domain of influence of each node is the union of the MLS domains of all points in the node's local domain.

Finally, local meshfree formulations use a node-by-node stiffness calculation to generate, in the domain of influence of the local node, the respective rows of the global stiffness matrix.

In the absence of body forces, the local form of the work theorem eqn (6), can be written simply as

$$\int_{\Gamma_Q - \Gamma_{Qt}} \boldsymbol{t}\, d\Gamma = -\int_{\Gamma_{Qt}} \bar{\boldsymbol{t}}\, d\Gamma, \tag{9}$$

which represents mechanical equilibrium of the boundary tractions of the local domain Ω_Q, associated with the field node $Q \in \Omega_Q$.

General numerical methods can be effectively formulated through a reduced integration of the equilibrium eqn (9) which, in the simplest linear case, leads to a point-wise discrete, form that improves the accuracy and the computational efficiency, as numerical results clearly demonstrate.

Hence, when a linear variation of tractions is assumed along each boundary segment of the local domain, the local integral form of equilibrium can be exactly evaluated with a single quadrature point, centered on each segment of the boundary. Eqn (9) then simply leads to

$$\frac{L_i}{n_i}\sum_{j=1}^{n_i} \boldsymbol{t}_{x_j} = \frac{L_t}{n_t}\sum_{k=1}^{n_t} \bar{\boldsymbol{t}}_{x_k}, \tag{10}$$

in which n_i and n_t denote the total number of integration points, or segments, defined on, respectively the interior local boundary $\Gamma_Q = \Gamma_{Qi} \cup \Gamma_{Qt} \cup \Gamma_{Qu}$, with length L_i, and the local static boundary Γ_{Qt}, with length L_t. This integrated equation obviously represents a point-wise discrete form of mechanical equilibrium of boundary tractions, evaluated at a set of points on the boundary of the local domain Ω_Q.

Consider a meshfree discretization of the body. Then, the local meshfree method with linear reduced integration, symbolically referred to as ILMF, is used to compute the respective system of algebraic equations, in a node-by-node process, throughout traction evaluation at each central point of the boundary segments of the corresponding integrated local form (10) assigned to each node, with rectangular or circular local domains. Fig. 2 schematically represents these local domains with four segments and one integration point on each side, or quadrant, of the respective local domain.

Discretization of the integrated local form (10) is carried out with the MLS approximation, in terms of the unknown nodal parameters $\hat{\boldsymbol{u}}$, thus leading to the system of two linear algebraic equations

$$\frac{L_i}{n_i}\sum_{j=1}^{n_i} \boldsymbol{n}_{x_j}\boldsymbol{D}\boldsymbol{B}_{x_j}\hat{\boldsymbol{u}} = \frac{L_t}{n_t}\sum_{k=1}^{n_t} \bar{\boldsymbol{t}}_{x_k}, \tag{11}$$

that can be written as

$$\boldsymbol{K}_Q\hat{\boldsymbol{u}} = \boldsymbol{F}_Q, \tag{12}$$

in which $\boldsymbol{K}_Q$, the nodal stiffness matrix associated with the field node Q, is a $2x2n$ matrix (n is the number of nodes included in the domain of influence of the reference node Q that is the union of the MLS domains of definition of all integration points in the local domain Ω_Q) given by

$$\boldsymbol{K}_Q = \frac{L_i}{n_i}\sum_{j=1}^{n_i} \boldsymbol{n}_{x_j}\boldsymbol{D}\boldsymbol{B}_{x_j}, \tag{13}$$

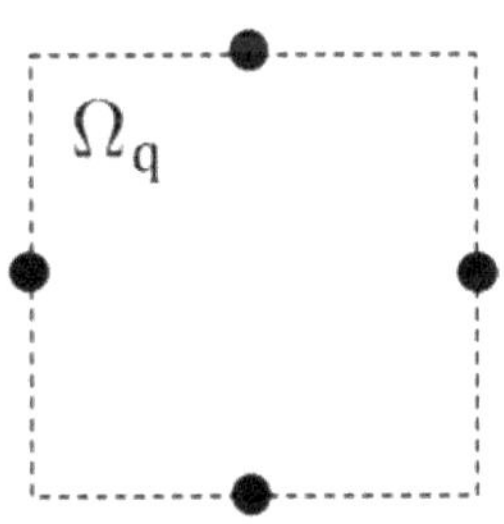

Figure 2: Schematic representation of rectangular domain, with one integration point on each edge of the local domain, for the computation of the local form of ILMF (10).

and $\boldsymbol{F}_Q$ is the respective force vector given by

$$\boldsymbol{F}_Q = \frac{L_t}{n_t}\sum_{k=1}^{n_t} \bar{\boldsymbol{t}}_{x_k}. \tag{14}$$

Consider that the problem has a total of N field nodes Q, each one associated with the respective local domain Ω_Q. Assembling eqn (9) for all M interior and static-boundary field nodes leads to the global system of $2M\ x\ 2N$ equations

$$\boldsymbol{K}\widehat{\mathbf{u}} = \boldsymbol{F}. \tag{15}$$

Finally, the remaining equations are obtained from the $N - M$ boundary field nodes on the kinematic boundary. For a field node on the kinematic boundary, a direct interpolation method is used to impose the kinematic boundary condition as

$$\boldsymbol{u}_k = \boldsymbol{\Phi}_k\widehat{\mathbf{u}} = \overline{\boldsymbol{u}}_k, \tag{16}$$

with $k = 1.2$, where $\overline{\boldsymbol{u}}_k$ is the specified nodal displacement component. Eqn (16) are directly assembled into the global system of eqn (15).

It can be easily anticipated high computational efficiency, with very accurate results, of this local formulation with linear reduced integration. As a matter of fact, the nodal stiffness matrix is effectively computed, in eqn (13), with only 4 integration points (1 integration point on each side of the local boundary), which basically implies a very short processing time to run the analysis. In addition, the reduced integration leads to high accuracy of the results, which plays a key role in the behavior of ILMF, since it implies a reduction of the nodal stiffness which, in turn, leads to an increase of the solution accuracy and, which is most important, presents no instabilities.

3 NUMERICAL EXAMPLES

This section presents some numerical results for Cantilever beam for three different nodal configurations. The effects regular and irregular nodal arrangement are analyzed and compared with exact solution and other popular meshfree method (MLPG).

For a generic node i, the size of the local support Ω_S and the local domain of integration Ω_q are respectively given by

$$r\Omega_S = \alpha_S d_i, \tag{17}$$

$$r\Omega_q = \alpha_q d_i, \tag{18}$$

in which d_i represents the distance of the node i, to the nearest neighboring node; for the analysis is performed for one value of the local support domain size ($\alpha_S = 2.13$), and the local quadrature domain size ($\alpha_q = 0.5$).

The nodal irregularity is generated by changing randomly the coordinates of the nodal regularity distribution by small distance, this movement can be calculated by

$$x'_{1i} = x_{1i} \pm c_n d_{x_{1i}}, \tag{19}$$

$$x'_{2i} = x_i \pm c_n d_{x_{2i}}, \tag{20}$$

in which C_n is a parameter that controls the nodal irregularity and **varies** randomly in the range of 0.0 and 0.4. For nodes located in the boundary there are restrictions that depend on the position of the node.

Displacement and energy norms can be used for error estimation and can be computed respectively as

$$||\boldsymbol{u}|| = \left[\int_{\Omega} \boldsymbol{u}^T \boldsymbol{u}\, d\Omega\right]^{\frac{1}{2}}, \tag{21}$$

$$||\varepsilon|| = \left[\frac{1}{2}\int_{\Omega} \varepsilon^T \mathbf{D} \varepsilon d\Omega\right]^{\frac{1}{2}}. \tag{22}$$

The relative error for $||u||$ and $||\varepsilon||$ is given, respectively by

$$r_u = \frac{||\boldsymbol{u}_{num} - \boldsymbol{u}_{exact}||}{||\boldsymbol{u}_{exact}||}, \tag{23}$$

$$r_\varepsilon = \frac{||\varepsilon_{num} - \varepsilon_{exact}||}{||\varepsilon_{exact}||}. \tag{24}$$

A cantilever beam, as shown in Fig. 3, is subjected to a parabolic traction at the free end. The main properties are tabulated in Table 1 and the problem is solved for plane stress case.

The parabolic traction and the moment of inertia are given by

$$\bar{t}_2(x_2) = -\frac{P}{2I}\left(\frac{D^2}{4} - x_2^2\right), \tag{25}$$

$$I = \frac{D^3}{12}. \tag{26}$$

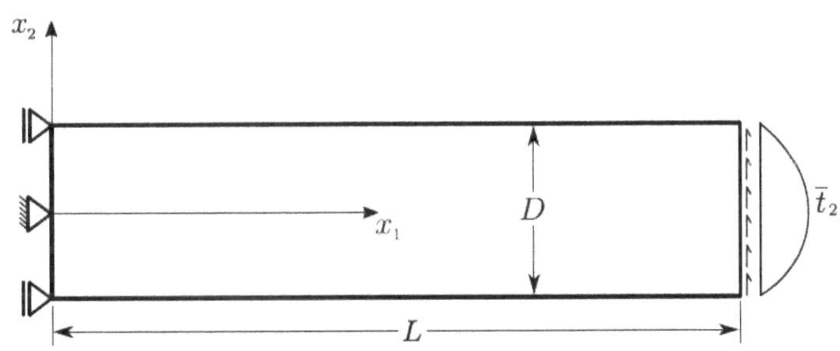

Figure 3: Cantilever beam.

Table 1: Properties of the cantilever beam.

Parameters	Values
Height, D	12 (m)
Length, L	48 (m)
Thickness, t	1 (m)
Load, P	1000 (N)
Modulus of elasticity, E	30 (GPa)
Poisson's ratio, ν	0.3

The equations for the exact displacement are:

$$u_1(x_1, x_2) = -\frac{Px_2}{6EI}\left[(6L - 3x_1) \cdot x_1 + (2 + \nu)\left(x_2^2 - \frac{D^2}{4}\right)\right], \tag{27}$$

$$u_2(x_1, x_2) = -\frac{Px_2}{6EI}\left[3\nu x_2^2(L - x_1) + (4 + 5\nu)\frac{D^2 x_1}{4} + (3L - x_1)x_1^2\right]. \tag{28}$$

And the exact stress components are given by

$$\sigma_{11}(x_1, x_2) = -\frac{P(L - x_1)x_2}{I}, \tag{29}$$

$$\sigma_{12}(x_1, x_2) = -\frac{P}{2I}\left(\frac{D^2}{4} - x_2^2\right), \tag{30}$$

$$\sigma_{22}(x_1, x_2) = 0. \tag{31}$$

The ILMF is used for solving this problem. Both a regular ($C_n = 0$) and irregular ($C_n = 0.1, 0.2, 0.3$ and 0.4) nodal distribution are employed with a discretization of 21 x 9 = 189 nodes.

In the first discretization, Fig. 4, the nodes located in the boundary have a regular distribution and nodes located inside the beam have irregular distribution for the solution using MLPG [15].

The energy relative error for two different methods are presented in Fig 6. The energy relative error for MLPG and ILMF have the same expressions presented in eqns (22), (24). These values for MLPG were obtained in [15].

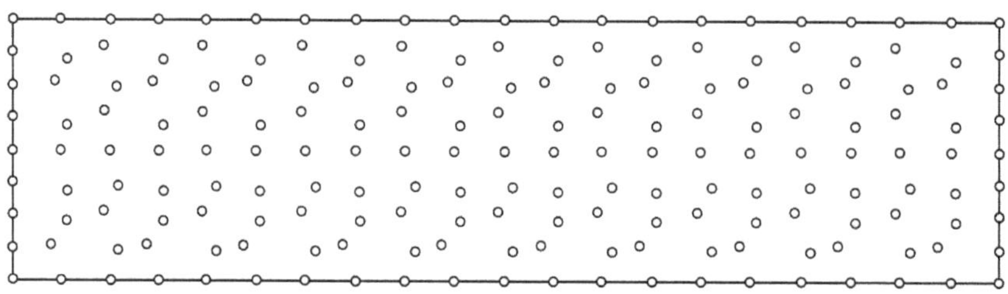

Figure 4: Irregular internal nodal arrangement for the cantilever beam (MLPG) [15].

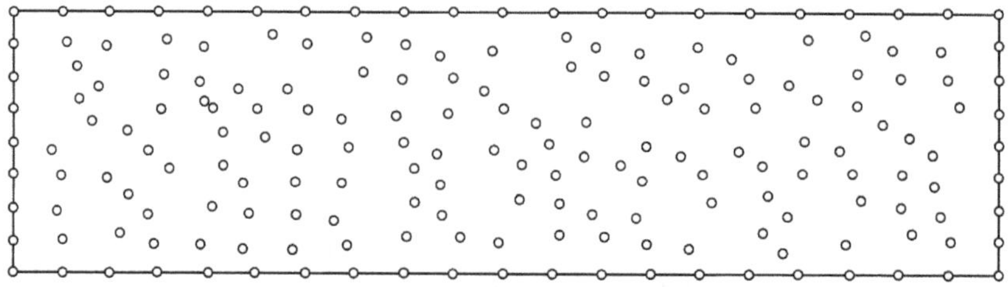

Figure 5: Irregular internal nodal arrangement (ILMF).

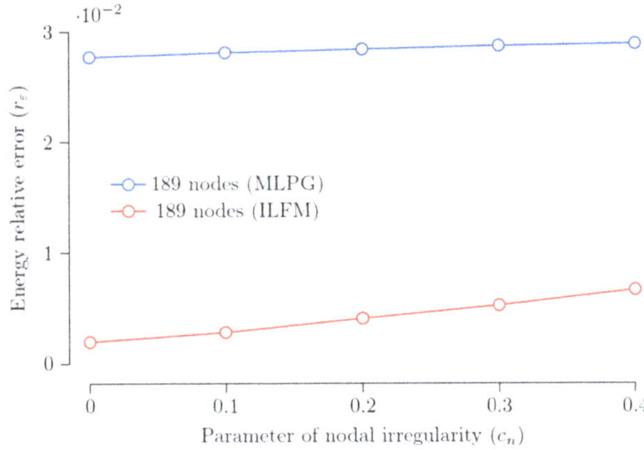

Figure 6: Effect of irregularity on the nodal arrangement on energy relative error for MLPG and ILMF.

Fig. 6 shows that the energy relative error for the irregular nodal discretization, using ILMF and MLPG presented similar behavior for the discretization, but the ILMF presented more accuracy.

Other different irregular nodal discretization was carried out to know the influence on the accuracy of the energy and displacement relatives errors. Two additional configurations are presented with 11 x 5 = 55 nodes and 33 x 17 = 561 nodes.

Figs 7 and 8 illustrates the variation of energy and displacement relative error as a function of the size of the nodal irregularity parameter, which varies between 0.0 to 0.4, with 0.1 increments. Results are presented for three nodal discretization using values fixed of the local support domain (α_s = 2.13) and the local quadrature domain (α_q = 0.5), using both nodal configurations.

Figs 7 and 8 show that the energy and displacement relative error decreases with finer nodal distributions for both configuration. This result evidences that the meshless methods with reduce integration presents an optimal behavior when the mesh is refined.

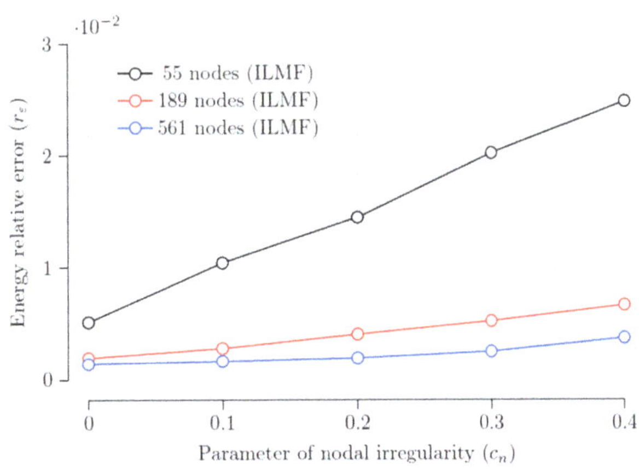

Figure 7: Effect of irregularity on the nodal arrangement on energy relative error for ILMF.

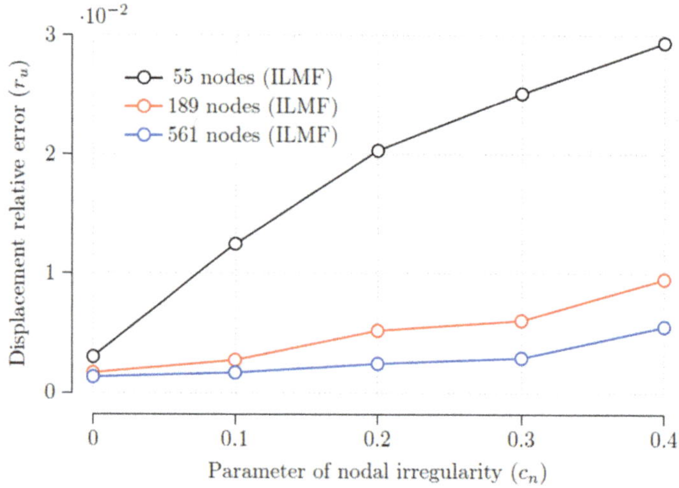

Figure 8: Nodal arrangement irregularity effect on displacement relative error for ILMF.

4 CONCLUSIONS

The effect of the nodal irregularity is irrelevant on energy and displacement relative errors for different configurations and discretization on ILMF. This fact reveals that the meshfree method with reduced integration (ILMF) and meshfree Local Petrov–Galerkin (MLPG) are stable for irregular nodal arrangements, but ILMF shows better accuracy.

The presented results prove that MLPG and ILMF are suitable for solving problems with irregular nodal arrangement. Both are accurate, but ILMF is more precise. Even with $C_n = 0.4$, the maximum irregularity used in this analysis, the ILMF presents a better precision than MLPG using the same discretization.

The results of the energy and displacement relative error for different nodal discretization for the cantilever beam, using ILMF and MLPG, confirm that the effect of the nodal irregularity is very small, and can be neglected.

The small errors obtained for different nodal discretization with a large parameter of irregular arrangement ($C_n = 0.4$), allow the application of this technique to realistic engineering problems involving inhomogeneous irregular geometries.

ACKNOWLEDGEMENTS

The program “PECC – Pós Graduação em Estruturas e Construção Civil”, Department of Civil and Environmental Engineering, Faculty of Technology, University of Brasília. CNPq – Brazilian National Counsel of Technological and Scientific Development for Wilber’s PhD scholarship. “FAP/DF – Fundação de Apoio à Pesquisa do Distrito Federal” for financial support.

REFERENCES

[1] Daxíni, S.D. & Prajapati, J.M., A review on recent contribution of the meshfree methods to structure and fracture mechanics applications. *The Scientific Word Journal*, **2014**, pp. 1–13, 2014. DOI: 10.1155/2014/247172.

[2] Finalyson, B.A., *The Method of Weighted Residuals and Vibrational Principles*, Academic Press, 1972.

[3] Chen, J.S., Hilhnan, M. & Chi, S.W., Meshfree Methods: Progress made after 20 years. *Journal of Engineering Mechanics,* **143**(4), p. 04017001, 2017. DOI: 10.1061/(asce)em.1943-7889.0001176.

[4] Nayroles, B., Touzot, G. & Villon, P., Generalized the Finite Element Method: Diffuse Approximation and Diffuse Elements. *Computational Mechanics*, **10**(5), pp. 307–318, 1992. DOI: 10.1007/bf00364252.

[5] Li, W.K., Jun, S. & Zhang, Y.F., Reproducing kernel particle methods. *International Journal for Numerical Methods in Engineering,* **20**, pp. 1081–1106, 2007.

[6] Belytshko, T., Lu, Y.Y. & Gu, L., Element free Galerkin methods. *International Journal for Numerical Methods in Engineering,* **37**(2), pp. 229–256, 1994.

[7] Atluri, S.N. & Zhu, T., A new Meshless Local Petrov–Galerkin (MLPG) approach in computational mechanics. *Computational Mechanics,* **22**(2), pp. 117–127, 1998. DOI: 10.1007/s004660050346.

[8] Atluri, S.N. & Shen, S., The meshless local Petrov–Galerkin (MLPG) method: A simple and less-costly alternative to the finite element and boundary element methods. *CMES: Computer Modelling in Engineering and Sciences,* **3**(1), pp. 11–15, 2002.

[9] Zhu, T., Zhang, J. & Atluri, S.N., A local boundary integral equation (LBJE) method in computational mechanics and a meshless discretization approach. *Computational Mechanics,* **21**(3), pp. 223–235, 1998. DOI: 10.1007/s004660050297.

[10] Liu, G.R. & Gu, Y.T., A local point interpolation method for stress analysis of two-dimensional solids. *Structural Engineering and Mechanics,* **11**(2), pp. 221–236, 2001. DOI: 10.12989/sem.2001.11.2.221.

[11] Liu, G.R., Yan, L., Wang, l. G. & Gu, Y.T., Point interpolation method based on local residual formulation using radial basis functions. *Structural Engineering and Mechanics,* **14**(6), pp. 713–732, 2002. DOI: 10.12989/sem.2002.14.6.713.

[12] Atluri, S.N., Han, Z.D. & Rajendran, A.M., A new implementation of the meshless finite volume method through the mlpg mixed approach. *CMES: Computer modelling in Engineering and Sciences,* **6**, pp. 491–513, 2004.

[13] Oliveira, T. & Portela, A., Weak–form-collocation: A local meshless method in linear elasticity. *Engineering Analysis with Boundary Elements,* **73**, pp. 144–160, 2016. DOI: 10.1016/j.enganabound.2016.09.010.

[14] Santana, E., Oliveira, T., Vélez, W. & Portela, A., Meshfree method with reduced integration and automatic parameter optimization. *Computers and Structures,* 2017.

[15] Liu, G.R., *Meshfree Methods Moving Beyond The Finite Element Method*, CRC Press, 2003.

MOVING FINITE ELEMENT METHOD

VLADIMIR SLADEK, MIROSLAV REPKA & JAN SLADEK
Institute of Construction and Architecture, Slovak Academy of Sciences, Slovakia

ABSTRACT

A novel discretization method is proposed and developed for numerical solution of boundary value problems governed by partial differential equations. The spatial variation of field variables is approximated by using Lagrange finite elements for interpolation without discretization of the analysed domain into the mesh of finite elements. Only the net of nodal points is used for discrete degrees of freedom on the analysed domain and its boundary. The governing equations are considered at interior nodal points while the boundary conditions at nodal points on the boundary. The finite elements are created for each nodal point properly instead of using fixed finite elements like in standard Finite Element Method. In this way, we can eliminate interfaces between elements as well as the difficulties with continuity of derivatives of field variables on such interfaces. Both the strong and weak formulations are implemented for governing equations. The reliability (accuracy and efficiency) of the new method has been verified in numerical simulations for 2D problems of heat conduction in solids with possible continuous gradation of the heat conduction coefficient.

Keywords: Lagrange finite element, bi-quadratic and bi-cubic approximation, strong and weak formulations.

1 INTRODUCTION

The Finite Element Method (FEM) is the most widely used computational method because of its universality and simplicity. Nevertheless, it suffers from discontinuities of derivatives on element interfaces as long as C^0 continuous finite elements (FE) are employed [1]. Another favourite subdomain method is the Finite Volume Method (FVM) [2], [3] enforcing the conservation (or balance) physical laws. Recently various mesh-free methods appeared which are universal. The price which should be paid for elimination of creation of discretization elements is time consuming evaluation of shape functions because of the loss of polynomial interpolation.

In 1981, Miller introduced the Moving Finite Element Method (MFEM) [5], [6] as an adaptive grid method in which the grid of finite elements deforms continuously in time with the solution. This method found wide application in vast diversity of problems in science and engineering involving time-dependent PDE describing solutions with pronounced spatial activity in some regions of space that move with time [7], [8].

In this paper, we propose the strong and weak formulations with developing the so called Moving Finite Element (MFE) approximation. In contrast to the MFEM [5]–[8], the grid of nodes is fixed (not adaptive) in the proposed method and the FE are created individually around each nodal point according to its position in the analysed domain.

2 HEAT CONDUCTION IN FGM SOLIDS: DISCRETIZATION WITH POLYNOMIAL INTERPOLATION

For the sake of simplicity, let us consider the stationary heat conduction in 2D domain of isotropic solid with the governing equation

$$\left(\lambda u_{,i}\right)_{,i} = 0 \ \text{ or } \ \lambda u_{,ii} + \lambda_{,i} u_{,i} = 0 \ \text{ in } \Omega \tag{1}$$

and the boundary conditions

WIT Transactions on Engineering Sciences, Vol 122, © 2019 WIT Press
www.witpress.com, ISSN 1743-3533 (on-line)
doi:10.2495/BE410111

$$u(\mathbf{x}) = T(\mathbf{x}) \text{ on } \partial\Omega_T\,;\; -\lambda(\mathbf{x})n_i(\mathbf{x})u_{,i}(\mathbf{x}) = f(\mathbf{x}) \text{ on } \partial\Omega_f\,;\; \partial\Omega = \partial\Omega_T \cup \partial\Omega_f\,. \quad (2)$$

The heat conduction coefficient λ is assumed to be continuous differentiable function of Cartesian coordinates. In general, the analytical solution of the mentioned boundary value problem is not available and discretization methods utilizing polynomial interpolation belong to widely spread numerical techniques because of their universality, simplicity and computational efficiency. The field variable is approximated in certain subdomain as

$$u(\mathbf{x}) = \sum_c u(\mathbf{x}^c)\varphi^c(\mathbf{x}),$$

where the unknown field variable is expressed in terms of finite number of nodal unknowns $u(\mathbf{x}^c)$ and known shape functions $\varphi^c(\mathbf{x})$. For calculation of the unknown nodal values, there are mostly used two approaches: (i) the strong formulation with collocation of the governing equation at nodal points; (ii) the weak formulation when the governing equation is considered in integral sense with using appropriate test functions. The prescribed boundary conditions can also be considered either in the strong or weak sense.

2.1 Moving Finite Element (MFE) approximation

Instead of discretization of the analysed domain into finite elements, we consider the net of nodal points which can be used for creation of so called moving finite elements individually around particular nodes according to their localization in the analysed domain as illustrated in Fig. 1 for planar elements. There are shown the bi-quadratic Lagrange elements (with 9 nodes). If $\mathbf{x}^a$ is an interior node of the analysed domain, the FE is constructed from surrounding nodes in such a way that $\mathbf{x}^a$ is the centre of the FE. In the case of a corner node the associated FE is constructed uniquely. Finally, the FE associated with other boundary node involves this node as the mid-side node of the FE. Obviously, a uniform mesh of nodes is the most appropriate one for generation of moving finite elements around each node. Having defined the FE around particular nodes $\mathbf{x}^a$, one can simply approximate the spatial variation of a field variable within the associated moving finite element E^a

$$u(\mathbf{x})\big|_{E^a} \approx \sum_{\gamma=1}^{n} u^{a_\gamma} N^\gamma(\xi)\,,\; u^{a_\gamma} = u(\mathbf{x}^{a_\gamma})\,, \quad (3)$$

where ξ is the shortcut for intrinsic coordinates (ξ_1, ξ_2), $N^\gamma(\xi)$ stand for standard interpolation shape functions in Lagrange FE, and a_γ is the global number of the node on the element E^a with the local number $\gamma \in \{1,2,...,n\}$. In order to account also the 2nd order derivatives of field variables, it is necessary to use finite elements with complete higher order polynomial interpolation, at least bi-quadratic Lagrange FE defined with 9 nodes [1].

If $\mathbf{x}^a$ is an interior node of the analysed domain, the FE is constructed from surrounding nodes in such a way that $\mathbf{x}^a$ is the centre of the FE. In the case of a corner node the associated FE is constructed uniquely. Finally, the FE associated with other boundary node involves this node as the mid-side node of the FE. Obviously, a uniform mesh of nodes is the most

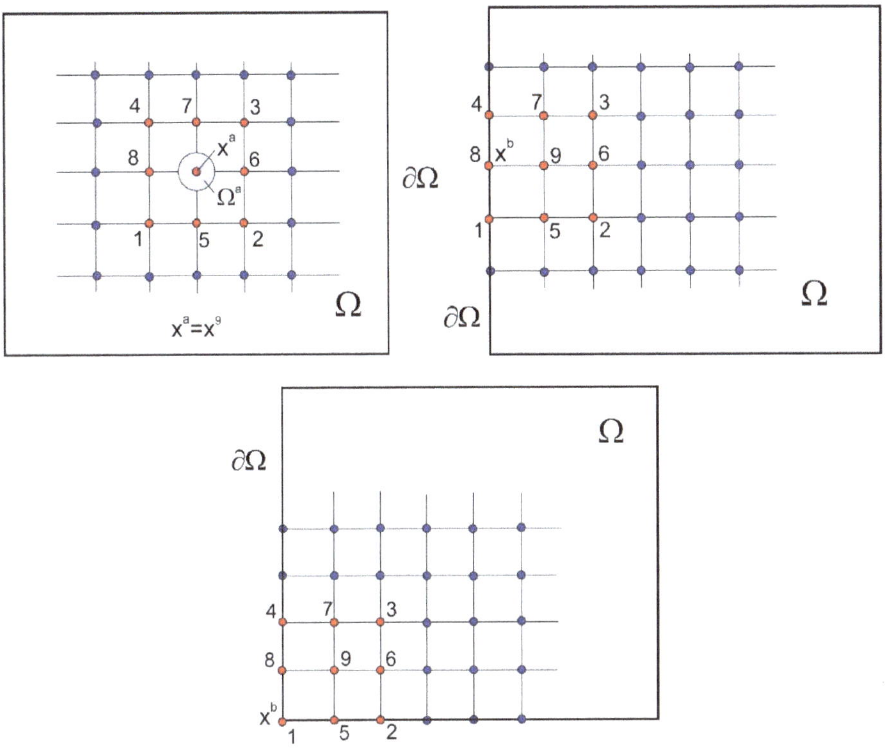

Figure 1: Creation of moving FE around the interior node $\mathbf{x}^a$ as well as the boundary node and corner nodes $\mathbf{x}^b$.

appropriate one for generation of moving finite elements around each node. Having defined the FE around particular nodes $\mathbf{x}^a$, one can simply approximate the spatial variation of a field variable within the associated moving finite element E^a

$$u(\mathbf{x})\big|_{E^a} \approx \sum_{\gamma=1}^{n} u^{a_\gamma} N^\gamma(\xi), \; u^{a_\gamma} = u(\mathbf{x}^{a_\gamma}), \tag{3}$$

where ξ is the shortcut for intrinsic coordinates (ξ_1, ξ_2), $N^\gamma(\xi)$ stand for standard interpolation shape functions in Lagrange FE, and a_γ is the global number of the node on the element E^a with the local number $\gamma \in \{1, 2, ..., n\}$. In order to account also the 2nd order derivatives of field variables, it is necessary to use finite elements with complete higher order polynomial interpolation, at least bi-quadratic Lagrange FE defined with 9 nodes [1]. The gradient of the field variable can be approximated on the FE as

$$u_{,i}(\mathbf{x})\big|_{E^a} = \left.\frac{\partial u(\mathbf{x})}{\partial x_i}\right|_{E^a} \approx \sum_{\gamma=1}^{n} u^{a_\gamma} b_i^{\gamma a}(\xi), \; b_i^{\gamma a}(\xi) = \mathrm{Y}_{ik}^a \frac{\partial N^\gamma(\xi)}{\partial \xi_k}, \; \mathrm{Y}_{ik}^a = \left.\frac{\partial \xi_k}{\partial x_i}\right|_{E^a}, \tag{4}$$

Recall that the matrix $[Y^a]$ is the inverse of the Jacobi matrix, i.e. $[Y^a]=[J^a]^{-1}$, and

$$\left[J^a\right]=\begin{pmatrix} \partial x_1/\partial\xi_1 & \partial x_2/\partial\xi_1 \\ \partial x_1/\partial\xi_2 & \partial x_2/\partial\xi_2 \end{pmatrix}\Bigg|_{E^a}=\sum_{\gamma=1}^{n}\begin{pmatrix} x_1^{a_\gamma}\partial N^\gamma/\partial\xi_1 & x_2^{a_\gamma}\partial N^\gamma/\partial\xi_1 \\ x_1^{a_\gamma}\partial N^\gamma/\partial\xi_2 & x_2^{a_\gamma}\partial N^\gamma/\partial\xi_2 \end{pmatrix},$$

$$J_{ki}^a=\frac{\partial x_i}{\partial\xi_k}\bigg|_{E^a}=\sum_{\gamma=1}^{n}x_i^{a_\gamma}\frac{\partial N^\gamma}{\partial\xi_k}, \tag{5}$$

since we use the isoparametric elements and the Cartesian coordinates are approximated as

$$x_i\big|_{E^a}\approx\sum_{\gamma=1}^{n}x_i^{a_\gamma}N^\gamma(\xi). \tag{6}$$

Apparently,

$$\left[Y^a\right]=\left[J^a\right]^{-1}=\frac{1}{\left|J^a\right|}\sum_{\beta=1}^{n}\begin{pmatrix} x_2^{a_\beta}\partial N^\beta/\partial\xi_2 & -x_2^{a_\beta}\partial N^\beta/\partial\xi_1 \\ -x_1^{a_\beta}\partial N^\beta/\partial\xi_2 & x_1^{a_\beta}\partial N^\beta/\partial\xi_1 \end{pmatrix},$$

$$\left|J^a\right|=\left(\frac{\partial x_1}{\partial\xi_1}\frac{\partial x_2}{\partial\xi_2}-\frac{\partial x_1}{\partial\xi_2}\frac{\partial x_2}{\partial\xi_1}\right)\Bigg|_{E^a}=\sum_{\alpha,\beta=1}^{n}x_1^{a_\alpha}x_2^{a_\beta}\left(\frac{\partial N^\alpha}{\partial\xi_1}\frac{\partial N^\beta}{\partial\xi_2}-\frac{\partial N^\alpha}{\partial\xi_2}\frac{\partial N^\beta}{\partial\xi_1}\right). \tag{7}$$

As regards the 2nd order derivatives, we can write

$$u_{,ij}(\mathbf{x})\big|_{E^a}=\frac{\partial^2u(\mathbf{x})}{\partial x_i\partial x_j}\bigg|_{E^a}\approx\sum_{\gamma=1}^{n}u^{a_\gamma}\frac{\partial\xi_m}{\partial x_j}\frac{\partial}{\partial\xi_m}b_i^{\gamma a}(\xi)=$$

$$=\sum_{\gamma=1}^{n}u^{a_\gamma}Y_{jm}^a\frac{\partial}{\partial\xi_m}\left(Y_{ik}^aN_{,k}^\gamma(\xi)\right)=\sum_{\gamma=1}^{n}u^{a_\gamma}Y_{jm}^a\left[Y_{ik}^aN_{,km}^\gamma(\xi)+Y_{ik,m}^aN_{,k}^\gamma(\xi)\right]=$$

$$=\sum_{\gamma=1}^{n}u^{a_\gamma}Y_{jm}^a\left[Y_{ik}^aN_{,km}^\gamma(\xi)-Y_{is}^aJ_{sl,m}^aY_{lk}^aN_{,k}^\gamma(\xi)\right]=$$

$$=\sum_{\gamma=1}^{n}u^{a_\gamma}Y_{jm}^a\left[Y_{ik}^aN_{,km}^\gamma(\xi)-Y_{is}^aJ_{sl,m}^ab_l^{\gamma a}(\xi)\right]=\sum_{\gamma=1}^{n}u^{a_\gamma}b_{ij}^{\gamma a}(\xi) \tag{8}$$

since

$$\left(Y_{ik}^aJ_{kl}^a\right)_{,m}=0=Y_{ik,m}^aJ_{kl}^a+Y_{ik}^aJ_{kl,m}^a\ \Rightarrow\ Y_{ik,m}^a=-Y_{is}^aJ_{sl,m}^aY_{lk}^a$$

$$\text{with } J^a_{sl,m} = \sum_{\gamma=1}^{n} x_l^{a_\gamma} \frac{\partial^2 N^\gamma}{\partial \xi_s \partial \xi_m} = J^a_{ml,s}$$

$$b_{ij}^{\gamma a}(\xi) := Y^a_{jm}\left[Y^a_{ik} N^\gamma_{,km}(\xi) - Y^a_{is} J^a_{sl,m} b_l^{\gamma a}(\xi) \right] = b_{ji}^{\gamma a}(\xi) . \tag{9}$$

Note that in the case of standard FEM with C^0 continuous elements, one can expect discontinuities of both the 1st and 2nd order derivatives on interfaces of finite elements. On the other hand, in the case of moving finite elements approximation, there are no element interfaces because each node is either interior node of the associated FE, or it is a mid-side node or a corner of the associated FE. Thus, the C^0-continuity troubles of standard FEM disappear in formulations based on MFE approximation.

2.2 Strong formulation based of MFE approximation

The strong formulation seems to be computationally the most efficient formulation because the only thing, needed for creation of discretized equations for calculation of nodal unknowns, is approximation of derivatives at nodal points, since the discretized equations take the form

$$\lambda(\mathbf{x}^c) \sum_{\gamma=1}^{n} u^{c_\gamma} b_{ij}^{\gamma c}(\xi^{\alpha_c}) + \lambda_{,i}(\mathbf{x}^c) \sum_{\gamma=1}^{n} u^{c_\gamma} b_i^{\gamma c}(\xi^{\alpha_c}) = 0 \text{ for each } \mathbf{x}^c \in \Omega , \tag{10a}$$

$$u(\mathbf{x}^b) = T(\mathbf{x}^b) \text{ for } \mathbf{x}^b \in \partial\Omega_T , \tag{10b}$$

$$\lambda n_i(\mathbf{x}^b) \sum_{\gamma=1}^{n} u^{b_\gamma} b_i^{\gamma b}(\xi^{\alpha_b}) = -f(\mathbf{x}^b) \text{ for } \mathbf{x}^b \in \partial\Omega_f , \tag{10c}$$

where α_c is the local number (and $\xi_1^{\alpha_c}$, $\xi_2^{\alpha_c}$ are intrinsic coordinates) of the global node $\mathbf{x}^c$ on the associated finite element E^c . The formulation resembles the Finite Difference Method (FDM), but the approximation of field variable used here is different.

2.3 Weak formulation based of MFE approximation

The weak form of the governing eqn (1) is given as

$$\int_\Omega \left(\lambda(\mathbf{x}) u_{,i}(\mathbf{x}) \right)_{,i} \omega(\mathbf{x}) d\Omega = 0 , \tag{11}$$

where $\omega(\mathbf{x})$ is a weight function. Selecting the weight function as the compactly supported function

$$\omega^a(\mathbf{x}) = \begin{cases} 1, & \mathbf{x} \in \overline{\Omega}^a , \\ 0, & \text{otherwise} , \end{cases}$$

with $\bar{\Omega}^a = \Omega^a \cup \partial\Omega^a$ being the subdomain of the analysed domain $\bar{\Omega}$ and taken around a point $\mathbf{x}^a \in \bar{\Omega}^a \subset \bar{\Omega}$, eqn (11) can be rewritten as

$$\int_{\partial\Omega^a} n_i(\mathbf{x})\lambda(\mathbf{x})u_{,i}(\mathbf{x})d\Omega = 0\,, \tag{12}$$

where $n_i(\mathbf{x})$ is the unit outward normal vector at $\mathbf{x} \in \partial\Omega^a$.

From the physical point of view, eqn (12) expresses the energy balance in local subdomain $\bar{\Omega}^a$ with absent volume heat sources. From the mathematical point of view, it is the formulation for solution of the considered problem by the Finite Volume Method (FVM) [4] without specifying the approximation of field variable. In this paper, we employ the MFE for approximation. In contrast to the strong formulation, now it is sufficient to approximate only the 1st order derivatives of the field variable, but we need to integrate over the boundary of local subdomain. Recall that the subdomain $\bar{\Omega}^a$ should involve only one nodal point $\mathbf{x}^a$ in the assumed MFE approximation. Since there are no other restrictions for selection of $\bar{\Omega}^a$, it is reasonable to select such a shape of $\bar{\Omega}^a$ that the integration over $\partial\Omega^a$ would be as simple as possible. Furthermore, the integration is carried out in the intrinsic space and Ω^a is the image of $\tilde{\Omega}^a$ resulting from the transformation: $\tilde{\Omega}^a \ni \boldsymbol{\xi} \to \mathbf{x} \in \Omega^a$ according to (6). Therefore we select $\tilde{\Omega}^a$ as the circle centred at $\left(\xi_1^{\alpha_a}, \xi_2^{\alpha_a}\right)$, i.e. $\tilde{\Omega}^a = \left\{\forall\, \boldsymbol{\xi} \in \tilde{E}^a;\, |\boldsymbol{\xi} - \boldsymbol{\xi}^{\alpha_a}| < 1\right\}$. Denoting the radius of the circular contour $\partial\tilde{\Omega}^a$ as ξ^a, the intrinsic coordinates of a point on $\partial\tilde{\Omega}^a$ can be expressed in terms of one angular parameter $\varphi \in [0,\, 2\pi]$ as

$$\partial\tilde{\Omega}^a = \left\{\forall\, \boldsymbol{\xi} \in \tilde{E}^a;\, \xi_k = \xi_k^{\alpha_a} + \xi^a\left(\delta_{k1}\cos\varphi + \delta_{k2}\sin\varphi\right)\right\}. \tag{13}$$

Since ξ^a is constant on $\partial\tilde{\Omega}^a$, we have $d\xi_k\big|_{\partial\tilde{\Omega}^a} = \xi^a\tilde{\tau}_k d\varphi$, $\tilde{\tau}_k = \delta_{k2}\cos\varphi - \delta_{k1}\sin\varphi$ and the infinitesimal length arc of circular contour $\partial\tilde{\Omega}^a$ is $ds\big|_{\partial\tilde{\Omega}^a} = \tilde{\tau}_k d\xi_k = \xi^a d\varphi$. Then,

$$d\Gamma\big|_{\partial\Omega^a} = \tau_i dx_i\big|_{\partial\Omega^a} = \tau_i \frac{dx_i}{ds}ds\bigg|_{\partial\Omega^a} = \frac{g_i}{g}g_i ds\bigg|_{\partial\Omega^a} = g\big|_{\partial\Omega^a}\,\xi^a d\varphi\,, \tag{14}$$

where the unnormalized tangent vector on $\xi_2^{\alpha_a}$ $\partial\Omega^a$ is defined as $g_i := dx_i / ds$ and it can be expressed as

$$g_i\big|_{\partial\Omega^a} = \frac{1}{\xi^a}\frac{dx_i}{d\varphi}\bigg|_{\partial\Omega^a} = \frac{1}{\xi^a}\frac{\partial x_i}{\partial\xi_j}\frac{\partial\xi_j}{d\varphi}\bigg|_{\partial\Omega^a} = \frac{1}{\xi^a}\left(\frac{\partial x_i}{\partial\xi_2}\cos\varphi - \frac{\partial x_i}{\partial\xi_1}\sin\varphi\right)\bigg|_{\partial\Omega^a}.$$

Hence,

$$g_i^{\partial\Omega^a} := g_i\big|_{\partial\Omega^a} = \frac{1}{\xi^a}\sum_{\gamma=1}^{n} x_i^{a_\gamma}\left(N_{,2}^{\gamma}(\xi^{\partial\Omega^a})\cos\varphi - N_{,1}^{\gamma}(\xi^{\partial\Omega^a})\sin\varphi\right), \tag{15}$$

with $\xi^{\partial\Omega^a} := \left(\xi_1^{\partial\Omega^a}, \xi_2^{\partial\Omega^a}\right) = \left(\xi_1^{\alpha_a}, \xi_2^{\alpha_a}\right) + \xi^a\left(\cos\varphi, \sin\varphi\right)$.

The unit normal vector on $\partial\Omega^a$ can be defined as $n_i\big|_{\partial\Omega^a} = \varepsilon_{ik3} g_k^{\partial\Omega^a} / g^{\partial\Omega^a}$. Thus, eqn (12) can be rewritten as

$$\varepsilon_{ik3}\xi^a \sum_{\gamma=1}^{n} u^{a_\gamma} \int_0^{2\pi} \lambda(\mathbf{x}^{\partial\Omega^a}) b_i^{\gamma a}(\xi^{\partial\Omega^a}) g_k^{\partial\Omega^a}\, d\varphi = 0, \tag{16}$$

with the Cartesian coordinates of the integration point on $\partial\Omega^a$ are given by

$$x_i^{\partial\Omega^a} := x_i\big|_{\partial\Omega^a} = \sum_{\gamma=1}^{n} x_i^{a_\gamma} N^{\gamma}(\xi^{\partial\Omega^a}).$$

The discretized prescribed boundary conditions can be considered in the same way as in the strong formulation. Then, the complete set of discretized equations in the weak formulation is given by eqns (16), (10b) and (10c).

3 NUMERICAL EXAMPLES

In order to test the applicability of the MFE approximation, we compared accuracy and computational efficiency achieved by both the strong and weak formulations with MFE approximation and by the standard FEM in solving simple problem of stationary heat conduction in a rectangular domain of solids with constant and/or functionally graded heat conduction coefficient λ. The dimensionless formulation is used with $L_1 \times L_2 = 3\times 3$, prescribed temperature on the bottom and top $u(x_1, x_2 = 0) = 0$, $u(x_1, x_2 = 3) = 100$; the lateral sides are thermally isolated $n_i u_{,i}\big|_{x_1=0} = 0 = n_i u_{,i}\big|_{x_1=3}$; the functional gradation is considered as power-law gradation in the vertical direction $\lambda = \lambda_0\left(1 + 2x_2 / L_2\right)^2$. In numerical solution, we employed uniform net of nodal points with n_1, n_2, $N = n_1 \times n_2$ being the number of nodes in horizontal and vertical directions, and the total number of nodes, respectively. Since we shall use also the standard FEM, the net of nodes is designed to be applicable also to FEM discretization. Furthermore, the Lagrange finite element is composed of $\sqrt{n}\times\sqrt{n}$ nodes. Therefore, $n_1 = e_1\left(\sqrt{n}-1\right)$ and $n_2 = e_2\left(\sqrt{n}-1\right)$, where e_k is the number of FE along the x_k-direction. Since the problem is actually 1-dimensional, we have chosen $e_1 = 1$ and e_2 (or n_2) to be variable. Finally, we have used quadrilateral bi-quadratic ($n = 9$) as well as bi-cubic ($n = 16$) Lagrange finite elements. The accuracy is characterized by average relative error defined as

$$\frac{1}{N}\sum_{a=1}^{N}\left|\frac{f^{c}(\mathbf{x}^{a})}{f^{e}(\mathbf{x}^{a})}-1\right|,$$

with the superscripts c and e denoting the computed and exact values, respectively.

Firstly, consider the results for heat conduction in a homogeneous solid. We achieved very precise numerical results for the temperature $u^{c}(\mathbf{x})$ as well as temperature gradients $u^{c}_{,2}(\mathbf{x})$. In this case, the exact solution is given by linear distribution of temperature and almost exact solution is achieved by using only one FE. Any other increase of degrees of freedom in numerical calculation yields increasing numerical error. Nevertheless, the best accuracy is achieved by strong formulations with MFE approximation. Since $u^{e}_{,22}(\mathbf{x})\equiv 0$, we have used average absolute error instead of average relative error. Now, the accuracy by standard FEM is significantly worse than that by using formulations with MFE approximation and the bi-cubic FE results into one order better accuracy than the bi-quadratic FE.

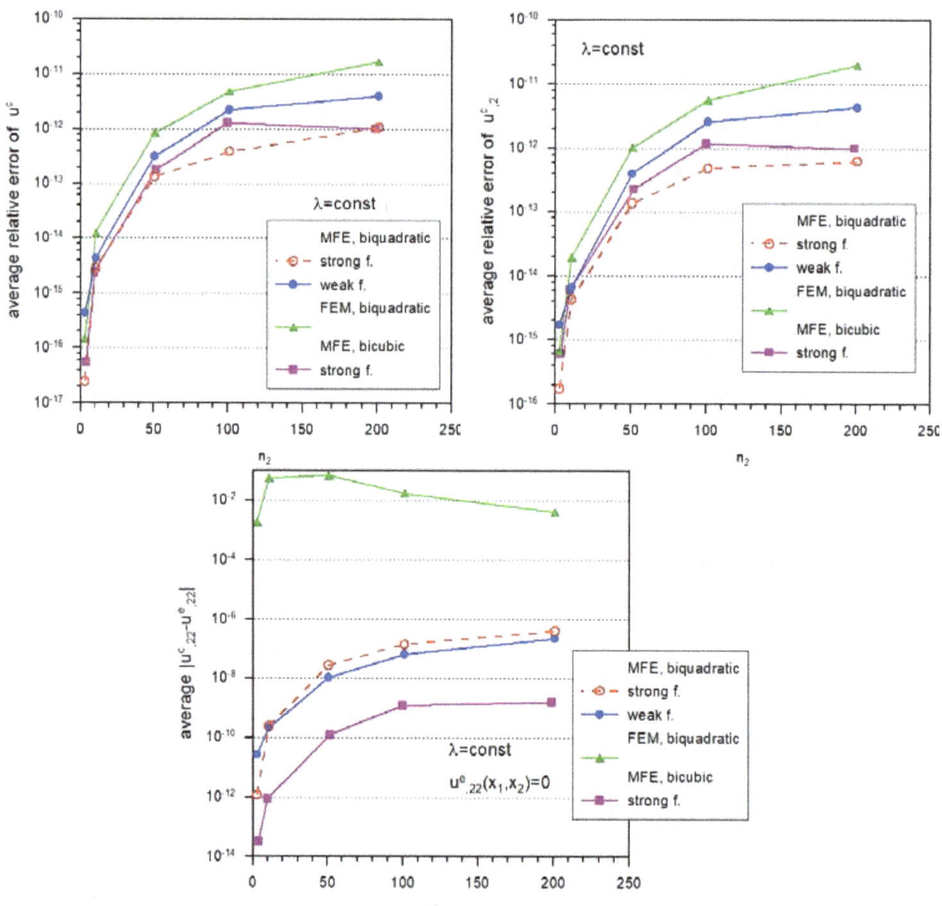

Figure 2: Comparison of accuracies for u^{c}, $u^{c}_{,2}$, and $u^{e}_{,22}$ in a homogeneous solid.

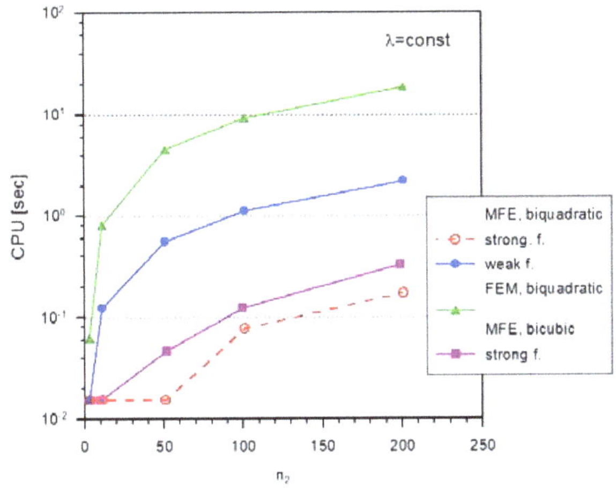

Figure 3: Comparison of computational efficiency in case of homogeneous solid.

There is significant difference between the computational times by using standard FEM and formulations with MFE as well as between the strong and weak formulations.

Substantially different accuracy is achieved for problems in FGM solid, when the exact solution exhibits a non-linear variation. Unexpectable accurate are numerical results for u^c obtained by strong formulation with MFE approximation. The other three techniques exhibit convergence to exact solution with increasing the amount of nodes and the convergence of u^c by the standard FEM is faster than by other two MFE approaches. Nevertheless, the best accuracy of $u^c_{,2}$ is achieved by the MFE with bi-cubic FE, while in the case of bi-quadratic FE there is no difference in accuracy achieved by the weak and strong formulations with MFE and the accuracy by standard FEM is the worst. Similar conclusion can be drawn also for the accuracy of $u^e_{,22}$ but the differences in achieved values are more remarkable. As regards the computational times, these are practically the same as in the case of homogeneous solid. The main shortcoming of the standard FEM approach consists in discontinuity of derivatives on FE interfaces. The differences in relative errors of $u^c_{,2}$ vary between 3% (on the top of bottom FE) and 0.5% (on the bottom of the top FE) if $e_2 = 5$, while in the case of $u^e_{,22}$ these differences are 28% and 6%.

4 CONCLUSIONS

The Moving Finite Element (MFE) approximation is proposed and developed. In contrast to the standard FEM, there is no mesh of FE created, but the polynomial interpolation is successfully utilized within MFE created individually around each nodal point of the net of nodes. Thus, the difficulty of discontinuous derivatives on FE interfaces is eliminated and C^0 continuous Lagrange FE are applicable to both the weak and strong formulations for solution of boundary value problems. In numerical test examples, the reliability of the method is verified and comparative study accomplished with using quadrilateral bi-quadratic and bi-cubic Lagrange FE.

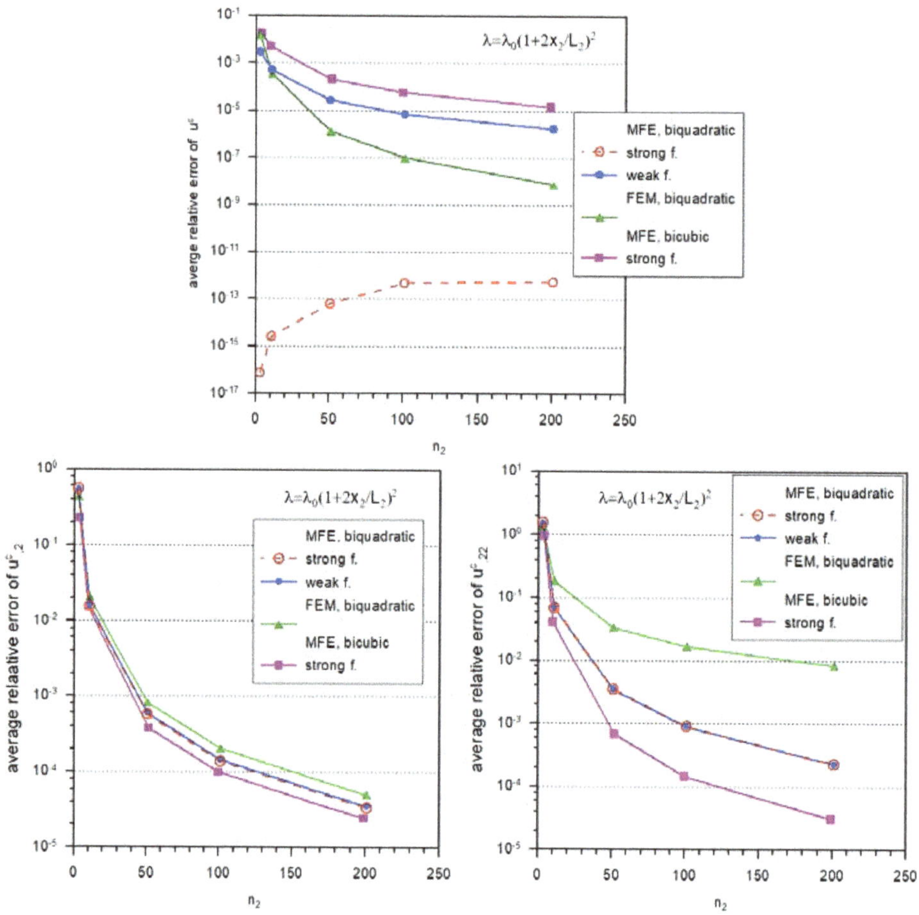

Figure 4: Comparison of accuracies for u^c, $u^c_{,2}$, and $u^e_{,22}$ in the FGM solid.

ACKNOWLEDGEMENT

This work was partially supported by the Slovak Research and Development Agency under the contract No. APVV-14-0440.

REFERENCES

[1] Hughes, T.J.R., *The Finite Element Method, Linear Static and Dynamic Finite Element Analysis*, Prentice-Hall, Inc.: Englewood Cliffs, 1987.

[2] McDonald, P.W., The computation of transonic flow through two-dimensional gas cascades. *ASME 1971 International Gas Turbine Conference and Products Show*, pp. V001T01A089, 1971. DOI:10.1115/71-GT-89.

[3] MacCormack, R.W. & Paullay, A.J., Computational efficiency achieved by time splitting of finite difference operators. *AIAA paper,* pp. 72–154, 1972.

[4] Kolditz, O., *Computational Methods in Environmental Fluid Mechanics, Chapter 8: Finite Volume Method*, Springer-Verlag: Heidelberg, pp. 173–190, 2002.

[5] Miller, K. & Miller, R., Moving finite elements I. *SIAM Journal Numerical Analysis,* **18**(6), pp. 1019–1032, 1981.
[6] Miller, K., Moving finite elements II. *SIAM Journal Numerical Analysis,* **18**(6), pp. 1033–1057, 1981.
[7] Coimbra, M., Rodrigues, A., Rodrigues, J., Robalo, R. & Almeida, R., *Moving Finite Element Method, Fundamentals and Applications in Chemical Engineering*, CRC Press: Boca Raton, 2016.
[8] Baines, M., Hubbard, M., Jimack, P. & Mahmood, R., A moving-mesh finite element method and its application to the numerical solution of phase-change problems. *Communications in Computational Physics*, **6**, pp. 595–624, 2009.

MULTIDOMAIN SINGULAR BOUNDARY METHOD FOR 2D LAMINAR VISCOUS FLOW

JURAJ MUŽÍK & ROMAN BULKO
Department of Geotechnics, Faculty of Civil Engineering, University of Zilina, Slovakia

ABSTRACT
In this paper, a numerical algorithm is developed for the solution of two-dimensional isothermal laminar viscous flow. The proposed numerical method implementation, which is quadrature-free, is based on the stream-velocity formulation of Navier–Stokes equations, the method of the particular solutions and the singular boundary method (SBM). The steady stream-velocity formulation of NS equation is 4th order biharmonic non-homogeneous nonlinear equation type which is solved using the proposed method with the non-homogeneous and nonlinear terms approximated using the method of particular solutions with multiquadrics RBF function. The accuracy of the method is proven using 2D backward-facing step and lid-driven cavity test cases.
Keywords: singular boundary method, dual reciprocity, Navier–Stokes equation, biharmonic equations.

1 INTRODUCTION

The numerical codes based on the theory of boundary integral equations are represented mainly by the boundary element method (BEM) [1], [2], that is one of the very important numerical methods used for scientific research and engineering practice.

However, the BEM is a very efficient numerical method from the mathematical and engineering point of view it brings some drawbacks caused mainly by time-consuming process of numerical integration. On the other side the boundary collocation methods, like the most-known method of fundamental solution (MFS) avoids the process of numerical integration and creation of a computational mesh in the sense of BEM. The singularity problem of the fundamental solution in MFS is overcome by construction of a fictitious boundary. The optimal position of the nodes that make up fictitious this boundary remains the open problem. To avoid the need for the fictitious boundary creation Sarler [5] proposed the modified method of fundamental solutions (MMFS). The MMFS uses the same set of boundary nodes as source and response nodes, but it requires a computationally intensive calculation of the characteristic matrix diagonal elements.

Chen et al. [6] and Chen and Gu [7] presented a numerical alternative to MMFS called a singular boundary method (SBM). Like MMFS, the SBM is using the nodes of regular boundary only and regularizes the singularities of fundamental solutions (diagonal elements of the characteristic matrix) using the values of so-called origin intensity factors (OIFs). There are several rigorous or empirical techniques used to determine the OIFs values, namely, inverse interpolation technique (IIT), subtracting and adding-back regularization and formulas [8], [9] based on empirical values. In this article, the regularized boundary integral equation is used to derive a formula for Dirichlet origin intensity factors.

This paper describes the possibility of the steady laminar isothermal flow solution, which is governed by the biharmonic (velocity-stream) form of Navier–Stokes (NS) equation using SBM with dual reciprocity (DR) scheme.

Section 2 of our paper introduces NS equations in velocity-stream form. Section 3 presents the basal concepts of the SBM and the connection with the DRM. Sections 4 and 5 are dedicated to two numerical examples and Section 6 contains conclusions.

2 GOVERNING EQUATIONS

An unsteady incompressible flow is governed by Navier–Stokes equations, which can be written in the non-dimensional form using primitive variables as

$$\frac{\partial u}{\partial t}+u\frac{\partial u}{\partial x}+v\frac{\partial u}{\partial y}=-\frac{\partial p}{\partial x}+\frac{1}{Re}\left(\frac{\partial^2 u}{\partial x^2}+\frac{\partial^2 u}{\partial y^2}\right), \tag{1}$$

$$\frac{\partial v}{\partial t}+u\frac{\partial v}{\partial x}+v\frac{\partial v}{\partial y}=-\frac{\partial p}{\partial y}+\frac{1}{Re}\left(\frac{\partial^2 v}{\partial x^2}+\frac{\partial^2 v}{\partial y^2}\right), \tag{2}$$

$$\frac{\partial v}{\partial x}+\frac{\partial v}{\partial y}=0, \tag{3}$$

where u and v are horizontal and vertical velocity components respectively, p represents the pressure, Re is the non-dimensional Reynolds number and x, y are the spatial coordinates and t is the time.

Though this formulation accurately represents the fluid flow, its direct solution has been difficult to obtain due to the mixed formulation in eqns (1), (2) and lack of the pressure development equation. To avoid the direct solution of the pressure field, a formulation using stream function and vorticity has been used as an alternative for several decades [10]. This formulation introduces the stream function ψ and vorticity ω:

$$\frac{\partial^2 \psi}{\partial x^2}+\frac{\partial^2 \psi}{\partial y^2}=-\omega, \tag{4}$$

$$\frac{\partial \omega}{\partial t}+u\frac{\partial \omega}{\partial x}+v\frac{\partial \omega}{\partial y}=\frac{1}{Re}\left(\frac{\partial^2 \omega}{\partial x^2}+\frac{\partial^2 \omega}{\partial y^2}\right). \tag{5}$$

This formulation has been very successful and has been used by many researchers over the past several decades to test new methods for the numerical solutions of a variety of fluid flow problems. Typical difficulty with this formulation consists of the specification of vorticity values at the no-slip boundaries; the vorticity ω is defined through the Poisson eqn (4) which needs to be solved discretely on the boundaries so that boundary values of the vorticity can be specified for the vorticity transport eqn (5) when this formulation is utilized. However, the values of vorticity ω on the boundaries are generally unspecified and one must carry out a variety of numerical approximations in order to specify the boundary values of vorticity [10].

To avoid the process of determination of boundary vorticity values the vorticity can be eliminated from the eqns (4), (5) at the price of 4th order non-linear, non-homogeneous, biharmonic partial differential eqn (6) [11]:

$$\frac{\partial^4 \psi}{\partial x^4}+2\frac{\partial^4 \psi}{\partial x^2 \partial y^2}+\frac{\partial^4 \psi}{\partial y^4}-Re\left[u\left(\frac{\partial^3 \psi}{\partial x^3}+\frac{\partial^3 \psi}{\partial x \partial y^2}\right)+v\left(\frac{\partial^3 \psi}{\partial y^3}+\frac{\partial^3 \psi}{\partial x^2 \partial y}\right)\right]=0, \tag{6}$$

where ψ represents stream function, u and v are velocity components that can be expressed in the terms of stream function as

$$u=\frac{\partial \psi}{\partial y},\quad v=-\frac{\partial \psi}{\partial x}. \tag{7}$$

The boundary conditions of eqn (6) are divided into two basic groups

- two boundary conditions of the Dirichlet type, where the value of variable ψ or its Laplacian $\Delta\psi$ on the part of boundary Γ_1 or Γ_3 is given, i.e.

$$\psi=\mathrm{U}_0(\mathbf{x})\ \mathbf{x}\in\Gamma_1,\quad \Delta\psi=\mathrm{V}_0(\mathbf{x})\ \mathbf{x}\in\Gamma_3 \tag{8}$$

- two boundary conditions of the Neumann type, where the value of the normal derivative of stream function or its Laplacian on the boundary Γ_2 or Γ_4 is given, i.e.

$$\frac{\partial \psi}{\partial \mathrm{n}} = \mathrm{q}_0(\mathbf{x})\ \mathbf{x} \in \Gamma_2, \quad \frac{\partial \Delta \psi}{\partial \mathrm{n}} = \mathrm{p}_0(\mathbf{x})\ \mathbf{x} \in \Gamma_4 \tag{9}$$

where n is the outward normal to the boundary Γ_2 or Γ_4.

For the numerical solution of the biharmonic eqn (6) the two boundary conditions need to be specified to obtain a solution [11].

3 SINGULAR BOUNDARY METHOD

The principle of the MFS is up to some level used by the singular boundary method (SBM) The main difference and superiority of SBM over MFS is no need of fictitious boundary. The SBM uses the same set of nodes as source and response points positioned on regular the boundary Γ. The solution of the homogeneous biharmonic equation is approximated by the linear combination of Laplace (11) and biharmonic (10) fundamental solutions (see eqn (12)) [12]

$$G_0 = -\frac{\|\mathbf{x}-\mathbf{s}\|^2}{2\pi}(ln\|\mathbf{x}-\mathbf{s}\| - 1), \tag{10}$$

$$G_1 = -\frac{1}{2\pi} ln\|\mathbf{x}-\mathbf{s}\|, \tag{11}$$

$$\psi_h(\mathbf{x}) = \sum_{j=1}^{N} a_j G_0(\mathbf{x}, \mathbf{s}_j) + \sum_{j=1}^{N} b_j G_1(\mathbf{x}, \mathbf{s}_j), \quad \mathbf{x} \in \Omega \cup \Gamma. \tag{12}$$

The diagonal term that arises when the source and response nodes are at the same position [7]–[9] can't be evaluated directly and needs to be determined using one of the aforementioned procedures and this process represents the crucial part of the SBM. To elaborate the solution (12) the singular terms that cannot be directly computed are moved outside the sum

$$\psi_h(\mathbf{x}_i) = \sum_{j=1, j\neq i}^{N} a_j G_0(\mathbf{x}_i, \mathbf{s}_j) + a_i U_{ii}^0 + \sum_{j=1, j\neq i}^{N} b_j G_1(\mathbf{x}_i, \mathbf{s}_j) + b_i U_{ii}^1, \quad \mathbf{x}_i \in \Gamma, \tag{13}$$

and similarly, for the boundary normal derivative at location $\mathbf{x}_i$

$$\frac{\partial \psi_h(\mathbf{x}_i)}{\partial n_{x_i}} = \sum_{j=1, j\neq i}^{N} a_j \frac{\partial G_0(\mathbf{x}_i, \mathbf{s}_j)}{\partial n_{x_i}} + a_i Q_{ii}^0 + \sum_{j=1, j\neq i}^{N} b_j \frac{\partial G_1(\mathbf{x}_i, \mathbf{s}_j)}{\partial n_{x_i}} + b_i Q_{ii}^1, \quad \mathbf{x}_i \in \Gamma, \tag{14}$$

the unknown coefficient $U_{ii}^0, Q_{ii}^0, U_{ii}^1, Q_{ii}^1$ are defined as OIFs. OIFs for a biharmonic fundamental solution can be handled very easy using limit to eliminate the singularity as the collocation point approaches the source point. So, for the OIFs that arises from biharmonic fundamental solution one can write

$$U_{ii}^0 = \lim_{\mathbf{x}_i \to \mathbf{s}_i} G_0(\mathbf{x}_i, \mathbf{s}_i) = 0, \tag{15}$$

$$Q_{ii}^0 = \lim_{\mathbf{x}_i \to \mathbf{s}_i} \frac{\partial G_0(\mathbf{x}_i, \mathbf{s}_i)}{\partial n_{x_i}} = 0. \tag{16}$$

The OIFs associated with the Laplace fundamental solution can be determined using subtracting and adding-back technique for the normal derivative (Neumann) boundary condition [7], [13]

$$Q_{ii}^1 = -\frac{1}{L_i} \sum_{j=1, j\neq i}^{N} L_j \frac{\partial G_1(\mathbf{x}_i, \mathbf{s}_j)}{\partial n_{s_j}}, \tag{17}$$

where L_j is the length of the appropriate part of boundary around point $\mathbf{x}_i$ (see Fig. 1). The value of the OIF for prescribed solution value (Dirichlet BC) on the boundary, the formula based on the regularized boundary integral equation is used [13]

$$U_{ii}^1 = -\frac{1}{L_i}\sum_{j=1,j\neq i}^{N} L_j \left[\mathbf{n}_{\mathbf{s}_i} \cdot \mathbf{n}_{\mathbf{x}_j} G_1(\mathbf{x}_i, \mathbf{s}_j) - \mathbf{n}_{\mathbf{s}_i} \cdot (\mathbf{x}_j - \mathbf{s}_i)\frac{\partial G_1(\mathbf{x}_j,\mathbf{s}_i)}{\partial n_{\mathbf{x}_j}}\right]. \tag{18}$$

3.1 Dual reciprocity method

The biharmonic (stream – velocity) form of the Navier–Stokes equation represents nonlinear, non-homogeneous boundary value problem. The governing eqn (6) can be rearranged

$$\frac{\partial^4\psi}{\partial x^4} + 2\frac{\partial^4\psi}{\partial x^2\partial y^2} + \frac{\partial^4\psi}{\partial y^4} = \Delta^2\psi = Re\left[u\left(\frac{\partial^3\psi}{\partial x^3} + \frac{\partial^3\psi}{\partial x\partial y^2}\right) + v\left(\frac{\partial^3\psi}{\partial y^3} + \frac{\partial^3\psi}{\partial x^2\partial y}\right)\right]. \tag{19}$$

The solution of (19) can now be defined as the sum

$$\psi = \psi_h + \psi_p, \tag{20}$$

where ψh is a solution that satisfies the homogeneous biharmonic equation and prescribed boundary conditions. The ψp represents a particular solution of the non-homogeneous eqn (19). The homogeneous part of the solution was calculated using the singular boundary method (SBM) and the particular part of the solution is obtained adopting dual reciprocity method. The particular solution ψp is approximated as a linear combination of radial basis (RBF) kernels as

$$\psi_p = \sum_{j=1}^{M} \beta_j \Phi(r_j), \tag{21}$$

where $\Phi(r_j)$ are radial basis functions, β_j are constants to be determined and M is the number of the internal points. The function $\Phi(r_j)$ represents a solution of the following equation

$$\Delta^2\Phi(r_j) = \varphi(r_j), \tag{22}$$

where $\varphi(r_j)$ are also radial basis kernels. The multiquadrics RBF (MQ see Table 1) have been tested for computations presented in this paper. Other appropriate possibilities can be found in [2], [13].

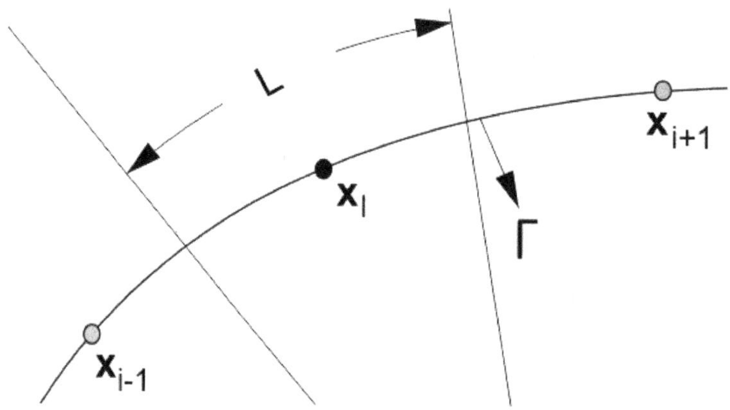

Figure 1: Part of boundary L around point $\mathbf{x}_i$.

Table 1: Integrated MQ radial basis functions tested in this paper.

$\varphi(r)$	$\Delta\Phi(r)$	$\Phi(r)$
$\sqrt{r^2+c^2}$	$\frac{1}{9}(r^2+4c^2)\sqrt{r^2+c^2}-\frac{c^3}{3}\ln\left(c\sqrt{r^2+c^2}+c^2\right)$	$\frac{1}{900}\left(\begin{array}{c}75c^3r^2-61c^4\sqrt{r^2+c^2}+48c^2r^2\sqrt{r^2+c^2}\\+4r^4\sqrt{r^2+c^2}+(30c^5-75c^3r^2)\ln\left(c+\sqrt{r^2+c^2}\right)\end{array}\right)$

According to the above definitions and the eqn (19), we can write

$$\sum_{j=1}^{M}\beta_j\varphi(r_j) = Re\left[\frac{\partial\psi}{\partial y}\left(\frac{\partial^3\psi}{\partial x^3}+\frac{\partial^3\psi}{\partial x\partial y^2}\right)-\frac{\partial\psi}{\partial x}\left(\frac{\partial^3\psi}{\partial y^3}+\frac{\partial^3\psi}{\partial x^2\partial y}\right)\right], \tag{23}$$

because the nonlinear character of eqn (17) iterative procedure is adopted to solve eqn (23) while the terms on the right-hand side of eqn (23) will be evaluated using MQ (see Table 1) interpolation functions with coefficients β and fundamental solutions with coefficients a and b knew from the previous iteration. The explicit iterative form of eqn (23) one can write as

$$\begin{aligned}\sum_{j=1}^{M}\beta_j^{n+1}\varphi(x,s_j) = Re\Bigg[&\left(\sum_{j=1}^{M}\beta_j^{n}\frac{\partial\Phi(x,s_j)}{\partial y}+\sum_{i=1}^{N}a_i^{n}\frac{\partial G0(x,s_i)}{\partial y}+\right.\\ &\left.\sum_{i=1}^{N}b_i^{n}\frac{\partial G1(x,s_i)}{\partial y}\right)\left(\sum_{j=1}^{M}\beta_j^{n}\left(\frac{\partial^3\Phi(x,s_j)}{\partial x^3}+\frac{\partial^3\Phi(x,s_j)}{\partial x\,\partial y^2}\right)+\sum_{i=1}^{N}a_i^{n}\left(\frac{\partial^3 G0(x,s_i)}{\partial x^3}+\frac{\partial^3 G0(x,s_i)}{\partial x\,\partial y^2}\right)+\right.\\ &\left.\sum_{i=1}^{N}b_i^{n}\left(\frac{\partial^3 G1(x,s_i)}{\partial x^3}+\frac{\partial^3 G1(x,s_i)}{\partial x\,\partial y^2}\right)\right)-\left(\sum_{j=1}^{M}\beta_j^{n}\frac{\partial\Phi(x,s_j)}{\partial x}+\sum_{i=1}^{N}a_i^{n}\frac{\partial G0(x,s_i)}{\partial x}+\right.\\ &\left.\sum_{i=1}^{N}b_i^{n}\frac{\partial G1(x,s_i)}{\partial x}\right)\left(\sum_{j=1}^{M}\beta_j^{n}\left(\frac{\partial^3\Phi(x,s_j)}{\partial y^3}+\frac{\partial^3\Phi(x,s_j)}{\partial x^2\,\partial y}\right)+\sum_{i=1}^{N}a_i^{n}\left(\frac{\partial^3 G0(x,s_i)}{\partial y^3}+\frac{\partial^3 G0(x,s_i)}{\partial x^2\,\partial y}\right)+\right.\\ &\left.\sum_{i=1}^{N}b_i^{n}\left(\frac{\partial^3 G0(x,s_i)}{\partial y^3}+\frac{\partial^3 G0(x,s_i)}{\partial x^2\,\partial y}\right)\right)\Bigg].\end{aligned} \tag{24}$$

Eqn (24) will be formed for every internal node (M – the number of internal nodes) which gives M equations. To close the linear system, we must impose $2N$ boundary condition values in all boundary nodes (two for each node). These $2N$ equations are obtained imposing the first type of Dirichlet and Neumann boundary conditions

$$\sum_{i=1}^{N}a_i^{n+1}G0(\mathbf{x},\mathbf{s}_i)+\sum_{i=1}^{N}b_i^{n+1}G1(\mathbf{x},\mathbf{s}_i)+\sum_{j=1}^{M}\beta_j^{n+1}\Phi(\mathbf{x},\mathbf{s}_i)=\bar{\psi}, \tag{25}$$

$$\sum_{i=1}^{N}a_i\frac{\partial G0(\mathbf{x},\mathbf{s}_i)}{\partial n}+\sum_{i=1}^{N}b_i\frac{\partial G1(\mathbf{x},\mathbf{s}_i)}{\partial n}+\sum_{j=1}^{M}\beta_j\frac{\partial\Phi(\mathbf{x},\mathbf{s}_i)}{\partial n}=\bar{q}, \tag{26}$$

The closed system of eqns (24)–(26) is then iteratively solved until the overall difference between two successive steps is smaller than $\varepsilon=1\times10^{-5}$.

The DRM technique is applied to approximate nonlinear and non-homogeneous terms of eqn (19) over the whole computational domain. However, as has been remarked by several researchers, the DRM yields accurate stable solutions for a Reynolds number up to 200.

This limitation shows that special domain treatments are required to obtain the solution for higher Reynolds numbers [14].

In the BEM focused articles, one can often find references to the division of computational domain into smaller regions – sub-domains [14]. When the convection effect becomes stronger, the sub-domain technique offers the possibility of emulating a computational mesh in homogeneous medium by dividing it into sub-domains and then applying the complete SBM with DRM to each of these sub-domains. Such a method used to be called *Multiple Domain Singular Boundary Method with Dual Reciprocity* – MD-SBMDR.

The detail of single simple square domain is shown in Fig. 2 with the detail of global computational domain "mesh".

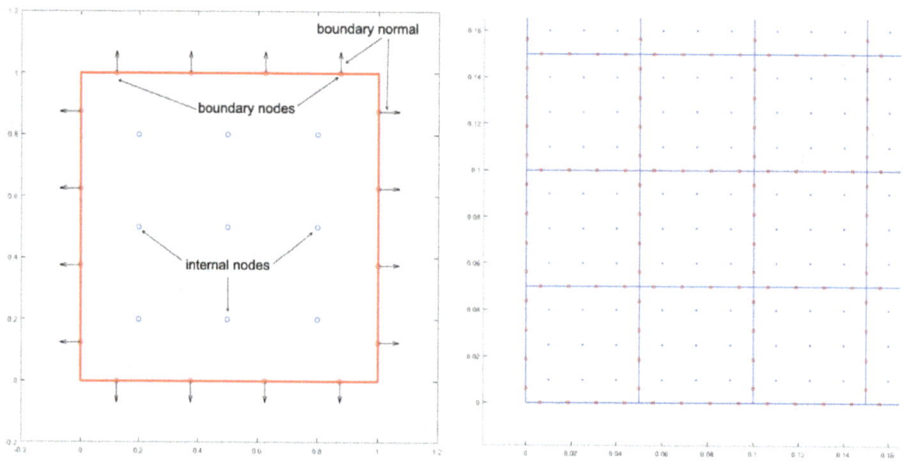

Figure 2: The sub-domain used in the MD-SBMDR formulation and "mesh" of multiple sub-domains covering the global computational domain.

4 SIMULATION OF THE LID-DRIVEN CAVITY PROBLEM

The standard benchmark problem for the numerical solver of incompressible, isothermal and laminar flow based on the Navier–Stokes equation is Lid-driven cavity flow. The top boundary of the cavity imposes a generation of the velocity $u=1$ to the flow moving the top lid, and no-slip impermeable boundary conditions are prescribed for the bottom, left and right wall. In a case of the stream function, the following boundary conditions are used

$$\psi(0,y) = \psi(1,y) = \psi(x,0) = \psi(x,1) = 0,$$

$$\frac{\partial\psi(0,y)}{\partial n_{left}} = \frac{\partial\psi(1,y)}{\partial n_{right}} = \frac{\partial\psi(x,0)}{\partial n_{bottom}} = 0,$$

$$\frac{\partial\psi(x,1)}{\partial n_{top}} = 1. \tag{27}$$

The geometrical configuration and prescribed boundary conditions are shown in Fig. 3. The challenging aspect of this benchmark problem lies in the presence of singularities of pressure and velocity field at the two upper corners of the cavity. Therefore, the size of computational sub-domains used decreases as their position is close to these corners (see Fig. 3) to capture high stream function gradients.

The steady solution using the MD-SBMDR is reached when the difference between two consecutive time steps (in the sense of the infinity norm) is lower than the prescribed, $\varepsilon_{tol} = 1\times10^{-6}$ is used for presented numerical computations.

The solutions obtained by the MD-SBMDR are compared with the solution presented in [15] for the Reynolds number Re=1000; the vertical velocity components along a horizontal line and the horizontal velocity components along a vertical line through the center of the cavity are plotted in Fig. 4. Figs 5 and 6 present stream-function ψ and pressure contours for Reynolds numbers Re=400 and 1000.

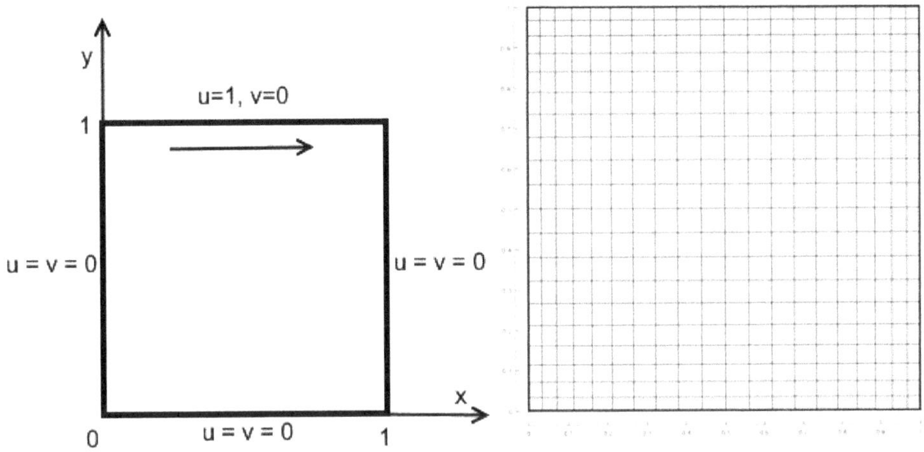

Figure 3: The lid-driven cavity problem geometry, boundary conditions, and computational domains distribution.

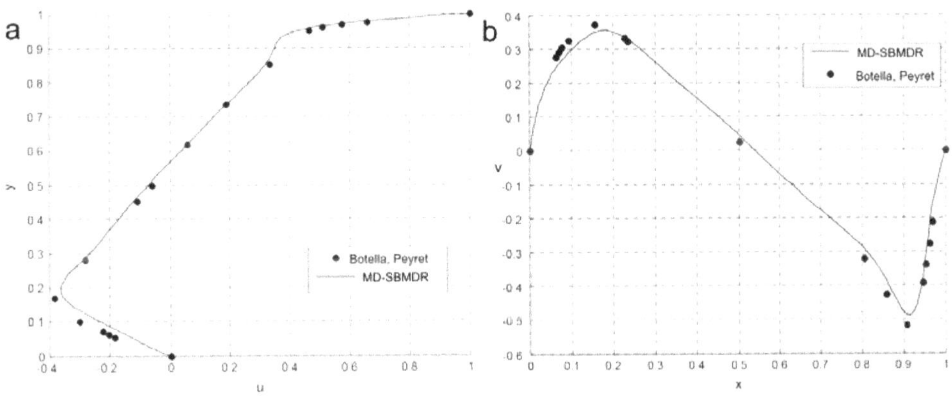

Figure 4: Lid-driven cavity, Re=1000. (a) Horizontal velocity components along vertical line x=0.5; (b) Vertical velocity components along horizontal line y=0.5.

5 SIMULATION OF THE BACKWARD-FACING STEP FLOW

The backward-facing step flow is another benchmark problem widely used to test the efficiency of the incompressible flow solver. Numerous numerical and experimental simulations of backward-facing step were performed [16]–[18]. The inlet velocity boundary condition is prescribed as fully developed flow, and maximum inflow velocity is u_{max}=1.5, and the average inflow velocity is u_a=1. The Reynolds number can then be defined as

$$Re = \frac{u_a H}{\nu}, \tag{28}$$

where H is the height of the channel (see Fig. 7).

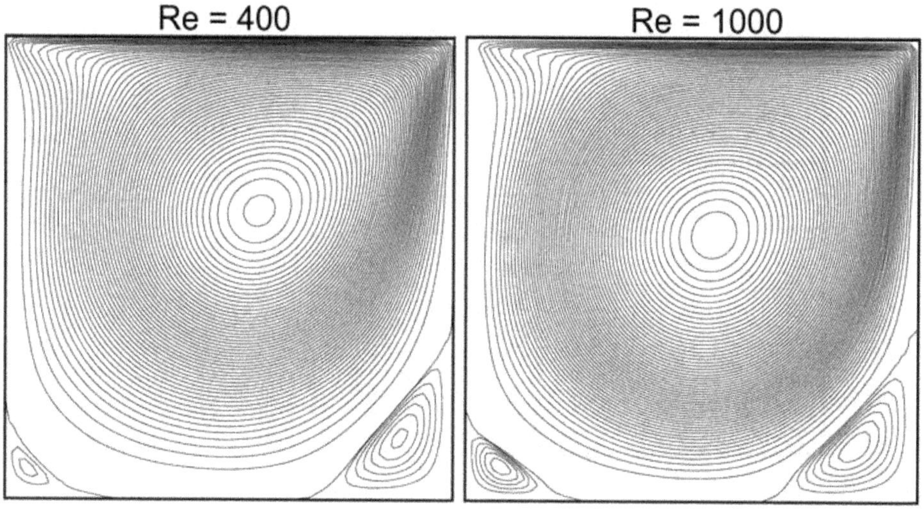

Figure 5: Stream function contours (streamlines) for Re=400 and 1000.

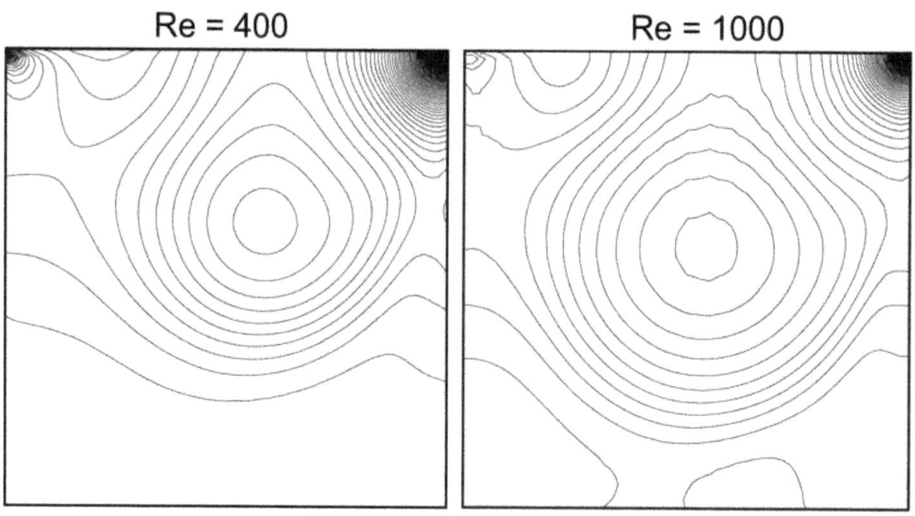

Figure 6: Pressure contours for Re=400 and 1000.

Stream function boundary conditions for a backward-facing step can be expressed in the following manner:

$$\begin{gathered}\psi(x,0) = 0 \;\; x \in [0, L],\\ \psi(0,y) = 0 \;\; y \in [0,0.5],\\ \psi(0,y) = 2y^2(3-4y) \;\; y \in [0.5,1],\\ \psi(L,y) = 0.25 + \frac{(3-4y^2)y}{4} \;\; y \in \; [0,1],\\ \psi(x,1) = 0 \;\; x \in [0, L].\end{gathered} \tag{29}$$

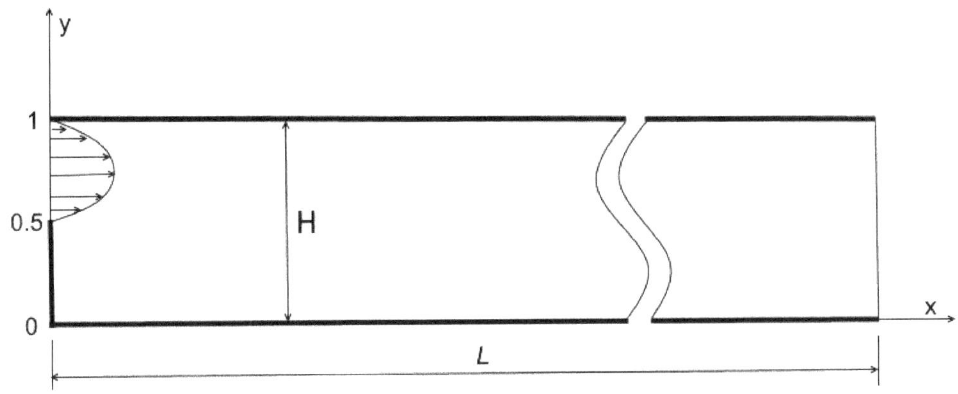

Figure 7: The geometry of the backward-facing step problem.

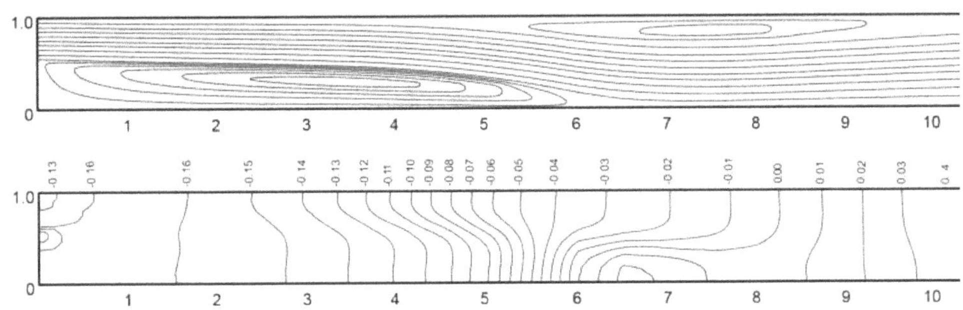

Figure 8: Stream function contours (streamlines) and pressure contours for backward-facing step, Re=800.

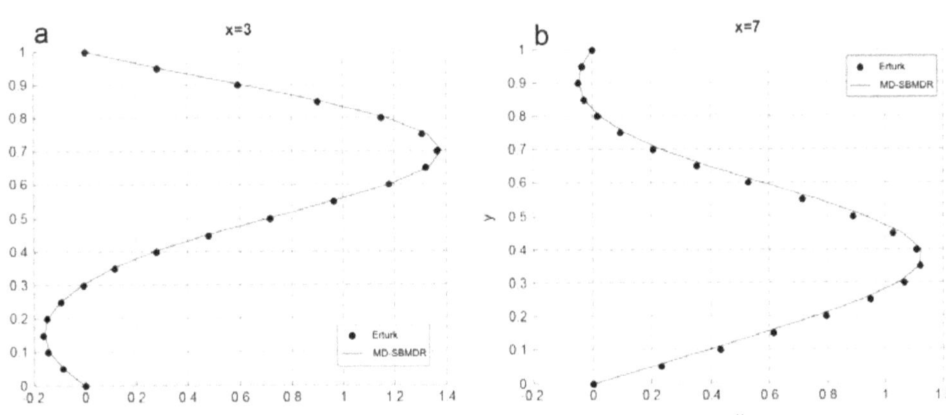

Figure 9: Backward-facing step, comparison of the horizontal velocities in profiles x=3 and x=7 for Re=800.

The computational domain of this problem is covered by uniform cell mesh that is made up of 161 cells in the horizontal and 11 in the vertical direction. The steady solution is reached when the difference between two consecutive stream function solutions is lower than a prescribed tolerance.

Fig. 8 shows streamlines and pressure contours for Re=800, and Fig. 9 shows the comparison of horizontal velocity values in vertical cuts positioned at x=3 and x=7, with the solution presented in [18]. Comparing results of the MD-SBMDR for Re=800 with results of other researchers is highly identical.

6 CONCLUSIONS

This article presents the application of multi-domain SBM and DRM (MD-SBMDR) for the solution the incompressible laminar flow described by a biharmonic stream-velocity form of Navier–Stokes equation. Two types of benchmarking problems, lid-driven cavity and backward-facing step, have been successfully tested. This type of boundary collocation method showed a potential to be a promising alternative to the solutions of incompressible fluid flow using the boundary based numerical schemes. The presented numerical code and numerical results need to be more analyzed in the sense of precision, convergence and computational demands.

ACKNOWLEDGEMENT

This contribution is the result of the project funded by the Scientific Grant Agency of Slovak Republic (VEGA) No. 1-0716-17.

REFERENCES

[1] Brebbia, C.A., Telles, J.C.F. & Wrobel, L.C., *Boundary Element Techniques*, Springer-Verlag: Berlin and New York, 1984.

[2] Partridge, P.W., Brebbia, C.A. & Wrobel, L.C., *The Dual Reciprocity Boundary Element Method*, CM Publications: Southampton, 1992.

[3] Golberg, M., The method of fundamental solutions for Poissons equations. *Engineering Analysis with Boundary Elements,* **16**, pp. 205–213, 1995.

[4] Golberg, M.A. & Chen, C.S., The method of fundamental solutions for potential. *Helmholtz and Diffusion Problems*, ed. M. A. Golberg, Boundary Integral Methods-Numerical and Mathematical Aspects, Computational Mechanics Publication: Southampton, pp. 103–176, 1998.

[5] Sarler, B., Solution of potential flow problems by the modified method of fundamental solutions: Formulations with the single layer and the double layer fundamental solutions. *Engineering Analysis with Boundary Elements*, **33**(12), pp. 1374–1382, 2009. DOI: 10.1016/j.enganabound.2009.06.008.

[6] Chen, W., Fu, Z. & Wei, X., Potential problems by singular boundary method satisfying moment condition. *CMES—Computer Modeling in Engineering and Sciences,* **54**, pp. 65–85, 2009.

[7] Chen, W. & Gu, Y., Recent advances on singular boundary method. *Joint International Workshop for Trefftz method*, **4**, pp. 543–558.

[8] Fu, Z.J., Chen, W., Chen, J.T. & Qu, W.Z., Singular boundary method: Three regularization approaches and exterior wave applications. *CMES—Computer Modeling in Engineering and Sciences*, **99**, pp. 417–443, 2014.

[9] Wei, X., Chen, W., Sun, L. & Chen, B., A simple accurate formula evaluating origin intensity factor in singular boundary method for two-dimensional potential problems

with Dirichlet boundary. *Engineering Analysis with Boundary Elements,* **58**, pp. 151–165, 2015. DOI: 10.1016/j.enganabound.2015.04.010.
[10] Kim, Y., Kim, D., Jun, S. & Lee, J., Meshfree point collocation method for the stream-vorticity formulation of 2D incompressible Navier–Stokes equations. *Computer Methods in Applied Mechanics and Engineering*, **196**(33–34), pp. 3095–3109, 2007. DOI: 10.1016/j.cma.2007.01.018.
[11] Gupta, M. & Kalita, J., A new paradigm for solving Navier–Stokes equations: streamfunction–velocity formulation. *Journal of Computational Physics,* **207**(1), pp. 52–68, 2005. DOI: 10.1016/j.jcp.2005.01.002.
[12] Yang, C. & Li, X., Meshless singular boundary methods for biharmonic problems. *Engineering Analysis with Boundary Elements,* **56**, pp. 39–48, 2015. DOI: 10.1016/j.enganabound.2015.02.001.
[13] Kovářík, K., Mužík, J., Masarovičová, S. & Sitányiová, D., Regularized singular boundary method for 3D potential flow. *Engineering Analysis with Boundary Elements,* **95**, pp. 85–92, 2018. DOI: 10.1016/j.enganabound.2018.06.022.
[14] Florez, W., Nonlinear flow using dual reciprocity. WIT Press: Southhampton, UK, 2001.
[15] Botella, O. & Peyret, R., Benchmark spectral results on the lid-driven cavity flow. *Computers and Fluids,* **27**(4), pp. 421–433, 1998. DOI: 10.1016/s0045-7930(98)00002-4.
[16] Armaly, B.F., Dursts, F., Pereira, J.C.F. & Schonung, B., Experimental and theoretical investigation of backward-facing step flow. *Journal of Fluid Mechanics*, **127**(1), pp. 473–496, 1983. DOI: 10.1017/s0022112083002839.
[17] Gartling, D.K., A test problem for outflow boundary conditions–flow over backward-facing step. *International Journal for Numerical Methods in Fluids*, **11**(7), pp. 953–967. DOI: 10.1002/fld.1650110704.
[18] Erturk, E., Numerical solutions of 2-D steady incompressible flow over a backward-facing step, Part I: high Reynolds number solutions. *Computers and Fluids*, **37**(6), pp. 633–655, 2008. DOI: 10.1016/j.compfluid.2007.09.003.

ELECTROMAGNETIC DOSIMETRY BASED ON EFIE FORMULATION AND RWG BASIS FUNCTION

MARIO CVETKOVIĆ & DRAGAN POLJAK
University of Split, Croatia

ABSTRACT

The paper deals with the electromagnetic dosimetry model for the lossy homogeneous biological body. The model is based on the use of electric field integral equation (EFIE) in the frequency domain derived using the surface equivalence theorem and the boundary conditions for the electric field. The scattered electric field is expressed in terms of the equivalent electric and magnetic current densities which are expanded using Rao–Wilton–Glisson (RWG) basis function, and a point wise orthogonal $\hat{n}\times$RWG basis function, respectively. Some illustrative results for the obtained equivalent surface current densities are given. The electromagnetic model can be useful in high frequency dosimetry of the human brain and also in modeling of medical technique such as transcranial magnetic stimulation (TMS).

Keywords: electric field integral equation, surface integral equation, equivalence principle, RWG basis functions, human exposure to electromagnetic fields, computational electromagnetics.

1 INTRODUCTION

In addition to the environmental exposure to artificially generated electromagnetic (EM) fields due to increasing number of power and communication equipment, the efficient applications of EM radiation in medical treatments and diagnosis also require the knowledge of the accurate distribution of these fields inside the biological tissues. As the direct measurement of these quantities is rather difficult, or even impossible, the computational methods have become necessary to determine the internal field distributions [1]–[4].

From the first days, the computational bioelectromagnetics models formulated using the differential equation approach became the standard, although the integral equation approach, using the Green integral representation, is more suitable in case of open boundary problems such as the human body exposed to incident EM field [5], [3]. Although the numerical methods based on the solution of integral equations in computational electromagnetics (CEM) were developed more than half a century ago [6], [7], only recently have they begun to attract more attention in the bioelectromagnetics community as seen from a number of published papers [2], [3], [5], [8], [9].

In order to formulate the bioelectromagnetics problem using the integral equation approach, one can carry out similar procedure as in a case of classical scattering problem [5]. In case of the homogeneous biological body of an arbitrary shape S representing the scatterer, placed in the incident electromagnetic field, the electric field integral equation (EFIE), based on the application of surface equivalence theorem can be used.

This paper features the application of EFIE to the problem of a biological body, i.e. the human brain exposed to high frequency electromagnetic radiation and also to the field due to a TMS stimulation coil.

The paper is organized as follows: following the introductory part, in the second part the surface integral equation set is derived using the surface equivalence theorem and the boundary conditions for the electric field. Some details on the expansion of the introduced equivalent surface current densities using the so called Rao–Wilton–Glisson (RWG) basis function are also given. This is followed by some numerical results for the equivalent current densities and the comparison with the commercial software. Furthermore, some

WIT Transactions on Engineering Sciences, Vol 122, © 2019 WIT Press
www.witpress.com, ISSN 1743-3533 (on-line)
doi:10.2495/BE410131

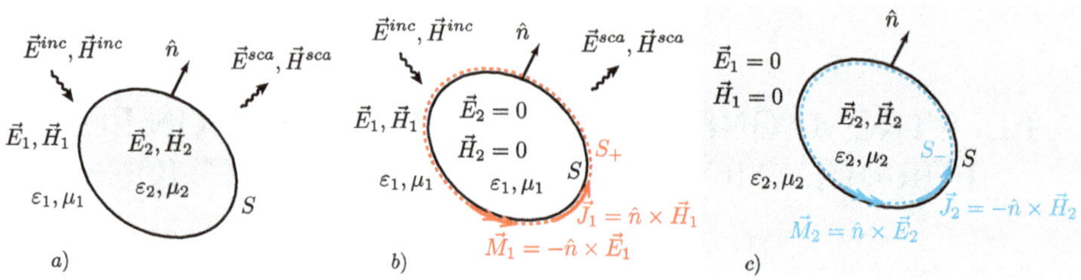

Figure 1: Biological body as a lossy homogeneous dielectric (ε_2, μ_2) placed in the incident field $(\vec{E}^{inc}, \vec{H}^{inc})$. (a) Original problem; (b) Equivalent problem for the outside; (c) Equivalent problem for the inside.

illustrative results are shown related to application of our model formulated using EFIE to the bioelectromagnetics problem. Finally, some concluding remarks are given.

2 THEORETICAL BACKGROUND

2.1 Surface integral equation formulation

The electromagnetic wave incident on the homogeneous biological body can be treated as a classical scattering problem. Applying the surface equivalence theorem and the boundary conditions for the electric field one can formulate the electromagnetic model as the surface integral equation (SIE), i.e. the electric field integral equation (EFIE).

To apply the equivalence theorem, the biological body is first replaced by a homogeneous dielectric body of arbitrary shape S placed in a free space with a given properties (ε_1, μ_1), as shown in Fig. 1(a). The properties of the biological object are represented by complex permitivity and permeability (ε_2, μ_2), where the complex permitivity is given by

$$\varepsilon_2 = \varepsilon_0 \varepsilon_r - j\frac{\sigma}{\omega}, \tag{1}$$

where ε_0 is permitivity of the free space, ε_r is relative permitivity, σ is electrical conductivity of the biological body, and $\omega = 2\pi f$ is the operating frequency. The value for the permeability can be taken as $\mu_0 = 4\pi \times 10^{-7}$ Vs/Am, as the biological tissues do not posses magnetic properties.

The lossy homogeneous object representing, e.g. the brain, is exposed to radiation of a known electromagnetic (EM) field $(\vec{E}^{inc}, \vec{H}^{inc})$. The incident EM field is present regardless of the presence of the material body. As the scattering object is present, there will also be the scattered field, denoted by $(\vec{E}^{sca}, \vec{H}^{sca})$. The fields exterior and interior to the surface S are, $(\vec{E}_1, \vec{H}_1)$ and $(\vec{E}_2, \vec{H}_2)$, respectively. Also, in Fig. 1, unit vector $\hat{n}$ placed on the surface S pointing from the region 2 into region 1, can be noticed.

Using the equivalence theorem for the surface S of the biological body, two equivalent problems can be formulated, in terms of the equivalent electric and magnetic current densities $\vec{J}$ i $\vec{M}$ placed on the surface S, one for the region 1 (outside) and other for the region 2 (inside) [10], [7], [6], [11]. The two equivalent problems are shown on Figs 1(b) and 1(c).

In the case of an external equivalent problem, shown in Fig. 1(b), field inside is assumed to be zero, $(\vec{E}_2 = 0, \vec{H}_2 = 0)$, hence allowing one to replace material properties of the region. Selecting the same properties as that of an exterior region, leads to the homogeneous domain of (ε_1, μ_1), allowing one to use the free space Green's function. The boundary conditions on the surface S are satisfied by introducing equivalent surface currents $\vec{J}_1$ and $\vec{M}_1$ at the surface S. Following the same procedure for the interior equivalent problem, yields another homogeneous domain of (ε_2, μ_2). Again, the equivalent surface currents $\vec{J}_2 = -\vec{J}_1$ and $\vec{M}_2 = -\vec{M}_1$ are introduced, as indicated in Fig. 1(c).

As both problems represent equivalent current densities radiating in a homogeneous medium, the following expressions for scattered fields can be used

$$\vec{E}_n^{sca}(\vec{J}, \vec{M}) = -j\omega \vec{A}_n - \nabla \varphi_n - \frac{1}{\varepsilon_n} \nabla \times \vec{F}_n, \tag{2}$$

$$\vec{H}_n^{sca}(\vec{J}, \vec{M}) = -j\omega \vec{F}_n - \nabla \psi_n + \frac{1}{\mu_n} \nabla \times \vec{A}_n, \tag{3}$$

where $n = 1, 2$ is index of the medium where equivalent surface currents radiate, and φ, $\vec{F}$, ψ i $\vec{A}$ are scalar and vector, electric and magnetic potentials, respectively, given in terms of integrals over the sources, i.e.

$$\vec{A}_n(\vec{r}) = \mu_n \iint_S \vec{J}(\vec{r}\,') G_n(\vec{r}, \vec{r}\,') \, \mathrm{d}S', \tag{4}$$

$$\vec{F}_n(\vec{r}) = \varepsilon_n \iint_S \vec{M}(\vec{r}\,') G_n(\vec{r}, \vec{r}\,') \, \mathrm{d}S', \tag{5}$$

$$\varphi_n(\vec{r}) = \frac{j}{\omega \varepsilon_n} \iint_S \nabla'_S \cdot \vec{J}(\vec{r}\,') G_n(\vec{r}, \vec{r}\,') \, \mathrm{d}S', \tag{6}$$

$$\psi_n(\vec{r}) = \frac{j}{\omega \mu_n} \iint_S \nabla'_S \cdot \vec{M}(\vec{r}\,') G_n(\vec{r}, \vec{r}\,') \, \mathrm{d}S', \tag{7}$$

where

$$G_n(\vec{r}, \vec{r}\,') = \frac{e^{-jk_n R}}{4\pi R}; \quad R = |\vec{r} - \vec{r}\,'|. \tag{8}$$

$G_n(\vec{r}, \vec{r}\,')$ is Green's function for the homogeneous medium n, R is the distance from observation point $\vec{r}$ to source point $\vec{r}\,'$, while k_n is wave number in medium n, $(n = 1, 2)$.

The electric and magnetic charges from (6) and (7), respectively, can be replaced with the divergence of the electric and magnetic currents, respectively, if the continuity equation is used.

Using (4)–(7), the scattered field from (2) and (3) can be expressed in terms of the equivalent surface currents. Applying the boundary conditions at the surface S being interface of the two equivalent problems, a set of four equations for electric and magnetic field, respectively, is obtained. Choosing the two for the electric field, yields the electric field integral equation (EFIE) formulation in the frequency domain for the lossy homogeneous biological object

$$\left[-\vec{E}_n^{sca}(\vec{J}, \vec{M})\right]_{tan} = \begin{cases} \left[\vec{E}^{inc}\right]_{tan}, & n = 1, \\ 0, & n = 2, \end{cases} \tag{9}$$

where $\vec{E}^{inc}$ is the known incident field, while $\vec{J}$ and $\vec{M}$ represent the unknown surface currents.

Inserting (4)–(7) into (2) and (3) we arrive at the coupled set of integral equations

$$j\omega\mu_n \iint_S \vec{J}(\vec{r}\,')G_n(\vec{r},\vec{r}\,')\,\mathrm{d}S' - \frac{j}{\omega\varepsilon_n}\nabla \iint_S \nabla'_S \cdot \vec{J}(\vec{r}\,')G_n(\vec{r},\vec{r}\,')\,\mathrm{d}S' + \\ + \nabla\times \iint_S \vec{M}(\vec{r}\,')G_n(\vec{r},\vec{r}\,')\,\mathrm{d}S' = \begin{cases} \vec{E}^{inc}, & n=1,\\ 0, & n=2, \end{cases} \tag{10}$$

where $\vec{J}$ and $\vec{M}$ are unknown equivalent surface electric and magnetic currents.

After some mathematical manipulations, details of which can be found in [8], the gradient operator outside the second integral of (10) and curl operator outside the third integral can be transferred to the Green's function

$$j\omega\mu_n \iint_S \vec{J}(\vec{r}\,')G_n(\vec{r},\vec{r}\,')\,\mathrm{d}S' - \frac{j}{\omega\varepsilon_n} \iint_S \nabla'_S \cdot \vec{J}(\vec{r}\,')\nabla G_n(\vec{r},\vec{r}\,')\,\mathrm{d}S' + \\ + \iint_S \vec{M}(\vec{r}\,')\times\nabla' G_n(\vec{r},\vec{r}\,')\,\mathrm{d}S' = \begin{cases} \vec{E}^{inc}, & n=1,\\ 0, & n=2, \end{cases} \tag{11}$$

where the following property for the Green's function gradient has been used

$$\nabla G_n(\vec{r},\vec{r}\,') = -\nabla' G_n(\vec{r},\vec{r}\,'). \tag{12}$$

2.2 Expansion of surface currents

In the case of a complex surface geometry, such as e.g. human brain, the coupled integral equations set (11) cannot be solved analytically, hence the numerical solution is necessary.

An efficient scheme of a method of moments (MoM) is implemented in this work. First, the equivalent electric and magnetic currents $\vec{J}$ and $\vec{M}$ in (11) are expanded in terms of a linear combination of known basis functions $\vec{f}_n$ and $\vec{g}_n$, respectively

$$\vec{J}(\vec{r}) = \sum_{n=1}^{N} J_n \vec{f}_n(\vec{r}), \tag{13}$$

$$\vec{M}(\vec{r}) = \sum_{n=1}^{N} M_n \vec{g}_n(\vec{r}), \tag{14}$$

where J_n and M_n are unknown coefficients, while N stands for total number of elements used to discretize the surface S.

If the surface of a biological body is discretized using the triangular elements, the so called Rao-Wilton-Glisson (RWG) basis functions [12] specially developed for triangular patches can be used.

RWG basis function $\vec{f}_n$, or, as sometimes called the *edge-element*, is defined on T_n^+ and T_n^- pair of triangles sharing a common edge, as shown on Fig. 2.

Namely, the function is given by

$$\vec{f}_n^{\pm}(\vec{r}) = \begin{cases} \dfrac{l_n}{2A_n^{\pm}}\vec{\rho}_n^{\,\pm}, & \vec{r}\in T_n^{\pm},\\ 0, & \vec{r}\notin T_n^{\pm}, \end{cases} \tag{15}$$

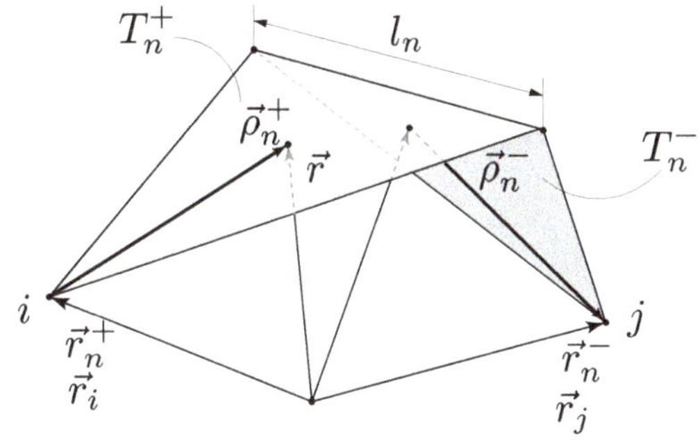

Figure 2: RWG basis function $\vec{f}_n(\vec{r})$ defined on a pair of triangles in $\mathbf{R}^3$ [12].

where l_n is the edge length at the interface between triangles T_n^+ and T_n^-, while A_n^+ and A_n^- denote the surface areas of those triangles. The vector $\vec{\rho}_n^+ = \vec{r} - \vec{r}_n^+$ is directed from the free vertex of T_n^+ and $\vec{\rho}_n^- = \vec{r}_n^- - \vec{r}$ is directed to the free vertex of T_n^-, as shown on Fig. 2.

While the surface electric current $\vec{J}$ is approximated by the RWG function $\vec{f}_n$, the surface magnetic current $\vec{M}$ is expanded by $\vec{g}_n = \hat{n} \times \vec{f}_n$, i.e. the function point wise orthogonal to the RWG function. Fig. 3 depicts the linear combination of a three basis functions (showed only on one triangle) obtained using the RWG and $\hat{n}\times$RWG function, respectively.

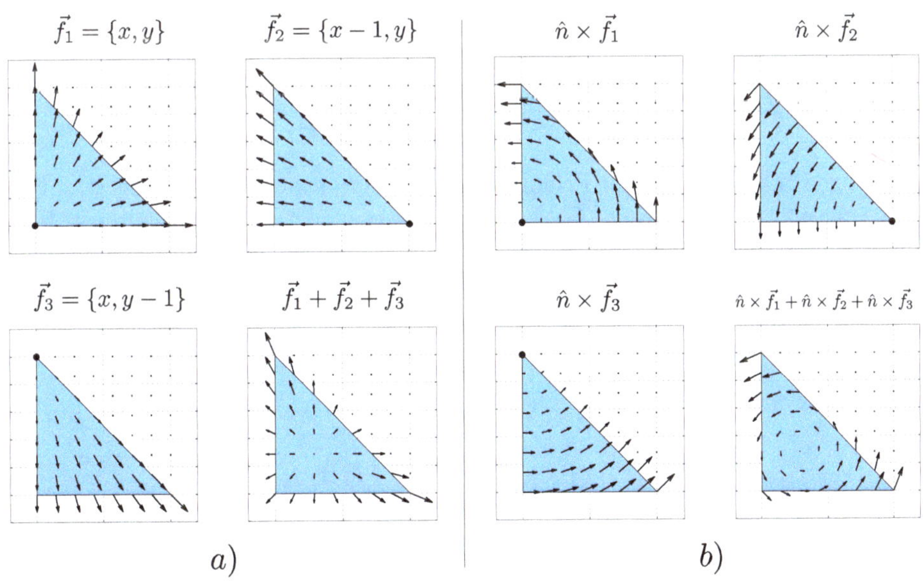

Figure 3: Three basis functions associated with three sides (edges) of a triangle and their linear combination, using: (a) RWG function; (b) $\hat{n}\times$RWG function.

The unknown equivalent currents $\vec{J}(\vec{r}')$ and $\vec{M}(\vec{r}')$ from (11) are substituted by (13) and (14). Eqn (11) is next multiplied by the set of a test functions $\vec{f}_m$, where $\vec{f}_m = \vec{f}_n$, and integrated over the surface S.

Undertaking some additional mathematical manipulations [5], [8], yields

$$\begin{aligned} & j\omega\mu_i \sum_{n=1}^{N} J_n \iint_S \vec{f}_m(\vec{r}) \cdot \iint_{S'} \vec{f}_n(\vec{r}\,')G_i \, dS' \, dS + \\ & + \frac{j}{\omega\varepsilon_i} \sum_{n=1}^{N} J_n \iint_S \nabla_S \cdot \vec{f}_m(\vec{r}) \iint_{S'} \nabla'_S \cdot \vec{f}_n(\vec{r}\,')G_i \, dS' \, dS + \\ & \pm \sum_{n=1}^{N} M_n \iint_S \vec{f}_m(\vec{r}) \cdot [\hat{n} \times \vec{g}_n(\vec{r}\,')] \, dS + \\ & + \sum_{n=1}^{N} M_n \iint_S \vec{f}_m(\vec{r}) \cdot \iint_{S'} \vec{g}_n(\vec{r}') \times \nabla' G_i \, dS' \, dS = \\ & = \begin{cases} \iint_S \vec{f}_m(\vec{r}) \cdot \vec{E}^{inc} \, dS, & i = 1, \\ 0, & i = 2, \end{cases} \end{aligned} \tag{16}$$

where subscript i stands for the region index. After extracting the two sums, (16) can be written as the following system of linear equations

$$\sum_{n=1}^{N} \left(j\omega\mu_i A_{mn,i} + \frac{j}{\omega\varepsilon_i} B_{mn,i} \right) J_n + \sum_{n=1}^{N} \left(C_{mn,i} + D_{mn,i} \right) M_n = \begin{cases} V_m, & i = 1, \\ 0, & i = 2. \end{cases} \tag{17}$$

Recasting the expression (17) in the matrix form, leads to

$$[\mathbf{Z}] \cdot \{\mathbf{I}\} = \{\mathbf{V}\}, \tag{18}$$

where $\mathbf{Z}$ is the $2N \times 2N$ matrix of the system, while $\mathbf{V}$ represents the source vector of a dimension $2N$.

Solving (18) results in vector $\mathbf{I}$ containing the unknown coefficients J_n and M_n. From these coefficients, using (13) and (14), the equivalent electric and magnetic currents, $\vec{J}$ and $\vec{M}$, respectively, could be determined.

From the equivalent surface currents, the electric field inside the biological body can be determined using the following:

$$\begin{aligned} \vec{E}_2(\vec{r}) = -j\omega\mu_2 \iint_S \vec{J}(\vec{r}\,')G_2(\vec{r}, \vec{r}\,') \, dS' - \frac{j}{\omega\varepsilon_2} \iint_S \nabla'_S \cdot \vec{J}(\vec{r}\,')G_2(\vec{r}, \vec{r}\,') \, dS' - \\ - \iint_S \vec{M}(\vec{r}\,') \times \nabla G_2(\vec{r}, \vec{r}\,') \, dS'. \end{aligned} \tag{19}$$

3 NUMERICAL RESULTS

The results obtained using the SIE formulation are compared to the results from the commercially available software FEKO. The first computational example is related to the dielectric sphere of a radius $r = 3$ cm, with parameters $\varepsilon_r = 35$ and $\sigma = 0.7$ S/m, placed in the field of a 918 MHz plane wave, directed in the negative z coordinate, polarized in x

direction, as shown on Fig. 4. The second example is the dielectric cube with side $a = 1$ m, $ka = 1$, frequency of incident wave $f = 47.74$ MHz, $\varepsilon_r = 54.5$, and $\sigma = 1.35$ S/m.

The incident electric field value in both cases is $E_0^{inc} = 86.83$ V/m, corresponding to power density of $P_0^{inc} = 1$ mW/cm^2. The surface of the sphere is discretized by $T = 152$ triangles, corresponding to $N = 228$ edge-elements, while the cube surface is discretized using $T = 768$ triangles, corresponding to $N = 1152$ edge-elements.

The distribution of the equivalent surface electric and magnetic currents obtained by the proposed model and using the FEKO are shown in Figs 4(a)–(b) and 4(c)–(d), respectively, while the maximum values obtained for J_{max} and M_{max} are also compared in Table 1.

It can be clearly seen that the maximum values for the equivalent surface currents obtained using EFIE formulation is similar to the results given by commercial software. Moreover, the obtained distribution of these equivalent surface currents are rather comparable.

As seen from Fig. 4, a much coarser mesh, compared to FEKO, has been used in the proposed method, in order to facilitate the solution process. Although the proposed method is applicable to an arbitrarily shaped biological object with more detailed geometry, the present implementation is limited to a roughly several thousand elements. The higher number of elements would consequently lead to a very large matrix system whose solution would

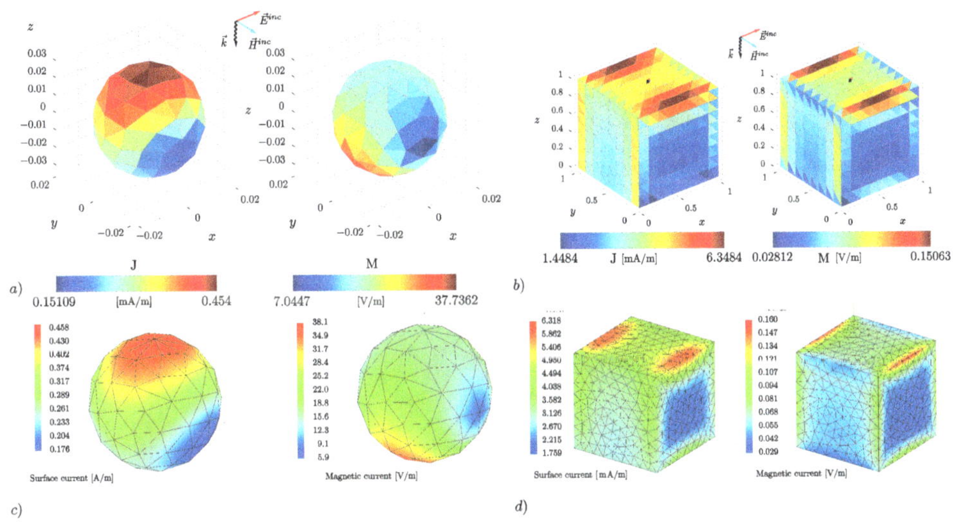

Figure 4: Distribution of the equivalent electric current $\vec{J}$ and magnetic current $\vec{M}$, respectively, on the surface S of a sphere and a unit cube due to vertically incident, horizontally polarized plane wave. Results obtained using EFIE formulation: (a) and (b). Results obtained using FEKO: (c) and (d).

Table 1: Comparison of maximum values for the surface electric and magnetic currents obtained by EFIE formulation and commercial software.

	Sphere		Cube	
	EFIE	FEKO	EFIE	FEKO
J_{max} [mA/m]	4.0209	3.947	6.348	6.318
M_{max} [V/m]	0.8706	0.843	0.151	0.160

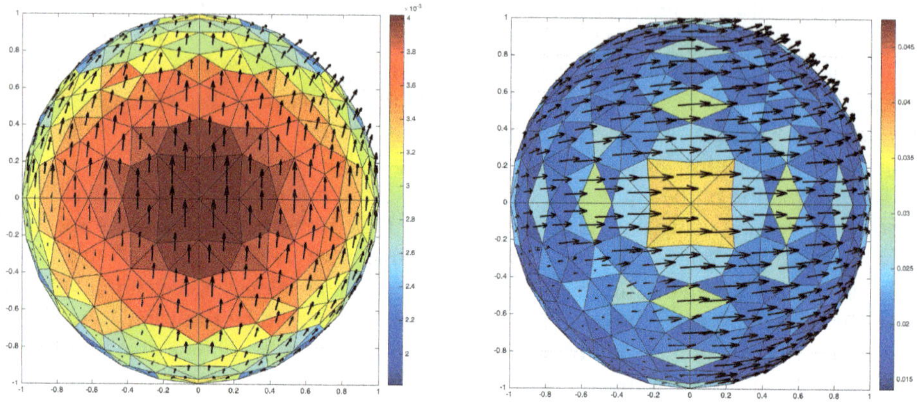

Figure 5: Distribution of the equivalent currents on the surface of a sphere due to vertically incident plane wave of frequency $f = 300$ MHz. Results for the equivalent electric current $\vec{J}$ (left), and the equivalent magnetic current $\vec{M}$ (right).

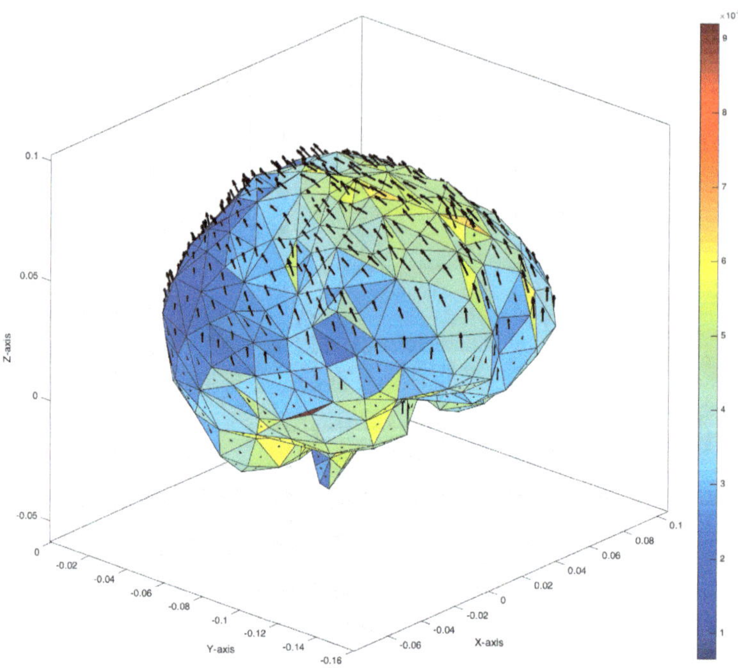

Figure 6: Distribution of the equivalent electric currents on the surface of a human brain model to vertically incident plane wave of frequency $f = 900$ MHz.

thus require the use of advanced techniques such as multilevel fast multipole method. The comparison of the proposed method to other numerical methods in case of low frequency field exposure could be found in [4].

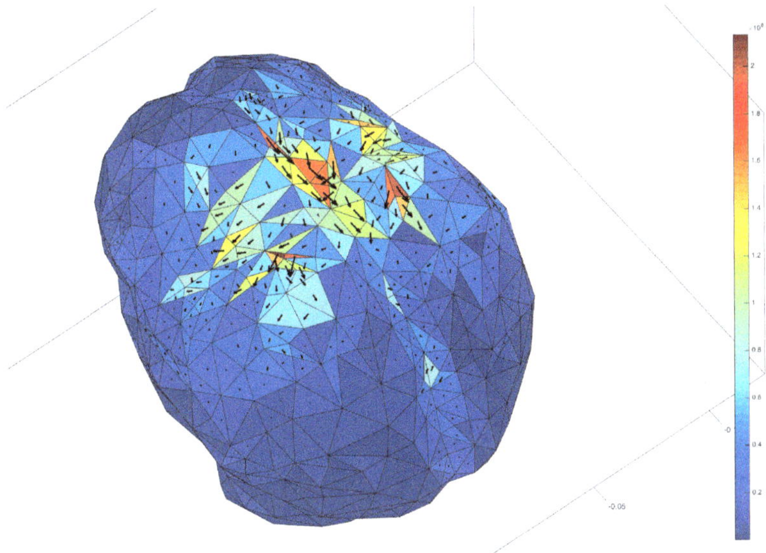

Figure 7: Distribution of the equivalent electric currents on the surface of a human brain model due to a transcranial magnetic stimulation coil positioned 1 cm over the brain surface.

Additionally, only as an illustrative results, Fig. 5 shows the distribution of the equivalent electric and magnetic currents, respectively, on a more finely discretized spherical surface due to incident wave of 300 MHz. One can note on each triangular element the electric current vector (expanded using RWG function) being piecewise orthogonal to the corresponding magnetic current vector (expanded using $\hat{n}\times$ RWG function).

Finally, some numerical results, related to the homogeneous human brain exposed to EM radiation, are given. The results for the equivalent surface electric current, obtained using the EFIE formulation (10), and the RWG basis functions (15), are shown on Figs 6 and 7.

Fig. 6, depicting the human brain exposed to electromagnetic wave incident from the above, corresponds to high frequency exposure scenario [13], while Fig. 7 is related to the human brain stimulated by a transcranial magnetic stimulation (TMS) figure-of-eight coil positioned 1 cm over the primary motor cortex [8]. As seen from 7, the distribution of the electric current densities over the brain surface is actually following the pattern similar to a so called figure-of-eight TMS coil, used as a low frequency EM source (2.4 kHz) in this case.

4 CONCLUSION

The paper presented the integral equation based alternative to the nowadays most frequently used differential equation based formulation of the bioelectromagnetics problem. The electromagnetic model for the homogeneous biological body of arbitrary shape, represents a powerful, accurate and efficient alternative for the treatment of open boundary type problems such as a human body exposed to electromagnetic radiation. The unknown equivalent surface electric and magnetic current densities, respectively, are expanded using Rao-Wilton-Glisson basis functions and the point wise orthogonal function set defined on the triangular elements. The numerical solution of the electric field integral equation is carried out using the method of moments technique. Some illustrative results for the equivalent current densities on the surface of the brain model are given. The proposed method could be found useful in the initial assessment of the human brain exposed to high frequency EM radiation, and also in modeling medical techniques such as TMS. The future work will be related to a more realistic scenarios featuring the use of a non-homogeneous models, and the corresponding integral equation based formulations.

REFERENCES

[1] Hand, J., Modelling the interaction of electromagnetic fields (10 MHz–10 GHz) with the human body: methods and applications. *Physics in Medicine and Biology*, **53**(16), p. R243, 2008.

[2] Poljak, D., Cavka, D., Dodig, H., Peratta, C. & Peratta, A., On the use of the boundary element analysis in bioelectromagnetics. *Engineering Analysis with Boundary Elements*, **49**, pp. 2–14, 2014.

[3] Poljak, D., Cvetković, M., Peratta, A., Peratta, C., Dodig, H. & Hirata, A., On some integral approaches in electromagnetic dosimetry. *BIOEM 2016*, 2016.

[4] Poljak, D., Cvetković, M., Bottauscio, O., Hirata, A., Laakso, I., Neufeld, E., Reboux, S., Warren, C., Giannopoulos, A. & Costen, F., On the use of conformal models and methods in dosimetry for nonuniform field exposure. *IEEE Transactions on Electromagnetic Compatibility*, **60**(2), pp. 328–337, 2018.

[5] Cvetković, M. & Poljak, D., An efficient integral equation based dosimetry model of the human brain. *Proceedings of 2014 International Symposium on Electromagnetic Compatibility (EMC EUROPE) 2014, Gothenburg, Sweden, 1–4 September 2014*, pp. 375–380, 2014.

[6] Poggio, A.J. & Miller, E.K., Integral equation solutions of three–dimensional scattering problems. *Computer Techniques for Electromagnetics*, 2nd ed., ed. R. Mittra, Hemisphere Publishing Corporation, Chapter 4, pp. 159–264, 1987.

[7] Chew, W.C., Tong, M.S. & Hu, B., *Integral Equation Methods for Electromagnetic and Elastic Waves*, Morgan & Claypol Publishers, p. 261, 2009.

[8] Cvetković, M., Poljak, D. & Haueisen, J., Analysis of transcranial magnetic stimulation based on the surface integral equation formulation. *Biomedical Engineering, IEEE Transactions on*, **62**(6), pp. 1535–1545, 2015.

[9] Wei, F. & Yilmaz, A., A More Scalable and Efficient Parallelization of the Adaptive Integral Method – Part II: BIOEM Application. *IEEE T Antenn Propag*, **62**(2), pp. 727–738, 2014.

[10] Harrington, R., Boundary integral formulations for homogeneous material bodies. *Journal of Electromagnetic Waves and Applications*, **3**(1), pp. 1–15, 1989.

[11] Umashankar, K., Taflove, A. & Rao, S., Electromagnetic scattering by arbitrary shaped three-dimensional homogeneous lossy dielectric objects. *IEEE Transactions on Antennas and Propagation*, **34**(6), pp. 758–766, 1986.

[12] Rao, S., Wilton, D.R. & Glisson, A., Electromagnetic scattering by surfaces of arbitrary shape. *IEEE Transactions on Antennas and Propagation*, **30**(3), pp. 409–418, 1982.

[13] Cvetković, M., Poljak, D. & Hirata, A., The electromagnetic-thermal dosimetry for the homogeneous human brain model. *Engineering Analysis with Boundary Elements*, **63**, pp. 61–73, 2016.

DETERMINATION OF SHAPE PARAMETER IN RBF APPROXIMATION

ANDREAS KARAGEORGHIS & PINELOPI TRYFONOS
Department of Mathematics and Statistics, University of Cyprus, Cyprus

ABSTRACT

We apply a radial basis function (RBF) collocation method for the approximation of functions in two dimensions. The solution is approximated by a linear combination of radial basis functions. The issue of determining the optimal value of the shape parameter is tackled by including it in the unknowns along with the coefficients of the RBFs in the approximation. The resulting nonlinear system of equations is solved by directly applying a standard non-linear solver. The results of some numerical experiments are presented and analysed.

Keywords: radial basis functions, function approximation, collocation.

1 INTRODUCTION

In the last few decades radial basis functions (RBFs) have been used extensively for solving problems in various fields of engineering and scientific computing. Initially, RBF methods were used primarily for multivariate approximation [2], [5], the theoretical foundations of which may be found in [11], [16]. RBF approximation methods are particularly attractive due to their accuracy, flexibility and ability to handle scattered data in complex domains. In addition, they are meshless and very easy to implement. However, they suffer from a number of unfavourable issues. One such issue is the illconditioning of the global matrix which means that as the number of collocation points increases, the condition number of the matrix increases and, eventually, the accuracy of the solution suffers. Another unresolved issue in RBF approximation is the determination of the optimal value of the shape parameter present in many RBFs, as will be explained later. In the past, various techniques have been proposed for the determination of an appropriate value of the shape parameter [4], [9], [10], [18]. More recently, RBF methods have also become popular for the solution of boundary and initial value problems for partial differential equations, in particular due to the method proposed by Kansa [14], which has become known as the Kansa method. The same advantages and disadvantages as those mentioned for RBF approximation still hold for the Kansa method. A recent review of Kansa RBF methods for the numerical solution of partial differential equations may be found in [3].

In the current work we attempt to find a way of resolving the shape parameter issue when RBF methods are used to approximate functions in two dimensional domains. In particular, we use an idea developed in [12], [13] for applications of the Kansa-RBF method [14] to nonlinear boundary value problems. In the RBF approximation of the function in question we let the shape parameter be one of the unknowns. Since the shape parameter appears nonlinearly in the RBF expansion, the collocation of the approximation at various points in the domain now yields a system of nonlinear equations. This system is then solved using standard software for the solution of nonlinear systems which yields, in addition to the unknown coefficients in the RBF expansion, a value for the shape parameter. It is shown, in several examples, that, provided the initial guess is taken within a certain range, this value is close to the optimal shape parameter value. The results of further numerical experiments may be found in [20].

The paper is organized as follows. In Section 2 we present the problem and describe the RBF formulation. The details regarding the solution of the resulting nonlinear system are

WIT Transactions on Engineering Sciences, Vol 122, © 2019 WIT Press
www.witpress.com, ISSN 1743-3533 (on-line)
doi:10.2495/BE410141

presented in Section 3. In Section 4 we present the results of two numerical examples. Finally, some concluding remarks and ideas for future work are provided in Section 5.

2 RADIAL BASIS FUNCTION INTERPOLATION

2.1 The problem

We consider the interpolation of a bivariate function $f : \overline{\Omega} \to \mathbb{R}$, where $\Omega \subset \mathbb{R}^2$, from a set of sample values $\{f(x_\mathsf{m}, y_\mathsf{m})\}_{\mathsf{m}=1}^{\mathsf{M}}$ on a discrete set $X = \{(x_\mathsf{m}, y_\mathsf{m})\}_{\mathsf{m}=1}^{\mathsf{M}} \subset \overline{\Omega}$.

2.2 RBF approximation

We approximate f by a linear combination of radial basis functions (RBFs)

$$f_\mathsf{N}(x,y) = \sum_{\mathsf{n}=1}^{N} a_\mathsf{n}\phi_\mathsf{n}(x,y), \qquad (x,y) \in \overline{\Omega}. \tag{1}$$

The RBFs $\phi(x,y)$, $\mathsf{n} = 1, \ldots, N$ can be expressed in the form

$$\phi_\mathsf{n}(x,y) = \Phi(r_\mathsf{n}), \qquad r_\mathsf{n}^2 = (x - x_\mathsf{n})^2 + (y - y_\mathsf{n})^2. \tag{2}$$

Thus each RBF ϕ_n is associated with a point $(x_\mathsf{n}, y_\mathsf{n})$. These points $\{(x_\mathsf{n}, y_\mathsf{n})\}_{\mathsf{n}=1}^{\mathsf{N}}$ are usually referred to as *centers*. We shall assume that we have N_int interior centers $\{(x_\mathsf{n}, y_\mathsf{n})\}_{\mathsf{n}=1}^{\mathsf{N}_\mathsf{int}}$ and N_bry boundary centers $\{(x_\mathsf{n}, y_\mathsf{n})\}_{\mathsf{n}=\mathsf{N}_\mathsf{int}+1}^{\mathsf{N}_\mathsf{int}+\mathsf{N}_\mathsf{bry}}$ and take $\mathsf{N} = \mathsf{N}_\mathsf{int} + \mathsf{N}_\mathsf{bry}$. Often, RBFs include *shape parameters*. In this paper, the shape parameter will be denoted by c. The determination of the optimal value of shape parameter remains a major challenge.

For example, we mention the following popular RBFs which include a shape parameter:

- the normalized multiquadric (MQ) basis function

$$\phi_\mathsf{n}(x,y) = \Phi(r_\mathsf{n}) = \sqrt{(cr_\mathsf{n})^2 + 1}, \qquad r_\mathsf{n}^2 = (x - x_\mathsf{n})^2 + (y - y_\mathsf{n})^2, \tag{3}$$

and

- the inverse multiquadric (IMQ) basis function

$$\phi_\mathsf{n}(x,y) = \Phi(r_\mathsf{n}) = \frac{1}{\sqrt{(cr_\mathsf{n})^2 + 1}}, \qquad r_\mathsf{n}^2 = (x - x_\mathsf{n})^2 + (y - y_\mathsf{n})^2. \tag{4}$$

In the case of shape parameter-dependent RBFs we shall denote $\Phi(r_\mathsf{n})$ in (2) by $\Phi(c, r_\mathsf{n})$. We therefore replace approximation (1) by

$$f_\mathsf{N}(x,y) = \sum_{\mathsf{n}=1}^{\mathsf{N}} a_\mathsf{n}\,\Phi(c, r_\mathsf{n}), \qquad (x,y) \in \overline{\Omega}. \tag{5}$$

The coefficients $\{a_\mathsf{n}\}_{\mathsf{n}=1}^{\mathsf{N}}$ in eqn (5) are determined from the collocation equations

$$f_\mathsf{N}(x_\mathsf{m}, y_\mathsf{m}) \; = \; f(x_\mathsf{m}, y_\mathsf{m}), \quad \mathsf{m} = 1, \ldots, \mathsf{M}, \tag{6}$$

where the points $\{(x_\mathsf{m}, y_\mathsf{m})\}_{\mathsf{m}=1}^{\mathsf{M}} \in \overline{\Omega}$ are the *collocation points*. Of these, we have M_int interior collocation points $\{(x_\mathsf{m}, y_\mathsf{m})\}_{\mathsf{m}=1}^{\mathsf{M}_\mathsf{int}}$ and M_bry boundary collocation points $\{(x_\mathsf{m}, y_\mathsf{m})\}_{\mathsf{m}=\mathsf{M}_\mathsf{int}+1}^{\mathsf{M}_\mathsf{int}+\mathsf{M}_\mathsf{bry}}$ and take $\mathsf{M} = \mathsf{M}_\mathsf{int} + \mathsf{M}_\mathsf{bry}$. Note that, in general, the collocation points are not the same as the centers. Clearly, we need to have $\mathsf{M} \geq \mathsf{N} + 1$.

2.3 Improved RBF approximation

Let $\mathbb{P}_{\mathsf{p}}$ be the set of bivariate polynomials of total degree up to p and $\{p_{\mathsf{k}}\}_{\mathsf{k}=1}^{\mathsf{K}}$ be a basis of $\mathbb{P}_{\mathsf{p}}$. It is known that the number of polynomial terms for degree p is $\mathsf{K} = (\mathsf{p}+1)(\mathsf{p}+2)/2$. The polynomial basis is thus

$$p_{\mathsf{k}}(x,y) = x^{i-j}y^{j}, \quad 0 \le j \le i,\ 0 \le i \le \mathsf{p}, \quad \text{for} \quad \mathsf{k} = 1,\ldots,\mathsf{K}. \tag{7}$$

In the RBF approximation with an augmented polynomial basis, the approximation (5) is replaced by

$$f_{\mathsf{N}}(x,y) = \sum_{\mathsf{n}=1}^{\mathsf{N}} a_{\mathsf{n}}\, \Phi(c, r_{\mathsf{n}}) + \sum_{\mathsf{k}=1}^{\mathsf{K}} a_{\mathsf{N}+\mathsf{k}}\, p_{\mathsf{k}}(x,y), \qquad (x,y) \in \overline{\Omega}. \tag{8}$$

Also, in addition to the collocation eqns (6), we have the insolvency conditions

$$\sum_{\mathsf{n}=1}^{\mathsf{N}} a_{\mathsf{n}}\, p_{\mathsf{k}}(\mathsf{x}_{\mathsf{n}}, \mathsf{y}_{\mathsf{n}}) = 0, \quad \mathsf{k} = 1,\ldots,\mathsf{K}. \tag{9}$$

Clearly, we still need to have $\mathsf{M} \ge \mathsf{N}+1$.

3 IMPLEMENTATIONAL DETAILS

Since the shape parameter appears nonlinearly in the collocation eqns (6), the system of $M+K$ eqns (6) and (9) is nonlinear and can be written in the form

$$\boldsymbol{F}(\boldsymbol{a},c) := \begin{bmatrix} F_1 \\ F_2 \\ \vdots \\ F_{\mathsf{M}+\mathsf{K}} \end{bmatrix} = \begin{bmatrix} f_{\mathsf{N}}(x_1,y_1) - f(x_1,y_1) \\ \vdots \\ f_{\mathsf{N}}(x_M,y_M) - f(x_M,y_M) \\ \sum_{\mathsf{n}=1}^{\mathsf{N}} a_{\mathsf{n}}\, p_1(x_{\mathsf{n}}, y_{\mathsf{n}}) \\ \vdots \\ \sum_{\mathsf{n}=1}^{\mathsf{N}} a_{\mathsf{n}}\, p_{\mathsf{K}}(\mathsf{x}_{\mathsf{n}}, \mathsf{y}_{\mathsf{n}}) \end{bmatrix} = \boldsymbol{0}, \tag{10}$$

where $\boldsymbol{a} = [a_1,\ldots,a_{\mathsf{N}+\mathsf{K}}]^T$ are the coefficients in approximation (8).

The nonlinear system (10) may be solved with the MATLAB© [15] optimization toolbox routine `fsolve`. The algorithm used is either a trust-region-dogleg/reflective or the Levenberg–Marquardt algorithm [17]. The routine `fsolve` does not require the user to provide the gradient. When the system is not square, i.e. the number of equations is larger than the number of unknowns, which is the case in the current problem, the routine uses the Levenberg–Marquardt algorithm. Alternatively, the solution of the nonlinear system (10) may be recast as a nonlinear least squares minimization problem by setting

$$S(\boldsymbol{a},c) := \sum_{\mathsf{m}=1}^{\mathsf{M}+\mathsf{K}} F_{\mathsf{m}}^2, \tag{11}$$

where the $\{F_{\mathsf{m}}\}_{\mathsf{m}=1}^{\mathsf{M}+\mathsf{K}}$ are defined in (10). The functional $S(\boldsymbol{a},c)$ is then minimized using the MATLAB© optimization toolbox routine `lsqnonlin` which solves nonlinear least squares problems using, like `fsolve`, either a subspace trust region method or the Levenberg–Marquardt algorithm. The routine `lsqnonlin` does not require the user to provide the gradient and, in addition, offers the option of imposing lower and upper bounds on the elements of the vector of unknowns $\boldsymbol{x} = [\boldsymbol{a}, c]^T$ through the vectors `lb` and `up`. We can thus easily impose the constraints on the values of the coefficients and in particular on the values of the shape parameter. It should be noted that both `lsqnonlin` and `fsolve` might give only local solutions.

4 NUMERICAL EXAMPLES

In the numerical examples considered in this section, we calculated the approximate solution f_{N} at L test points in Ω. We calculated the maximum relative error E defined by

$$E = \frac{||f - f_{\mathsf{N}}||_{\infty,\Omega}}{||f||_{\infty,\Omega}} \tag{12}$$

and the root mean square error (RMSE) $\mathcal{E}$ from

$$\mathcal{E} = \sqrt{\frac{1}{\mathsf{L}}\sum_{\ell=1}^{\mathsf{L}} [f(x_\ell, y_\ell) - f_{\mathsf{N}}(x_\ell, y_\ell)]^2}\,. \tag{13}$$

In Example 1 we used MQ (3) and IMQ (4) while in the other numerical examples we used only the MQ. For all examples considered the analytical solution is known. In each case, once the value of the shape parameter c was obtained, the solution was re-calculated for this value of c. The interior collocation points are taken to be the Halton points [7, Appendix A.1] while the boundary points are uniformly distributed. The corresponding boundary and interior centers are obtained in a similar way. We chose a set of $\mathsf{L} = 101$ interior Halton points to be the set of test points. These may be generated using the MATLAB© code `haltonseq.m`. It was observed that the addition of the bivariate polynomials did not improve the accuracy of the solution, and hence in all the examples presented we took $\mathsf{p} = 0$. In both numerical examples, the domain Ω was the unit square $(0, 1) \times (0, 1)$.

4.1 Example 1

We approximate Franke's function [7, Section 2.2]

$$f(x,y) = \frac{3}{4}\mathrm{e}^{-\left(\frac{(9x-2)^2+(9y-2)^2}{4}\right)} + \frac{3}{4}\mathrm{e}^{-\left(\frac{(9x+1)^2}{49}+\frac{(9y+1)^2}{10}\right)}$$
$$+\frac{1}{2}\mathrm{e}^{-\left(\frac{(9x-7)^2+(9y-3)^2}{4}\right)} - \frac{1}{5}\mathrm{e}^{-\left((9x-4)^2+(9y-7)^2\right)}$$

(see Fig. 1).

4.1.1 Experiments with MQ

In Fig. 2(a) we present the results obtained while fixing the value of the shape parameter for various such values in the case $\mathsf{M} = 336$, $\mathsf{N} = 278$. In Table 1 we present the corresponding results for the same numbers of collocation points and centers but with the value of the shape parameter c free with different initial values c_0. The obtained results for different c_0 are remarkably consistent and agree with the value of c giving the apparent minima in Fig. 2(a).

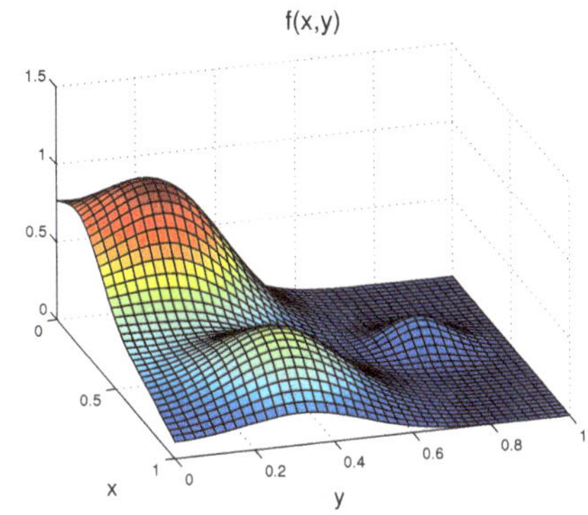

Figure 1: Example 1. The function $f(x, y)$.

Table 1: Example 1: Results for various c_0 using MQ with $\mathsf{M} = 336$, $\mathsf{N} = 278$ after 2000 iterations.

M	N	c_0	c	E	$\mathcal{E}$
336	278	1.0000	4.8689	9.4509(-5)	1.9220(-5)
336	278	2.0000	4.8629	9.4089(-5)	1.9352(-5)
336	278	3.0000	4.8663	9.4283(-5)	1.9025(-5)
336	278	4.0000	4.8902	9.9538(-5)	1.9664(-5)
336	278	5.0000	4.8768	9.7329(-5)	1.9331(-5)

4.1.2 Experiments with IMQ

In Fig. 2(b) we present the results obtained while fixing the value of the shape parameter for various such values in the case $\mathsf{M} = 336$, $\mathsf{N} = 278$. In Table 2 we present the corresponding results for the same numbers of collocation points and centers but with the value of the

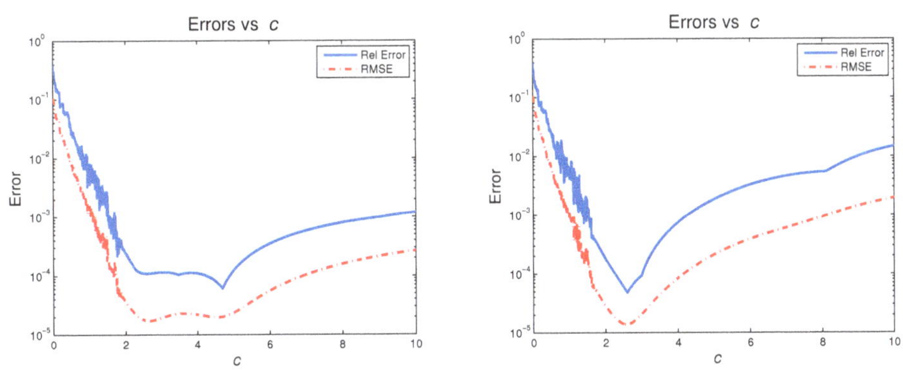

(a) MQ

(b) IMQ

Figure 2: Example 1. Results with MQ and IMQ for varying shape parameter for $\mathsf{M} = 336, \mathsf{N} = 278, \mathsf{p} = 0$.

Table 2: Example 1: Results for various c_0 using IMQ with $\mathsf{M} = 336$, $\mathsf{N} = 278$ after 2000 iterations.

M	N	c_0	c	E	$\mathcal{E}$
336	278	1.0000	3.5573	4.0088(-4)	4.8095(-5)
336	278	2.0000	3.5709	4.1049(-4)	4.8774(-5)
336	278	3.0000	3.5687	4.0895(-4)	4.8864(-5)
336	278	4.0000	3.5812	4.1749(-4)	4.9525(-5)
336	278	5.0000	3.5785	4.1542(-4)	4.9293(-5)

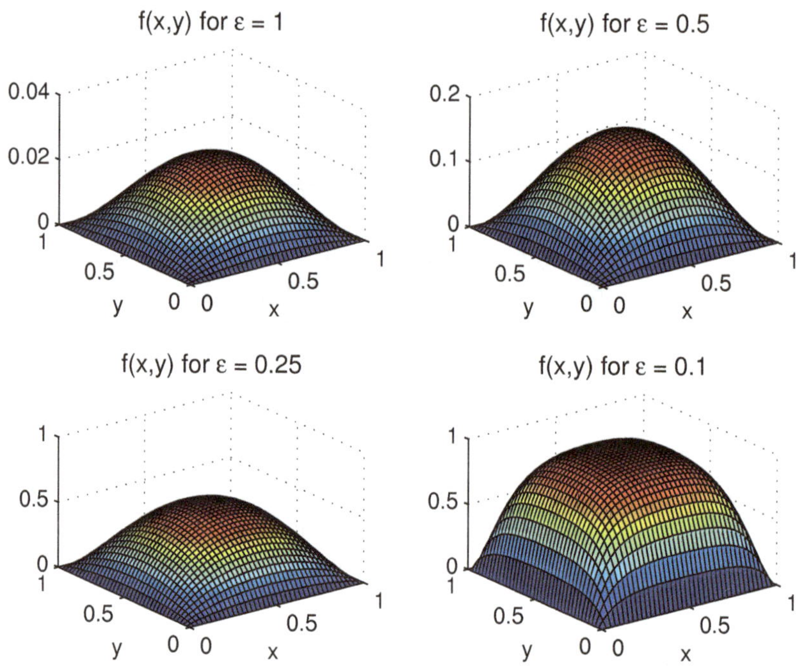

Figure 3: Example 2. The function $f(x, y)$.

shape parameter c free with different initial values c_0. The obtained results for different c_0 are remarkably consistent and agree well with the value of c giving the apparent minima in Fig. 2(b).

4.2 Example 2

We next approximate the function [6], [8], [13]

$$f(x, y) = \left(1 + \mathrm{e}^{-1/\varepsilon} - \mathrm{e}^{-x/\varepsilon} - \mathrm{e}^{(x-1)/\varepsilon}\right)\left(1 + \mathrm{e}^{-1/\varepsilon} - \mathrm{e}^{-y/\varepsilon} - \mathrm{e}^{(y-1)/\varepsilon}\right).$$

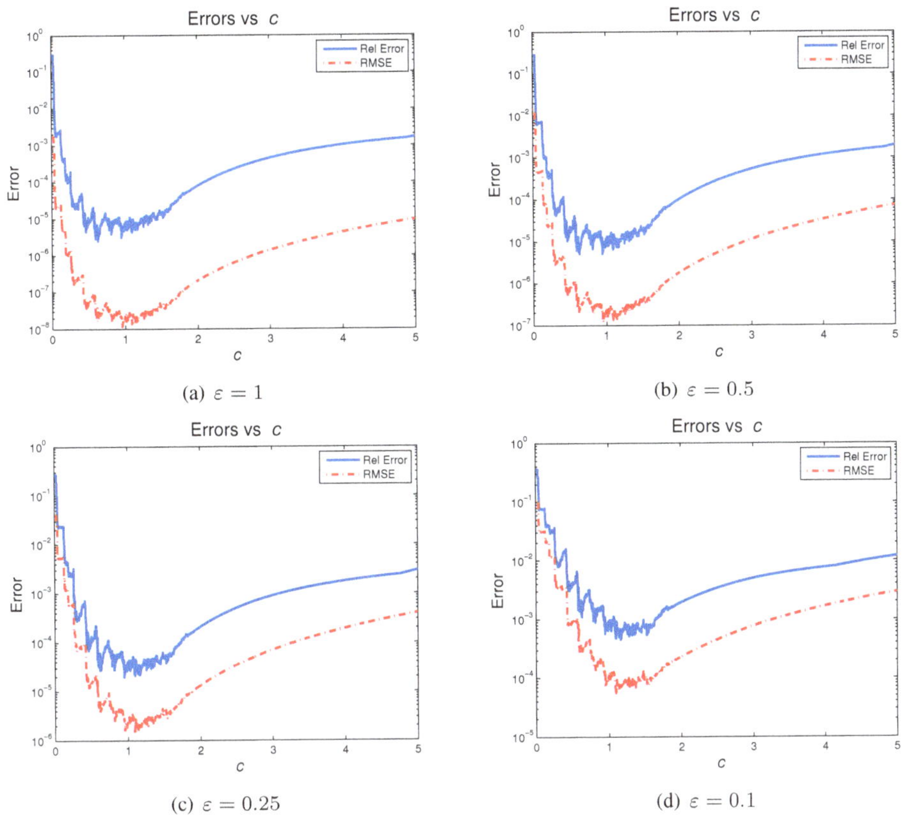

(a) $\varepsilon = 1$ (b) $\varepsilon = 0.5$

(c) $\varepsilon = 0.25$ (d) $\varepsilon = 0.1$

Figure 4: Example 2. Results for varying shape parameter for $\varepsilon = 1, 0.5, 0.25$ and 0.1, $\mathsf{M} = 336, \mathsf{N} = 278, \mathsf{p} = 0$.

As $\varepsilon > 0$ decreases, the problem becomes more difficult (see Fig. 3). In Figs 4(a)–(d) we present the results obtained while fixing the value of the shape parameter for various such values in the case $\mathsf{M} = 336$, $\mathsf{N} = 278$ and $\varepsilon = 1$, 0.5, 0.25 and 0.1. In Table 3 we present the corresponding results for the same numbers of collocation points and centers but with the value of the shape parameter c free with different initial values c_0.

4.2.1 Case $\varepsilon = 1$

In this case the solution is quite flat and convergence is rapid. In Table 3 we present typical results obtained using MQ RBFs with different initial values for c_0. The results show that the final value of c is close to the one giving a minimum in Fig. 4(a). The nonlinear system was solved with the MATLAB© optimization toolbox routine `fsolve`.

4.2.2 Case $\varepsilon = 0.5$

In this case the solution is not as flat and convergence is not as rapid as for $\varepsilon = 1$. In Table 3 we present typical results obtained using MQ RBFs. We used the routine `lsqnonlin` for different initial values c_0. The find value of the shape parameter c is close to the optimal value of Fig. 4(b).

Table 3: Example 2: Results for various c_0 using MQ with M = 336, N = 278 after 2000 iterations for $\varepsilon = 1, 0.5, 0.25$ and 0.1.

	M	N	c_0	c	E	$\mathcal{E}$
$\varepsilon = 1$						
	336	278	0.4000	1.4312	1.0344(-5)	3.5067(-8)
	336	278	0.5000	1.4384	1.0296(-5)	3.5667(-8)
	336	278	0.6000	1.4277	9.7104(-6)	3.3584(-8)
	336	278	0.7000	1.4171	1.1609(-5)	3.3984(-8)
	336	278	0.8000	1.4102	1.1596(-5)	3.4626(-8)
$\varepsilon = 0.5$						
	336	278	0.3000	0.6222	4.1773(-6)	2.2671(-7)
	336	278	0.4000	0.6301	4.9118(-6)	2.4898(-7)
	336	278	0.5000	0.6477	8.2606(-6)	2.8472(-7)
	336	278	0.6000	0.6310	4.7927(-6)	2.3990(-7)
	336	278	0.7000	0.6575	1.1300(-5)	3.1951(-7)
$\varepsilon = 0.25$						
	336	278	0.3000	0.8754	5.2289(-5)	3.7164(-6)
	336	278	0.4000	0.9115	5.5683(-5)	3.8877(-6)
	336	278	0.5000	0.9189	6.4466(-5)	4.2627(-6)
	336	278	0.6000	0.9159	6.9501(-5)	4.6381(-6)
	336	278	0.7000	0.9347	4.7612(-5)	3.0524(-6)
$\varepsilon = 0.1$						
	336	278	0.4000	0.9668	8.4964(-4)	9.5993(-5)
	336	278	0.5000	0.9857	7.2374(-4)	9.2681(-5)
	336	278	0.6000	0.9955	9.2337(-4)	1.1115(-4)
	336	278	0.7000	1.0219	7.3247(-4)	1.0153(-4)
	336	278	0.8000	1.0111	9.5063(-4)	1.2014(-4)

4.2.3 Case $\varepsilon = 0.25$

Again, the solution is not as flat and convergence is not as rapid as for $\varepsilon = 1$ and $\varepsilon = 0.5$. In Table 3 we present typical results obtained using MQ RBFs. We used the routine `lsqnonlin` for different initial values c_0. The find value of the shape parameter c is close to the optimal value of Fig. 4(c).

4.2.4 Case $\varepsilon = 0.1$

This is the most difficult case. Covergence is much slower as the solution is not flat. In Table 3 we present typical results obtained using MQ RBFs. We used the routine `lsqnonlin` for different initial values c_0. The find value of the shape parameter c is close to the optimal value of Fig. 4(d).

5 CONCLUSIONS

In this work we have considered the approximation of functions by RBF expansions. The coefficients in these expansions are obtained by collocation. The issue of the optimal value of the shape parameter in the RBFs used is addressed by including its value in the unknowns of the problem to be determined by collocation. The resulting system of nonlinear equations is solved using standard software. The results of several tests reveal that, provided the initial value is taken within a certain range, the value of the shape parameter obtained is close to the optimal value of the shape parameter. A criticism of the proposed approach is clearly the cost, as we are essentially turning a linear problem into a nonlinear one. This however has been the case with various other approaches proposed for the determination of a suitable shape parameter. For example, we mention the use of the so called leave-one-out cross validation (LOOCV) proposed by Rippa [18]. Future work could involve the use of different shape parameters for a each RBF in the approximation expansion. The idea of using variable shape parameters in RBF approximations has been previously used in [1], [19], [21]. Moreover, since both `fsolve` and `lsqnonlin` offer the user the option of providing the Jacobian of the nonlinear system under consideration, it would be interesting to study how this modification affects the convergence of the method.

REFERENCES

[1] Afiatdoust, F. & Esmaeilbeigi, M., Optimal variable shape parameters using genetic algorithm for radial basis function approximation. *Ain Shams Eng. J.*, **6** (2015), 639–647.

[2] Buhmann, M.D., *Radial Basis Functions*, Cambridge University Press: Cambridge, 2003.

[3] Chen, W., Fu, Z.-J. & Chen, C.S., Recent advances in radial basis function collocation methods. *Springer Briefs in Applied Sciences and Technology*, Springer: Heidelberg, 2014.

[4] Cheng, A.H.-D., Multiquadric and its shape parameter – a numerical investigation of error estimate, condition number, and round-off error by arbitrary precision computation. *Eng. Anal. Bound. Elem.*, **36** (2012), 220–239.

[5] Dyn, N., *Interpolation of scattered data by radial basis function. Topics in Multivariate Approximation*, eds. C.K. Chui, L.L. Schumarket & F.I. Utreras, Academic Press: New York, pp. 47–61, 1987.

[6] Fasshauer, G.E., Newton iteration with multiquadrics for the solution of nonlinear PDEs. *Comput. Math. Appl.*, **43** (2002), 423–438.

[7] Fasshauer, G.E., Meshfree approximation methods with MATLAB. *Interdisciplinary Mathematical Sciences*, vol. 6, World Scientific Publishing Co. Pte. Ltd: Hackensack, NJ, 2007.

[8] Fasshauer, G.E., Gartland, E.C. & Jerome, J.W., Newton iteration for partial differential equations and the approximation of the identity. *Numer. Algorithms*, **25** (2000), 181–195.

[9] Fasshauer, G.E. & Zhang, J.G., On choosing optimal shape parameters for RBF approximation. *Numer. Algorithms*, **45** (2007), 345–368.

[10] Fornberg, B. & Wright, G., Stable computation of multiquadric interpolants for all values of the shape parameter. *Comput. Math. Appl.*, **48**(5-6) (2004), 853–867.

[11] Franke, R., Scattered data interpolation: tests of some methods. *Math. Comp.*, **48** (1982), 181–200.

[12] Jankowska, M.A., Karageorghis, A. & Chen, C.S., Kansa RBF method for nonlinear problems. *Int. J. Comp. Meth. Exp. Meas.*, **6** (2018), 1000–1007.

[13] Jankowska, M.A., Karageorghis, A. & Chen, C.S., Improved Kansa RBF method for the solution of nonlinear boundary value problems. *Eng. Anal. Bound. Elem.*, **87** (2018), 173–183.

[14] Kansa, E.J., Multiquadrics—a scattered data approximation scheme with applications to computational fluid-dynamics. II. Solutions to parabolic, hyperbolic and elliptic partial differential equations. *Comput. Math. Appl.*, **19** (1990), 147–161.

[15] The MathWorks, Inc., 3 Apple Hill Dr., Natick, MA, *Matlab*.

[16] Micchelli, C.A., Interpolation of scattered data: distance matrices and conditionally positive definite functions. *Constr. Approx.*, **2** (1986), 11–12.

[17] Moré, J.J., The Levenberg–Marquardt algorithm: implementation and theory. *Numer. Analysis*, (1977), 105–116.

[18] Rippa, S., An algorithm for selecting a good value for the parameter c in radial basis function interpolation. *Adv. Comput. Math.*, **11** (1999), 193–210.

[19] Sarra, S.A. & Sturgill, D., A random variable shape parameter strategy for radial basis function approximation methods. *Eng. Anal. Bound. Elem.*, **33** (2009), 1239–1245.

[20] Tryfonos, P., Some numerical experiments on radial basis function approximation, MSc thesis, Department of Mathematics and Statistics, University of Cyprus, 2018.

[21] Xiang, S., Wang, K.-M., Ai, Y.-T., Sha, Y.-D. & Shi, H., Thrigonometric variable shape parameter and exponent strategy for generalized multiquadric radial basis function approximation. *Appl. Math. Model.*, **36** (2012), 1931–1938.

FAST GENERATION OF VARIABLE DENSITY NODE DISTRIBUTIONS FOR MESH-FREE METHODS

JURE SLAK & GREGOR KOSEC
Parallel and Distributed Systems Laboratory, "Jožef Stefan" Institute, Slovenia

ABSTRACT

Problems with pronounced differences in intensity, such as those that appear in contact mechanics, where locally concentrated high stresses are present, are usually attacked with a spatially variable nodal distribution. In a meshless context, such distribution has to be quasi-smooth with minimal spacing requirements to produce satisfactory results. To that end, development of fast discretization procedures, which distribute the nodes according to given (non-constant) spacing function, has become of interest. We improve a recently published algorithm for fast 2D meshless discretizations, by lowering its time complexity from $O(NS)$ to $O(N \log S)$ resulting in an algorithm that generates a million nodes per second. The proposed algorithm is independent of the dimensionality of the space, does not rely on a coordinate system and has provable minimal node spacing requirements. The original and new algorithms are compared in terms of node quality and execution time. The usability and robustness of the new algorithm is presented by solving PDE examples on irregular 3D domains with the RBF-FD method and by using it as a node generation algorithm in fully the automatic adaptive solver for linear elasticity.

Keywords: node generation, adaptive discretizations, mesh-free methods, RBF-FD, PDEs.

1 INTRODUCTION

One of the most attractive features of mesh-free methods is their ability to work on unstructured sets of nodes without any connectivity relations. This does not imply that any distribution of nodes will produce satisfying results, as especially many strong form methods have minimal spacing requirements that impact stability [2] and need locally regular distributions for good performance. An example of such a method is the Radial Basis Functions Finite Differences (RBF-FD) method [1]. Nonetheless, node positioning is an easier problem than mesh generation and specialized algorithms for fast generation of nodes on irregular domains have been developed [3], [4]. Often the nodes are required to be approximately distributed according to a spacing function, allowing for variable density node distributions to be generated. The need for such distributions arose from problems with pronounced differences in intensity, such as contact problems in linear elasticity. Furthermore, the high-intensity regions might not be known in advance and adaptive iteration is needed to solve the problem. In such cases, it is essential for the node positioning algorithm to be fast and robust, to be able to generate quality distributions as required by the error indicator automatically.

In this work, we improve a recently published algorithm by Fornberg and Flyer for 2D meshless discretizations [3] by improving its time complexity and generalize it to higher dimensions. We propose and alternative algorithm in Section 3 and compare the algorithms in terms of execution time and node quality in Section 4. In Section 5, the practical application of the new algorithm is demonstrated on irregular 3D domains and in the fully automatic adaptive setting.

2 IMPROVEMENTS OF 2D ALGORITHM BY FORNBERG AND FLYER

The algorithm by Fornberg and Flyer [3] fills a rectangle with nodes spaced according to the spacing function δr. The algorithm works as an advancing front algorithm that begins by

WIT Transactions on Engineering Sciences, Vol 122, © 2019 WIT Press
www.witpress.com, ISSN 1743-3533 (on-line)
doi:10.2495/BE410151

filling the bottom line of the rectangle with candidates for node positions. On every iteration, a point p with minimal y-coordinate is chosen and is added to the solution set. The point is then removed from the candidate list as well as other points being too close to it. A small constant number, e.g. 5, of new candidates are then computed, by spacing them uniformly by angle on an arc of radius $\delta r(p)$ spanning between the closest remaining left and right candidates, and are inserted into the list. The procedure finishes when the minimal y-coordinate is above the upper rectangle boundary.

A naïve implementation of the above algorithm using arrays has time complexity $O(NS)$, where N is the number of generated nodes and S is the maximal number of points in the candidate list array. For uniform nodal spacing h, it holds that $N = O(1/h^2)$ and $S = O(1/h)$, bringing the total time complexity to $O(1/h^3)$.

Additionally, if an irregular 2D domain is treated, the algorithm first fills its bounding box, discards all the nodes outside the domain and nodes too close to the boundary. The boundary discretization is then superimposed on the remaining nodes, and the distribution is then further regularized using a few iterations of a repel-type algorithm.

2.1 Time complexity improvements

The naïve implementation of the above algorithm can be improved by taking note of the required operations in each iteration. First, a point p with minimal y-coordinate is found, then all candidates in radius $\delta r(p)$ are removed from the candidate list and some new candidates are inserted instead. The insertions and removals can be sped up using a linked list instead of an array, performing them in constant time from a known position. The complexity of the search operation is not changed and as the candidates are ordered according to the x-coordinate, only a few candidates left and right from the known position must be checked. The only operation left is the minimum extraction, which can be done fast, if additionally, a data structure that allows fast minimum extraction (e.g. a heap) is used in the combination with a linked list. The heap contains references to the linked list elements and extracting the minimum is now an $O(\log S)$ operation. Introducing the heap requires some more bookkeeping at insertions and removals. Upon insertion of a candidate in the linked list, a reference to it must also be inserted in the heap. Upon removal of the candidate from the linked list, it is lazily marked as removed and al the removed elements are ignored at the next minimum extraction. This brings the complexity of one iteration down to a few constant time updates of the linked list and $O(\log S)$ time updates of the heap, making the overall time complexity $O(N \log S)$. For uniform nodal spacing this results in a time complexity $O(1/h^2 \log 1/h)$, improving the $O(1/h^3)$ complexity of the naïve algorithm.

Both proposed implementations were compared in practice using uniform nodal spacing h. They were implemented in C++11, compiled using g++v7 on Linux and run on a laptop computer with 4 Intel® Core™ i7-7700HQ CPU @ 2.80 GHz cores and 16 GB DDR4 RAM. The execution times are shown in Fig. 1.

As expected, the naïve implementation is faster for small N up to around $N = 10\,000$ due to less bookkeeping overhead. For large N the improved implementation with better asymptotic behaviour is faster and generates a million nodes under one second.

2.2 Generalization to higher dimensions

The algorithm and the efficient implementation can be extended to higher dimensions $d \geq 2$. Instead of a line, the advancing front forms a $d - 1$ dimensional plane. In each iteration, the

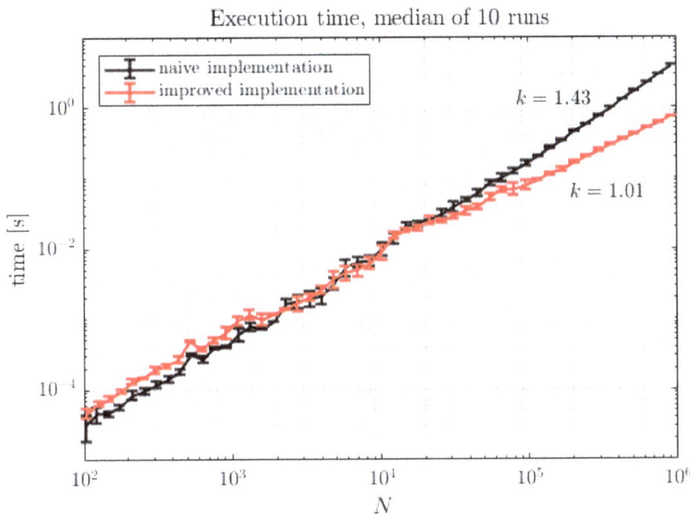

Figure 1: Execution time of naive and improved implementation. The error bars indicate standard deviation over 10 runs.

candidate p with the minimal d-th coordinate is extracted and its neighbourhood in radius $\delta r(p)$ is removed. New candidates are then inserted instead. It is not entirely obvious how to generalize the definition of new candidates in higher dimensions. One direct generalization would be to find the two closest remaining points to p along each of the $d-1$ dimensions, one in each direction. The angles between those points would then be split into uniform parts, thus defining the angle of the new candidate along each dimension. The scale is given by $\delta r(p)$ as in 1D case. In other words, angles of the new candidates form a tensor product mesh and the points are scaled to have radius equal to $\delta r(p)$.

Efficient implementation can also be generalized to higher dimensions: instead of a sorted doubly linked list with a heap, a spatial search structure, such a k-d tree must be used in combination with a heap to achieve similar time complexity. This makes the implementation somewhat more complex and out of the scope of this work.

2.3 Minimal spacing guarantees

It is often desirable for node generation algorithms to satisfy minimal node spacing guarantees, as small distances between nodes can cause ill conditioning in numerical methods [2]. If the nodes were generated using a constant density function $\delta r(p) = \delta r_0$, it is reasonable to expect that $\|p_i - p_j\| \geq \delta r_0$ holds for $i \neq j$. For variable density functions, some constraints were introduced by [5], were assumptions on the Lipschitz constant of the function must be made. Even if no such assumptions can be made (and none are needed by any of the algorithms in this work), a weak requirement $\|p_i - p_j\| \geq \min \delta r$, for $i \neq j$ can still be made. However, the algorithm by Fornberg and Flyer does not satisfy even that, as shown in Fig. 2.

A unit square was filled with uniform density $h = 0.01$ and marked points violate the proximity criterion. Most of the violations appear on the boundary, with some also appearing sporadically in the domain interior.

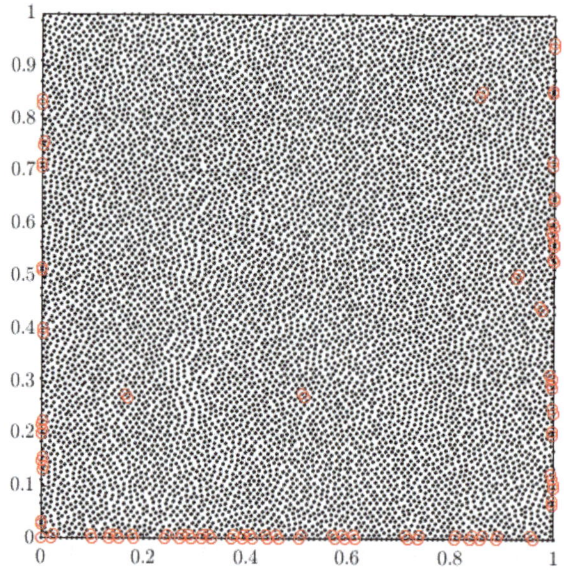

Figure 2: A sample run with marked proximity violations.

3 POISSON DISK SAMPLING BASED ALGORITHM

The algorithm proposed by Fornberg and Flyer has a few undesired properties. First, it has no minimal spacing guarantees, it is directionally dependent, and it does not work for arbitrary shapes without additional algorithms. Therefore, we propose an alternative algorithm for node generation, which is not directional, it is dimension independent, works with irregular domains, and satisfies minimal spacing criteria.

The algorithm starts with a given boundary discretization and works inwards. All boundary nodes and optionally some other starting user-supplied nodes are placed in a queue of *non-expanded* nodes. In each iteration, a new node p is popped from the queue and new candidates arising from it are considered. The candidates are considered one by one. If a candidate is outside of the domain, it is removed from consideration immediately. Otherwise, if it is not too close to any existing nodes, it is accepted and added to the queue and the result set. This iteration continues until the queue is empty.

The new candidates arising from each point can be generated in a few different ways: either chosen uniformly at random from a sphere with radius $\delta r(p)$ or generated in a deterministic way. One good way is such that hex packing is reproduced in 2D and 3D when using uniform spacing.

3.1 Time complexity

Each iteration consists of considering one node and adding a few new nodes in the queue. The most time-consuming operation is to check that new candidates are not too close to the existing nodes. There are different approaches to answering these types of queries, using e.g. background cell meshes or spatial trees. If k-d trees are used, then insertions and queries can be performed in $O(n \log n)$, where n is the number of points in the tree. As each candidate is inserted in the tree at most once and queries at most once, the overall time complexity is $O(N \log N)$.

4 COMPARISON OF ALGORITHMS

4.1 Node quality

Both algorithms are first compared in terms of node quality. A sample function

$$\delta r(x,y) = \frac{1}{8.5979}\left(3(x-1)^2 e^{-x^2-(y+1)^2} - 5e^{-x^2-y^2} + 4.2166\right)\cdot(H-h)+h \quad (1)$$

was used to compare and asses the quality of both algorithms. The function is such that H is the largest and h is the smallest node spacing used. Values $H = 0.1$ and $h = 0.01$ were used in our examples. The generated nodes are shown in Fig. 3.

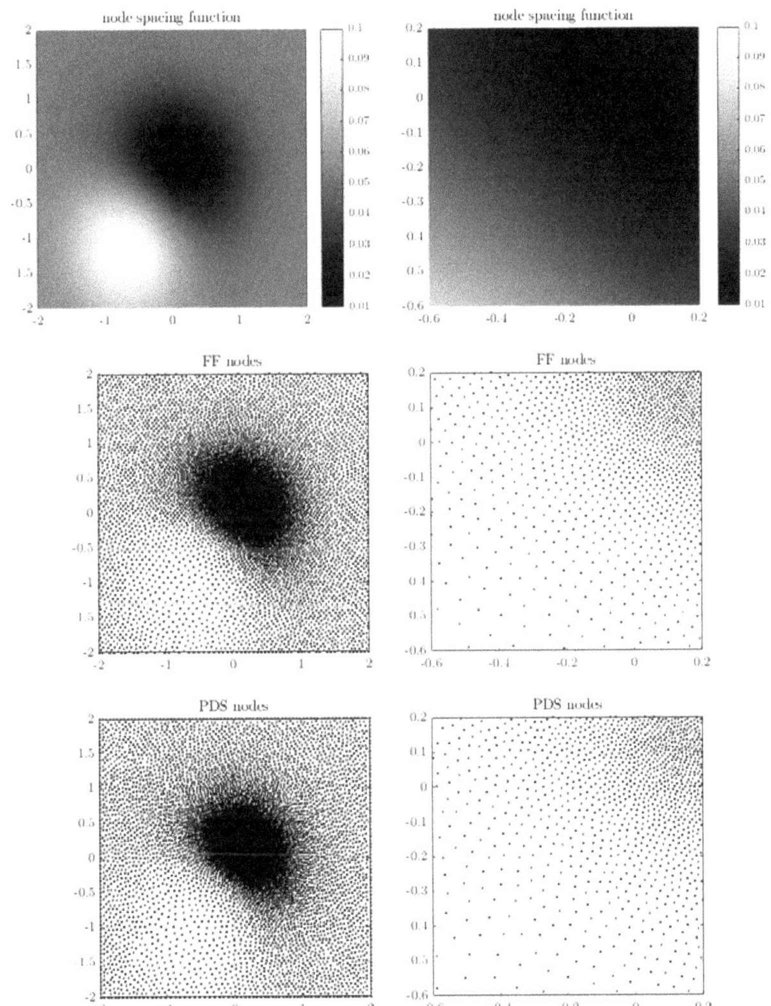

Figure 3: The left column represents the density function, the nodes generated by the algorithm published by Fornberg and Flyer (FF) and the nodes generated by the Poisson Disk Sampling (PDS) based algorithm, respectively. The right column shows zoomed in versions of the left column to better asses the node quality.

The node distributions produced by both algorithms do clearly conform to the specified nodal spacing function. Additionally, both distributions are locally regular as shown in the zoomed in plots. Further analysis of node quality can be done by comparing generated node distributions against the theoretical distribution, shown in Fig. 4.

Both algorithms produce a matching distribution of nodal spacing, with 9,277 and 8,693 nodes, respectively.

4.2 Execution time

Execution times of both algorithms are compared. The same density function as before (eqn (1)) was used. The procedure was run using $H = 10h$ with h varying in such a way that the number of generated nodes ranged approximately from 100 to 1,000,000. The domain of choice was an annulus, centred at $(0, 0)$ with inner radius 0.25 and outer radius 2. The proposed Poisson Disk Sampling based algorithm was applied directly, while the algorithm by Fornberg and Flyer was applied to the bounding box $[-2, 2] \times [-2, 2]$ and generated nodes were later filtered to keep only the nodes inside the domain and not too close to the domain boundary. The results are shown in Fig. 5.

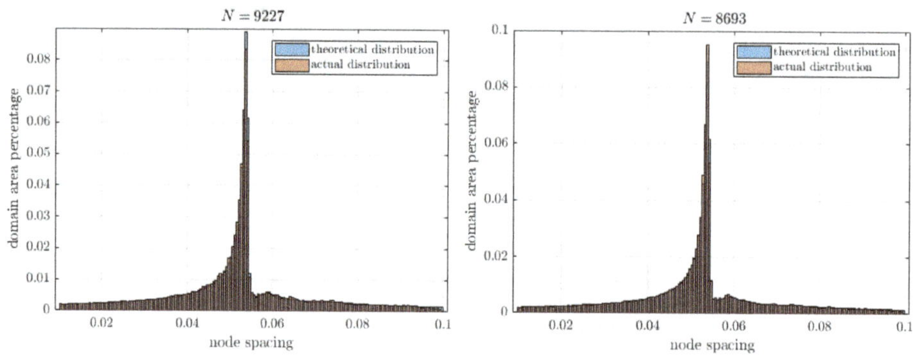

Figure 4: Spacing distribution produced by FF (left) and PDS algorithms (right).

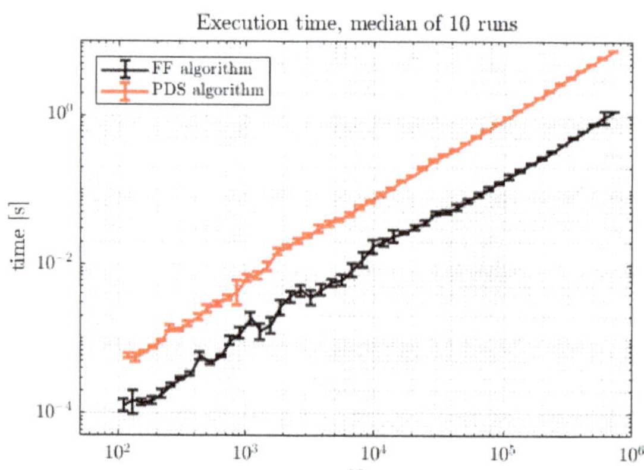

Figure 5: Comparison of execution time of both algorithms on an annulus domain. Shown execution time is the median of 10 runs, with error bars representing standard deviation.

An irregular domain was used on purpose to include a moderate overhead that the first algorithm has due to generating nodes within the whole bounding box. Nonetheless, it outperforms the Poisson Disk Sampling based algorithm for all tested N and exhibits very similar time complexity.

5 NUMERICAL EXAMPLES

The numerical examples in this work serve to demonstrate that nodes generated by the Poisson Disk Sampling based algorithm described in Section 3, can be used to solve PDEs using the RBF-FD method.

5.1 Electrostatics in 2D

Let us begin with a simple two-dimensional example originating from electrostatics. The electric field $\vec{E}$ satisfies $\nabla \cdot \vec{E} = \rho/\varepsilon$ and $\nabla \times \vec{E} = 0$. Introducing the electric potential ϕ, such that $\vec{E} = -\nabla\phi$, we get the Poisson's equation

$$\nabla^2 \phi = -\rho/\varepsilon. \tag{2}$$

In the absence of charge, this becomes the Laplace's equation.

The first case will computing the electric potential around 4 uniformly charged circles in a box. The Medusa framework [7] implementation of RBF-FD was used in the solution procedure, specifically; a neighbourhood of 9 support nodes and a basis consisting of 9 Gaussian functions with shape parameter $\sigma = 30$. The obtained solution is shown in Fig. 6.

The plotted values are values of potential ϕ obtained in computational nodes, which are distributed densely enough to create an illusion of a full plot.

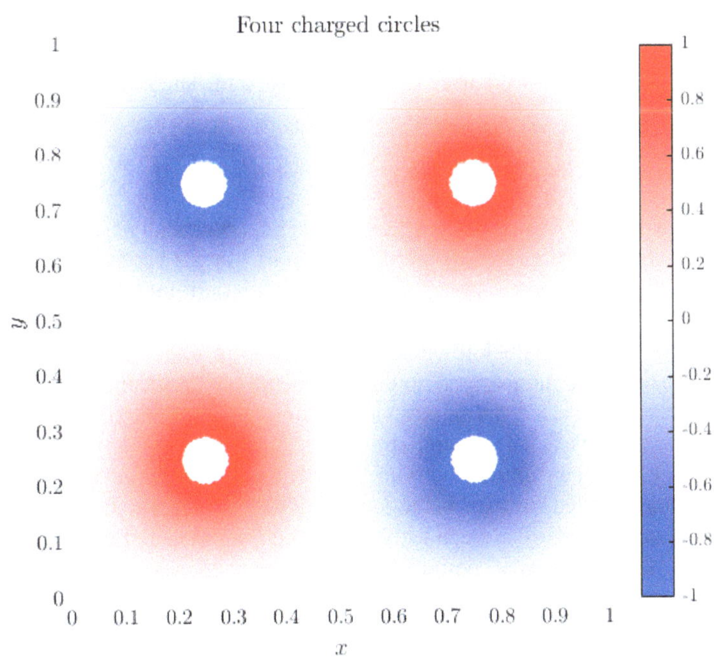

Figure 6: Electric potential around four uniformly charged circles in a box.

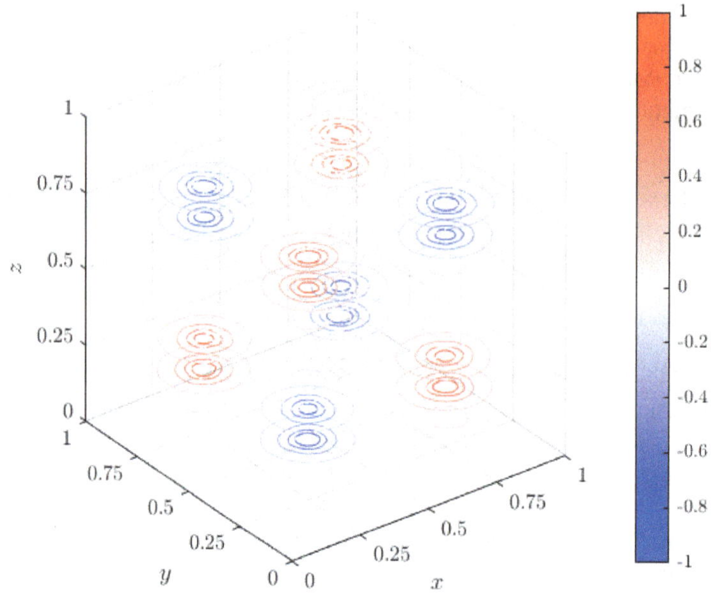

Figure 7: Electric potential around eight uniformly charged spheres in a box.

5.2 Electrostatics in 3D

The second case is a generalization of the first case to three dimensions. Eight uniformly charges spheres are put in a box, and the electric potential ϕ is computed using RBF-FD as before. A neighbourhood of 7 support nodes and a basis consisting of 7 Gaussian functions with shape parameter $\sigma = 30$ were used to solve the problem. The z cross-sections are shown in Fig. 7.

5.3 Adaptivity in linear elasticity

The final example deals with a linear elasticity problem, specifically fretting fatigue contact on the case described by Pereira et al. [6]. Two cylindrical pads act on a specimen of width W, length L and thickness t, creating a contact area, that is split into the stick and slip zones. Numerically, the problem is simplified to two dimensions and solved on the domain $\Omega = [-L/2, L/2] \times [-W/2, 0]$, with known top and right side tractions, symmetry boundary conditions on the bottom boundary and no displacement boundary conditions on the left. For further explanation and specific parameter values, the reader is referred to [9]. An important feature of this problem is that stick and slip contact zones cause high stress peaks on top boundary, which cannot be captured with uniform discretizations on current day computers. Refinement is thus needed and its behaviour in RBF-FD context has been investigated recently by Slak and Kosec [8]. Here, a fully automatic adaptive solution of the problem will be shown, with the purpose of demonstrating the capabilities of the node generation algorithm, as described in Section 3.

Fig. 8 shows top traction of the RBF-FD solution compared to a solution obtained using well-established FreeFem++ framework [10]. The solution is gradually improved over the course of 4 iterations, demonstrating that the node generator was generating distributions,

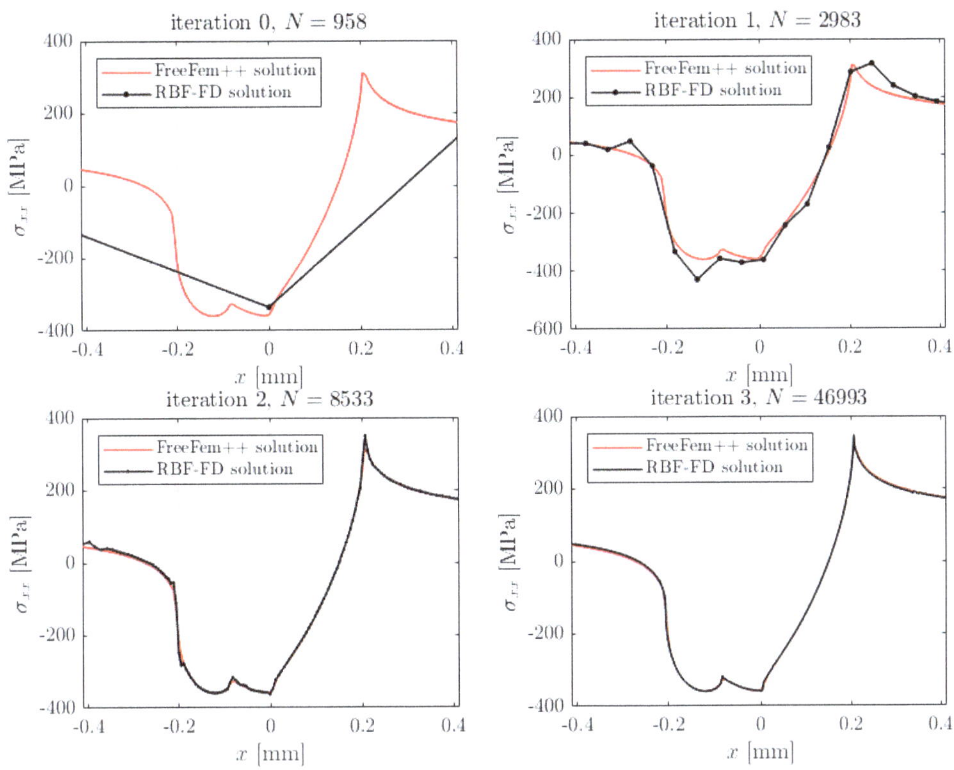

Figure 8: Adaptive solution of fretting fatigue contact problem. Top traction σ_{xx} is shown.

which were suitable for RBF-FD method. The generated distributions under the contact area arc shown in Fig. 9, coloured by von Mises stress. In the first iteration, only three nodes on the top of the domain are present. In the next iteration, the node distribution becomes denser, capturing the contact phenomenon well enough to separate high and low stress areas. Consequently, the third iteration produces a variable density distribution, and the final distribution is already well adapted to the solution behaviour.

6 CONCLUSIONS

A recently published algorithm for node positioning in 2D was improved, by lowering its theoretical time complexity and demonstrating the decrease in execution time in practice, generating 1 000 000 nodes per second. A new, slower, but dimension and domain shape independent algorithm with provable minimal spacing requirements and same time complexity is proposed. The algorithm is shown to generate node distributions conforming to an arbitrary nodal spacing function, and the produced nodal distributions can be used to solve problems with the RBF-FD method on irregular domains in two and three dimensions. Furthermore, it can also be used in the fully automatic adaptive setting.

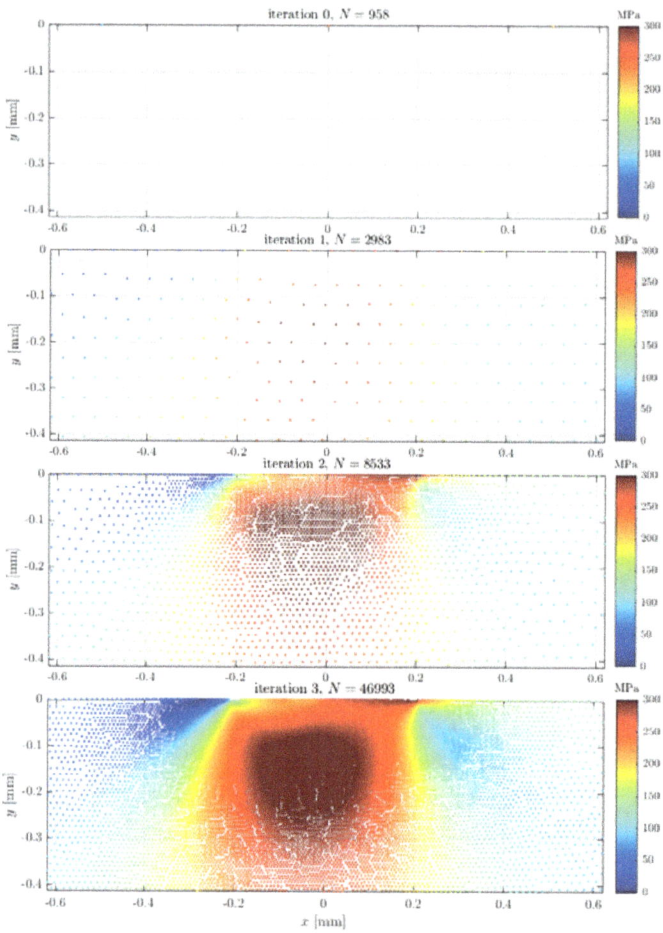

Figure 9: Nodes under contact in final iteration coloured by von Mises stress.

ACKNOWLEDGEMENTS

The authors would like to acknowledge the financial support of the Research Foundation Flanders (FWO), the Luxembourg National Research Fund (FNR) and Slovenian Research Agency (ARRS) in the framework of the FWO Lead Agency project: G018916N Multi-analysis of fretting fatigue using physical and virtual experiments, and the ARRS research core funding No. P2-0095.

REFERENCES

[1] Tolstykh, A.I. & Shirobokov, D.A., On using radial basis functions in a "finite difference mode" with applications to elasticity problems. *Computational Mechanics*, **33**(1), pp. 68–79, 2003. DOI: 10.1007/s00466-003-0501-9.

[2] Liu, G.R., *Mesh Free Methods: Moving Beyond the Finite Element Method*, CRC Press, 2002.

[3] Fornberg, B. & Flyer, N., Fast generation of 2-D node distributions for mesh-free PDE discretizations. *Computers and Mathematics with Applications*, **69**(7), pp. 531–544, 2015. DOI: 10.1016/j.camwa.2015.01.009.

[4] Shankar, V., Kirby, R.M. & Fogelson, A.L., Robust node generation for mesh-free discretizations on irregular domains and surfaces. *SIAM Journal on Scientific Computing*, **40**(4), pp. A2584–A2608, 2018. DOI: 10.1137/17m114090x.

[5] Mitchell, S.A., Rand, A., Ebeida, M.S. & Bajaj, C., Variable radii Poisson-disk sampling, extended version. *Proceedings of the 24th Canadian Conference on Computational Geometry*, **5**, 2012.

[6] Pereira, K. et al., On the convergence of stresses in fretting fatigue. *Materials*, **9**(8), p. 639, 2016.

[7] Medusa: coordinate free meshless method implementation. http://e6.ijs.si/medusa/. Accessed on: 28 Aug. 2018.

[8] Slak, J. & Kosec, G., Refined Meshless Local Strong Form solution of Cauchy–Navier equation on an irregular domain. *Engineering Analysis with Boundary Elements*, 2018. DOI: 10.1016/j.enganabound.2018.01.001.

[9] Trobec, R. & Kosec, G., Parallel scientific computing: theory, algorithms, and applications of mesh based and meshless methods. *Springer Briefs in Computer Science*, Springer, 2015.

[10] Hecht, F., New development in FreeFem++. *Journal of Numerical Mathematics*, **20**(3–4), pp. 251–266, 2012. DOI: 10.1515/jnum-2012-0013.

RBF-BASED LASER SPECKLE PATTERN DIGITAL IMAGE CORRELATION METHOD FOR SURFACE STRAIN MEASUREMENTS

EDUARDO DIVO[1], FAISAL MOSLEHY[2] & ALAIN KASSAB[2]
[1]Embry-Riddle Aeronautical University, Mechanical Engineering Department, USA
[2]University of Central Florida, Mechanical and Aerospace Engineering Department, USA

ABSTRACT

This paper introduces an innovative technique that integrates a genetic algorithm based digital image correlation with laser speckle photography (LSP) for the measurement of surface displacements in structures. The images (before and after deformation) are digitized using a digital camera, and the grayscale intensity matrices are read and processed by Matlab image processing toolbox. The two matrices of the images are then inputted into an iterative program based on the genetic algorithm that utilizes an advanced cross correlation technique to determine the surface displacements. Additionally, the strains are computed by radial basis function (RBF) differentiation. The computed displacements are compared with simulated results obtained by finite element analysis. Close agreement between the two results proved the validity of the developed non-contact technique for accurately measuring surface displacements. The experimentally measured displacements can be directly used in an inverse technique to detect and characterize subsurface cavities in structures.
Keywords: laser speckle pattern, RBF interpolation, genetic algorithm, surface strain.

1 INTRODUCTION

Optical methods such as holographic interferometry, shearography, laser speckle interferometry and Moiré were developed for measuring surface displacement and deformation. Some of these methods have been combined with the latest computer technology and imaging systems and developed into commercial scientific instruments. In practice, however, this equipment is expensive and requires a stable environment as well as laborious data reduction processes. References [1]–[4] are resourceful in literature and description of the laser speckle interferometry and photography techniques. Fraley et al. [5] proposed the use of the two-dimensional correlation method to measure the local displacement with laser speckle patterns. This approach has been applied in several special cases to determine the local displacement components due to uniform translation in pixel resolution.

A more straightforward technique using digital image correlation method was developed in the early 1980s for measuring surface displacements and deformation [6]. The method required a digital imaging system to optically record the images of the surface before and after deformation. The gray level functions of the images are then compared using advanced image correlation and processing algorithms in order to determine the displacement and deformation gradients. Pilch et al. [7] and Chu et al. [8] developed a new optimizing algorithm based on evolutionary techniques for digital image correlation. The optimizing process was tested for a rubber specimen under tensile load using speckle formed by spray painting the specimen surface.

In this paper, we developed an image domain decomposition approach to laser speckle photography (LSP) as applied to the vision-based method of deformation measurements. Here, a genetic algorithm is utilized to determine the motion of the center of pixel groupings by maximizing a simplified cross correlation expression. The whole specimen

WIT Transactions on Engineering Sciences, Vol 122, © 2019 WIT Press
www.witpress.com, ISSN 1743-3533 (on-line)
doi:10.2495/BE410161

deformation is subsequently reconstructed by means of radial basis function interpolation (RBF), and the strains are found by RBF differentiation.

2 THE DOMAIN DECOMPOSITION LASER SPECKLE TECHNIQUE

The application of the LSP technique involves the use of speckle pattern formed on the surface of a specimen impinged by a coherent source of light such as a laser beam. The procedure entails the application of a random speckle pattern to a surface of interest, then capturing a digital image of the surface, followed by capturing an image of the specimen after the surface has been displaced or deformed and then comparing the two images to compute the displacement for the surface. A schematic of the experimental setup is shown in Fig. 1.

Suppose a speckle pattern is created on a surface and an image is captured, as depicted in Fig. 2(a). A point of interest in this image undergoes motion in the second image (Fig. 2(b)) depending upon the loading conditions. A region (subset) around the point of interest may be selected. This subset may undergo a displacement and strain such that the new location and shape of the subset is depicted by the parallelogram in Fig. 2(b). The center of the parallelogram is the point of interest.

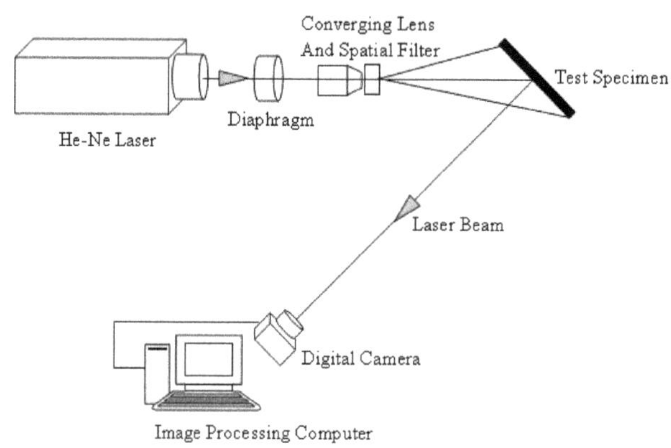

Figure 1: Schematic of the experimental setup.

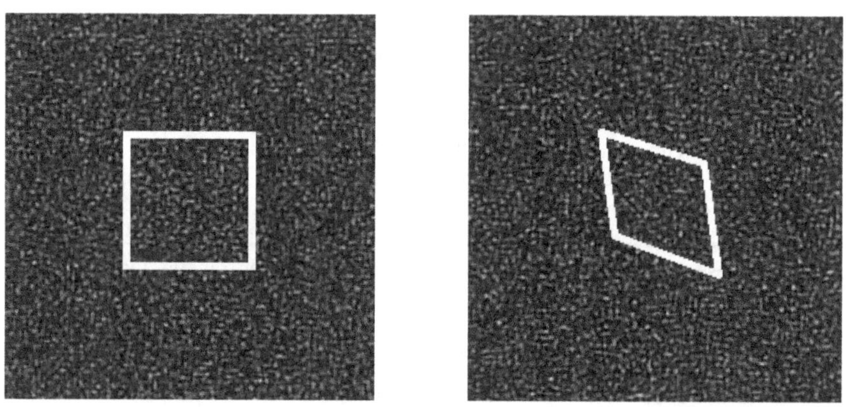

Figure 2: (a) Undeformed image; (b) Actual deformed image.

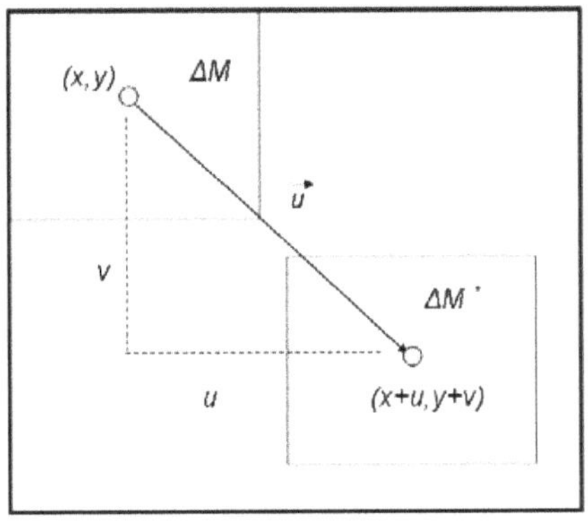

Figure 3: Model utilized in this study: motion of the center of a grouping of pixels.

The method for comparing the two subsets (to determine the displacement vector $\vec{u}$, see Fig. 3) is commonly given by the cross correlation coefficient, C:

$$C(u,v) = \frac{\int_{\Delta M^*} f(x,y) f^*(x+u, y+v) dA}{\sqrt{\left(\int_{\Delta M} f^2(x,y) dA \right) \left(\int_{\Delta M} f^{*2}(x+u, y+v) dA \right)}}, \tag{1}$$

where: ΔM = subset in the undeformed image;
ΔM^* = subset in the deformed image;
f(x,y) = grey level of speckle at a point (pixel).

The values of u and v which maximize C are the local deformation (displacement components) for the selected subset. The main objective of the image correlation process is to find these values for the subset under investigation, and then repeat it for all subsets in a given region to find the whole field deformation profile. In practice, the above is then implemented as:

$$C\,(u,v) = \frac{\left(\sum_{\Delta M} (grey\ levels) \right) \left(\sum_{\Delta M^*} (grey\ levels) \right)}{\left(\sum_{\Delta M} (grey\ levels)^2 \right) \left(\sum_{\Delta M^*} (grey\ levels)^2 \right)}. \tag{2}$$

Generally, the above correlation should be at least augmented by first derivatives to account for actual deformations not just translation. However, we since adopt a domain decomposition approach to the problem, the deforming regions that are followed are significantly reduced in size with each subdomain viewed with the full resolution of the camera as illustrated in Fig. 4. For each grouping of pixels, the values of u and v that maximize the correlation in eqn (1) are determined by a genetic algorithm (GA). The image of the whole specimen is then reconstructed via Radial Basis function (RBF) interpolation of the motion of each pixel grouping.

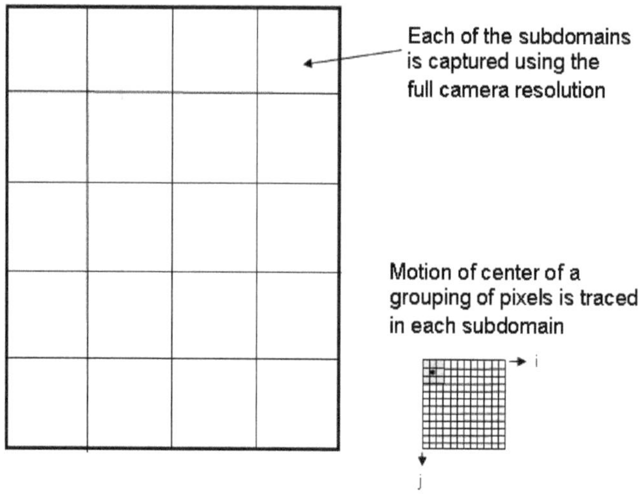

Figure 4: Domain decomposition of the image.

3 THE GENETIC ALGORITHM

The genetic algorithm (GA) optimization process begins by setting up a random set of possible solutions, called the population, with a fixed initial size or number of individuals. Each individual is defined by optimization variables and is represented as a bit string or a chromosome. An objective function, in our case $Z(u,v)=1/C(u,v)$, is evaluated for every individual in the current population defining their fitness or their probability of survival. At each of the iterations of the GA, the processes of selection, cross-over, and mutation operators are used to update the population of designs. A selection operator is first applied to the population in order to determine and select the individuals that are going to pass information in a mating process with the rest of the individuals in the population. This process allows the genetic information contained in the best individuals to be combined to form offspring. Additionally, a mutation operator randomly affects the information obtained by the mating of individuals. This is a crucial step for continuous improvement.

In nature, the properties of an organism are described by a string of genes in the chromosomes. Therefore, if one is trying to simulate nature using computers one must encode the design variable in a convenient way. We adopt a haploid model using a binary vector to model a single chromosome. The length of the vector is dictated by the number of design variables and the required precision of each design variable. Each design variable has to be bounded with a minimum and a maximum value and in the process the precision of the variable is determined. The number of divisions used in the discretization has to be integer power of two. This procedure allows an easy mapping from real numbers to binary strings and vice versa. The haploid GAs place all design variables into one binary string. In turn, each individual is equipped with a given set of design variables to which corresponds a value of the objective function. This value is the measure of "fitness" of the individual design. In GAs, poorly fit designs are not discarded; rather they are kept, as in nature, to provide genetic diversity in the evolution of the population. This genetic diversity is required to provide forward movement of the population during the mating, cross-over, and mutation processes which characterize the GA (see Goldberg [9]).

After initializing the population with random values the fitness of each individual is calculated by computing the value of the objective function. Then, the probability of being a selected individual for mating is calculated as the ratio between the value of the fitness function of each individual and the sum of all fitness function values. This is given by:

$$P_{selescted} = \frac{Fitness(v_i)}{\sum_{i=1}^{population-size} Fitness(v_i)}. \tag{3}$$

where v_i is *i*th member of the population, and *Fitness*(v_i) is the measure of the fitness of that member under its currently evolved parameter set configuration. A weighted roulette wheel is generated, where each member of the current population is assigned a portion of the wheel in proportion to its probability of selection. The wheel is spun as many times as there are individuals in the population to select which members mate. Obviously, some individuals could be selected more than once, where the best chromosomes get more copies, the average stay even, and the worst die off. Once selection has been applied, crossover and mutation occur to the resulting offspring. These operations further expand genetic diversity in the population. All other probabilities referred to in the description of the GA adopted in this paper are computed in an analogous fashion as the selection probability.

Following selection, crossover and mutation the new population is ready for its next evolution until the convergence criteria "fitness" is reached. It is the very nature of the binary representation of the design variables of the objective function and the random search process which provide yet another but implicit degree of regularization in this optimization process. The sensitivity of the objective function can be tuned depending on the size of each element of the chromosome. Thus, low bit representation is insensitive to large variations in input (regularized but may lead to poor solution due to low resolution), while high bit representation is sensitive to large variations in input (not regularized and therefore may lead to poor solution as well). There is a range of bit size which produces a regularized and sensitive response leading to stable solutions.

In the GA we used to generate the results presented in this paper, the following parameters are chosen: population size of individuals/generation, with a string of ten bits to define each parameter within each individual, two offspring per mating, a 1% probability of mutation, and a 70% probability of crossover. The population is not allowed to grow. This combination of parameters has been proven to yield efficient and accurate optimization results for different studies carried out by the authors [10].

4 RBF INTERPOLATION AND STRAIN COMPUTATION

The radial basis function (RBF) interpolation begins by defining a set of data centers, x_c, comprised of points distributed throughout the domain and the boundary which need not be uniformly distributed. These data centers serve as collocation points for the localized expansion of the variables of interest in the domain and on the boundary.

A localized expansion over a group or topology of influence points, *NF*, around each data center is sought such that:

$$\phi(x) = \sum_{j=1}^{NF} \alpha_j \chi_j(x) + \sum_{j=1}^{NP} \alpha_{j+NF} P_j(x). \tag{4}$$

The variable φ will take on the values of u or v depending on the application, x denotes the x–y pair of coordinates of a general point, α_j denote the unknown expansion coefficients, χ_j are the RBF expansion functions, NP is the number of additional polynomial functions, $P_j(x)$, added to the expansion to guarantee that constant and linear fields can be retrieved exactly. The RBF we utilize belong to the family of Hardy Multiquadrics [11]:

$$\chi_j(x) = \left[r_j^2(x) + c^2 \right]^{n-\frac{3}{2}}, \tag{5}$$

where, n is a positive exponent, c is a shape parameter, and $r_j(x)$ is the Euclidean distance from x to x_i. We utilize n = 1, a choice that yields the well-established inverse Multiquadrics, whose behavior has been extensively studied in the literature [12], [13]. The shape parameter, c, controls the behavior of the interpolation and the accuracy of its derivative. For a specific expansion over a set of data centers, the larger the shape parameter, the smoother is the computed derivative field. However, the magnitude of the shape parameter cannot be increased without bound as the expansion functions become flatter and, hence, the collocation coefficient matrix becomes ill-conditioned. For this reason, a simple optimization search is employed to determine every value for the shape parameter used in every expansion over the different local topologies that cover the entire field. An initial guess for c is based on the ratio of the average distance between data centers in a topology to the number of points in the topology.

The localized expansion approach [14], [15] reduces the computational burden global RBF interpolation methods by expanding the field variable locally around each data center to interpolation the function and to obtain its derivatives. This approach yields the generation of a different but small interpolation matrix for each data center rather than the large and fully-populated global interpolation matrix of the standard global interpolation methods.

The selection of an influence region or localized topology of expansion around each data center is easily accomplished by a circular (spherical in 3D) search around each data center. The search is automated to guarantee that a minimum number of points will be included and additional criteria, such as including all directions around internal data centers, are met. The collocation of the field variable at the points within the localized topology, leads to the following in matrix-vector form:

$$\{\phi\} = [C]\{\alpha\} \Rightarrow \{\alpha\} = [C]^{-1}\{\phi\}, \tag{6}$$

where the matrix [C] and the vector { φ } are composed as:

$$[C] = \begin{bmatrix} \chi_1(x_1) & \dots & \chi_{NF}(x_1) & P_1(x_1) & \dots & P_{NP}(x_1) \\ \cdot & \cdot & \cdot & \cdot & \cdot & \cdot \\ \chi_1(x_{NF}) & \dots & \chi_{NF}(x_{NF}) & P_1(x_{NF}) & \dots & P_{NP}(x_{NF}) \\ P_1(x_1) & \dots & P_1(x_{NF}) & 0 & \dots & 0 \\ \cdot & \cdot & \cdot & \cdot & \cdot & \cdot \\ P_{NP}(x_1) & \dots & P_{NP}(x_{NF}) & 0 & \dots & 0 \end{bmatrix} \text{ and } \{\varphi\} = \begin{Bmatrix} \varphi(x_1) \\ \cdot \\ \varphi(x_{NF}) \\ 0 \\ \cdot \\ 0 \end{Bmatrix}, \tag{7}$$

and, evaluating eqn (4) for field variable at the data center x_c leads to the matrix form:

$$\phi(x_c) = \{\chi(x_c)\}^T [C]^{-1}\{\phi\} = \{\psi(x_c)\}^T \{\phi\}. \tag{8}$$

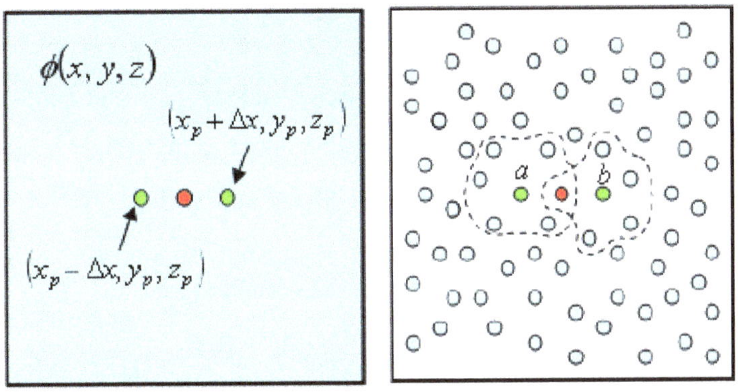

Figure 5: Differentiation stencil and local topology for RBF-FD.

To compute variable derivatives at the data center there are two possibilities: (1) RBF native differentiation (RBF-ND) we apply any linear differential operator, for example, $L_x = \partial / \partial x$, over the localized expansion equation, or (2) RBF-enhanced finite difference (RBF-FD) where we utilize the RBF to interpolate the field variable at locations on a finite difference stencil (virtual points) centered at the data center, and subsequently apply a standard finite difference on that stencil. In the first case of RBF-ND, the approach leads to:

$$L\phi(x_c) = \sum_{j=1}^{NF} \alpha_j L\chi_j(x_c) + \sum_{j=1}^{NP} \alpha_{j+NF} LP_j(x_c), \tag{9}$$

where x_c is the data center of the topology. Thus, in matrix-vector form:

$$L\phi_c = \{L_c\}^T \{\alpha\} \Rightarrow L\phi_c = \{L_c\}^T [C]^{-1} \{\phi\} \Rightarrow L\phi_c = \{L\}^T \{\phi\}, \tag{10}$$

where the vector $\{L_c\}$ is composed by:

$$\{L_c\} = \begin{pmatrix} L\chi_1(x_c) \\ \vdots \\ L\chi_{NF}(x_c) \\ LP_1(x_c) \\ \vdots \\ LP_{NP}(x_c) \end{pmatrix}. \tag{11}$$

In the second case of RBF-FD, illustrated for the second order central difference evaluation of the first derivative at the data center x_p, the RBF is utilized to generate the values for the field variable at the virtual finite difference stencil locations *a* and *b*. Local topologies are constructed for each virtual data center and these do not include the virtual points themselves, see Fig. 5. Eqn (6) is applied to the set of points surrounding the virtual points *a* and *b*, leading to:

$$\phi(x_a) = \{\psi(x_a)\}^T \{\phi\}_a, \tag{12}$$

$$\phi(x_b) = \{\psi(x_b)\}^T \{\phi\}_b , \tag{13}$$

where $\{\psi(x_a)\}$ is the interpolating weight vector and $\{\phi\}_a$ is the nodal value vector associated with virtual point *a*, and $\{\psi(x_b)\}$ is the interpolating weight vector and $\{\phi\}_b$ is the nodal value vector associated with virtual point *b*.

Therefore, substituting these virtual point expressions into a second order accurate central difference of the first x-derivative at the data center x_c leads to:

$$\frac{\partial \phi}{\partial x} = \left\{ \frac{1}{2\Delta x}\{\psi(x_a)\}^T \ \frac{-1}{2\Delta x}\{\psi(x_b)\}^T \right\}^T \left\{ \begin{matrix} \{\phi_a\} \\ \{\phi_b\} \end{matrix} \right\} = \{L_{c,x}\}^T \{\phi\} . \tag{14}$$

This is readily extended for any other derivative of interest. Therefore, using either method, RBF-ND or RBF-FD, the evaluation of the field variable derivatives at every one of the data centers x_c is provided by a simple inner product of two small vectors: $\{L_c\}$ which can be pre-built and stored and $\{\phi\}$ which is the field variable values at the surrounding RBF points within the topology of the data center x_c. In this paper, we compute all derivatives utilizing the RBF-FD. With the derivatives of the displacements now readily available, the strains can be computed as:

$$\varepsilon_x = \frac{\partial u}{\partial x},\ \varepsilon_y = \frac{\partial v}{\partial y},\ \gamma_{xy} = \frac{\partial u}{\partial y} + \frac{\partial v}{\partial x} . \tag{15}$$

5 EXPERIMENTAL PROCEDURE AND DATA COLLECTION

A Plexiglas test specimen of length = 250 mm, width = 168 mm, and thickness = 4.8 mm, with two circular holes was fabricated and all the optical instruments were clamped firmly onto an optical table (see Fig. 6). The translation stage holding the spatial filter was adjusted till a clean divergent laser beam is obtained to illuminate the test specimen. The test specimen was then clamped onto the loading frame and a tensile load of 500 lb was applied. An image of the specimen (covered by laser speckles) in the deformed state was captured by a digital camera. Subsequently the specimen was unloaded to 100 lb and another digital image was captured.

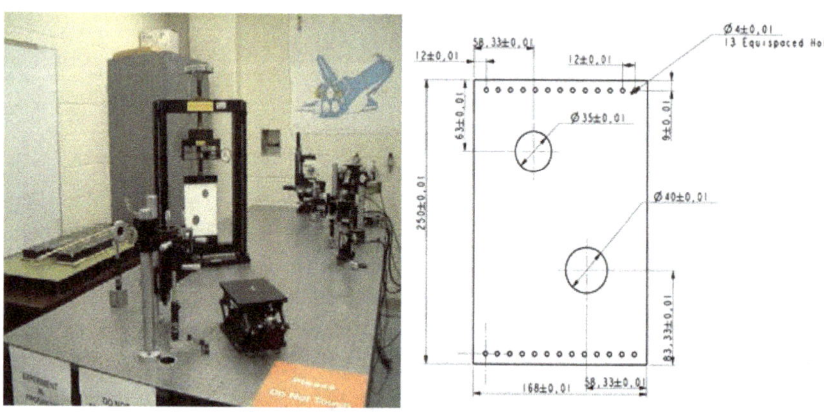

Figure 6: Photograph of the experimental setup and details of the test specimen.

The digital camera at highest resolution and full zoom can give resolution of 2832 x 2128 pixels. Once the digital images are cropped to size, the resolution for the images was calculated to be 0.125 mm/pixel. Fig. 7 depicts the gray level distribution. Preliminary finite element analysis indicated that the maximum deformation would be in the order of 0.3 mm, or in terms of pixels the maximum expected deformation would be less than 3 pixels. Preliminary runs of the image correlation using the genetic algorithm proved that such low-resolution images gave poor output for surface displacements.

To overcome the low-resolution problem it was then decided to subdivide the entire test specimen into blocks and capture the image of each block separately at highest resolution and full zoom. As shown in Fig. 8(a) and (b) the specimen surface was divided into an array of 20 blocks using a soft-tipped pen and each block was identified by its own numbering scheme.

The camera was mounted on a translation stage firmly clamped onto the isolation mount and placed 250 mm away from the test specimen. The camera was set at the highest resolution and zoomed to focus on the first block of the specimen. The specimen was

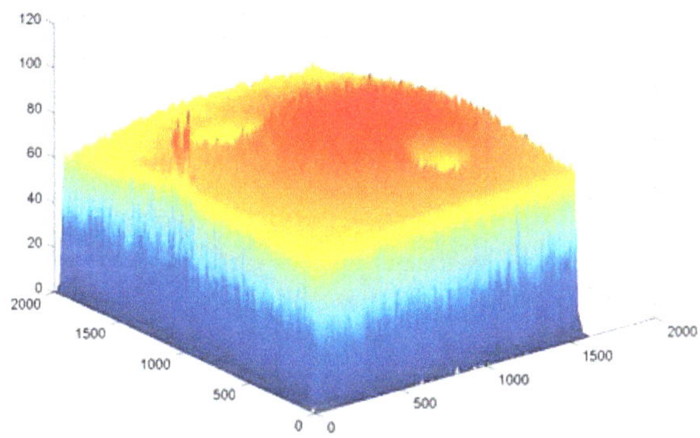

Figure 7: Grey level distribution of specimen covered by laser speckles.

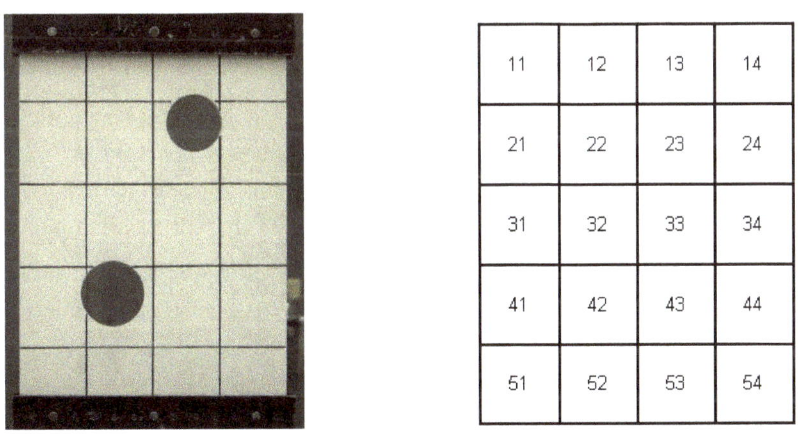

Figure 8: (a) Subdivided specimen; and (b) Numbering scheme.

loaded to 500 lb; an image captured and subsequently the specimen was unloaded to 100 lb and another image captured. The resolution obtained by this technique was found to be 0.027 mm/pixel. This procedure was repeated for the twenty blocks of the specimen. The image intensity matrix for each set was correlated using the genetic algorithm-based correlation technique. The input parameters of the algorithm were continuously tuned or calibrated to give a deformation gradient similar to the gradient obtained from preliminary finite element analysis (mesh shown in Fig. 9).

6 RESULTS AND DISCUSSIONS

The results from each block were then assembled and the final deformation gradient was obtained and shown in Fig. 10.

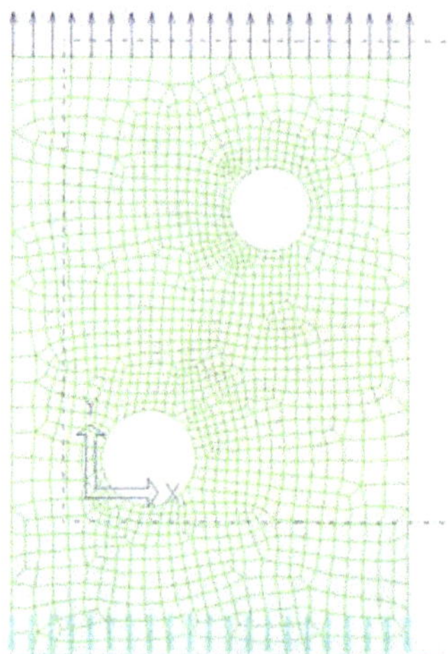

Figure 9: Finite element model of the test specimen.

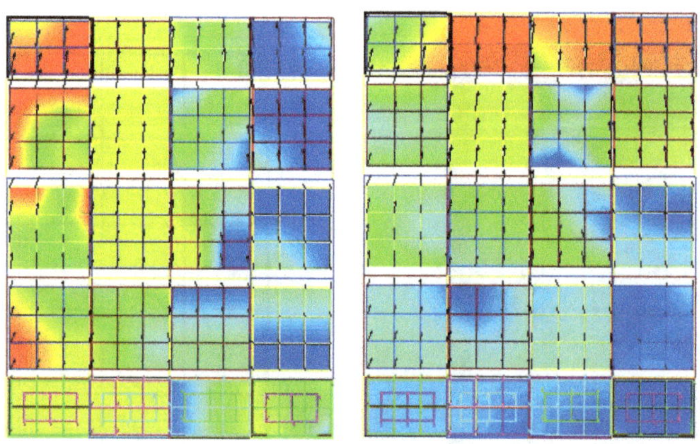

Figure 10: Displacement distribution of displacement components u and v from the genetic algorithm.

Referring to Fig. 10, it is evident that the information at the interface between the different blocks is yet unknown and the elastostatic compatibility equations are not satisfied. This is achieved by computationally integrating the various blocks together by averaging the intensity of the nodes at the edges between neighboring blocks along the median coordinates between the edges, using the RBF interpolation technique. This resulted in a displacement distribution of the overall specimen as a single entity and not a compilation of several entities. Figs 11 and 12 exhibit the deformation distribution plots magnified by a factor of 10 as compared to the solution obtained from preliminary finite element analysis.

The strain components $\varepsilon_x, \varepsilon_y$ and γ_{xy} were computed by the RBF-FD differentiation scheme and their distribution is shown in Fig. 13.

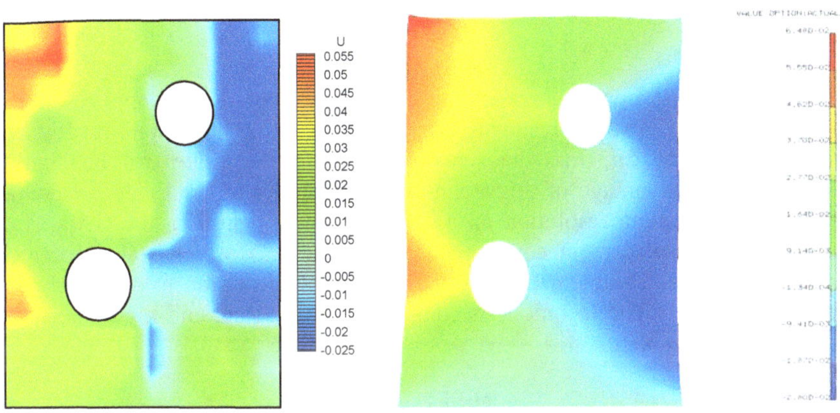

Figure 11: Displacement distribution of *u* at 10X Magnification for the integrated genetic algorithm data as compared to solution obtained from FEA.

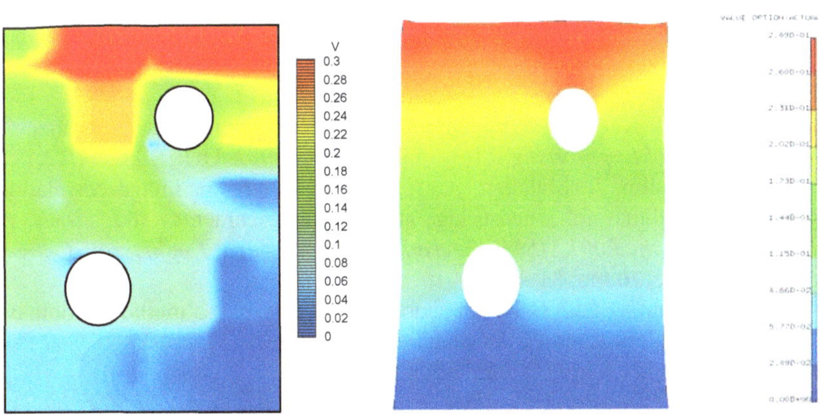

Figure 12: Displacement distribution of *v* at 10X Magnification for the integrated genetic algorithm data as compared to solution obtained from FEA.

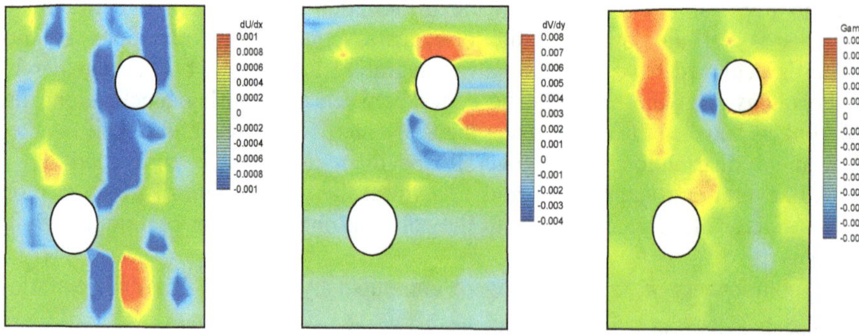

Figure 13: Strains components ε_x, ε_y and γ_{xy} distribution obtained by RBF interpolation.

7 CONCLUSIONS

The GA-based digital image correlation technique is successfully applied to measure surface displacements in structures using laser speckle photography. With readily available high-resolution digital cameras, the sensitivity of this technique can be significantly increased. A major advantage of this technique is its ability to subdivide the domain and hence use the full camera resolution on a small region. In addition, the GA was used to compute only the displacement components (two variables), which makes the technique more efficient. This is complemented by the RBF implementation to accurately compute the strain components. The technique is completely non-contact and highly accurate even for very small displacements.

REFERENCES

[1] Erf, R.K., *Speckle Metrology*, Academic Press: New York, 1978.

[2] Vest, C.M., *Holographic Interferometry*, John Wiley: New York, 1979.

[3] Kobayashi, A.S., *Handbook on Experimental Mechanics*, VCH Publishers: New York, 1993.

[4] Dainty, J.C., *Laser Speckle and Related Phenomena*, Springer-Verlag: New York, 1975.

[5] Fraley, J.E., Hamed, M.A., Peters, W.H. & Ranson, W.F., Experimental boundary integral equation application in speckle interferometry. *Proceedings 1981, SESA Spring Conference*, pp. 68–71, 1981.

[6] Peters, W.H. & Ranson, W.F., Digital imaging techniques in experimental stress analysis. *Optical Engineering*, **21**(3), pp. 427–431, 1982. DOI: 10.1117/12.7972925.

[7] Pilch, A., Maudlin, J., Mahajan, A. & Chu, T., Intelligent image correlation using genetic algorithms for measuring surface displacements and strain profiles. *Proceedings of the 2001, ASME International Mechanical Engineering Congress and Exposition, DSC* **70**, pp. 81–88, 2001.

[8] Chu, T., Mahajan, & Liu, C.T., An economical vision-based method to obtain whole-field deformation profiles. *Experimental Techniques*, **26**(6), pp. 25–28. DOI: 10.1111/j.1747-1567.2002.tb00087.x.

[9] Goldberg, D.E., *Genetic Algorithms in Search, Optimization and Machine Learning*, Addison-Wesley: Reading, MA, 1989.

[10] Divo, E.A., Kassab, A.J. & Rodríguez, F., An efficient singular superposition technique for cavity detection and shape optimization. *Numerical Heat Transfer, Part B*, **46**(1), pp. 1–30, 2004.

[11] Hardy, R.L., Multiquadric equations of topography and other irregular surfaces. *Journal of Geophysical Research*, **176**, pp. 1905–1915.

[12] Buhmann, M.D., *Radial Basis Functions: Theory and Implementation*, Cambridge University Press: Cambridge, 2003.

[13] Kansa, E.J. & Hon, Y.C., Circumventing the Ill-conditioning problem with multiquadric radial basis functions: Applications to elliptic partial differential equations. *Computers and Mathematics with Applications*, **39**(7–8), pp. 123–137, 2000. DOI: 10.1016/s0898-1221(00)00071-7.

[14] Pepper, D., Kassab, A. & Divo, E., *An Introduction to Finite Element, Boundary Element, and Meshless Methods: with Applications to Heat Transfer and Fluid Flow*, ASME Press, 2014.

[15] Sarler, B., Tran-Cong, T. & Chen, C.S., Meshfree direct and indirect local radial basis function collocation formulations for transport phenomena. *Boundary Elements XVII*, eds A. Kassab, C.A. Brebbia & E. Divo, WIT Press: Southampton and Boston, pp. 417–428, 2005.

IMPLICIT MODELLING OF GEOLOGICAL STRUCTURES: A CARTESIAN GRID METHOD HANDLING DISCONTINUITIES WITH GHOST POINTS

JULIEN RENAUDEAU[1,2], MODESTE IRAKARAMA[2], GAUTIER LAURENT[2],
FRANTZ MAERTEN[1] & GUILLAUM CAUMON[2]
[1]Schlumberger Petroleum Services, Montpellier, France
[2]Université de Lorraine, CNRS, GeoRessources, Nancy, France

ABSTRACT

In geology, implicit structural modelling constructs the geometry of geological structures (e.g. layers) by interpolating between sparse field data. A model is represented by a volumetric scalar field which is discontinuous on structural discontinuities such as faults or stratigraphic unconformities. The management of such discontinuities may involve boolean operations on several scalar fields or the creation of conformal meshes. Instead, we propose a ghost cell technique for the cartesian grid together with a set of relations between the ghost points, the regular nodes, and the discontinuities. Consequently, poor quality meshes are avoided and only the resolution of the grid has an impact on the solution. The modelling problem is posed as a least square's minimization of a bending energy penalization on data mismatch functions and approximated by finite difference. As all relationships in the grid are implicitly defined, except close to the discontinuities, this algorithm is computationally efficient. We provide some benchmarks of the method on two-dimensional examples with folds, faults, and erosions.

Keywords: 2D structural modelling, mesh reduction methods, energy penalization, ghost points.

1 INTRODUCTION

Implicit structural modelling consists in interpreting available field data into a model representing subsurface geological structures [1]–[8]. The model is represented by a scalar function defined on the entire volume of interest; this scalar function is called the implicit function. Horizons (interfaces between stratigraphic layers) are given by iso-values of the function, and structural discontinuities are given by discontinuous jumps in the function. Input data are represented by different types of modelling objects, such as points, vectors, polylines and surfaces [9]. The output model should represent natural geological structures while honouring the data.

The discrete smooth interpolation (DSI) [1], [4], [10] is an algorithm that discretises the implicit function on a background volumetric mesh where the mesh nodes are duplicated on either side of the discontinuities. The construction of such a mesh can represent a challenge depending on the geometrical complexity of the region being studied [11], [12]. The implicit function is then obtained by a spatial regression of data points sorted by horizons. DSI poses this problem as a least square's minimization of both this data regression and a roughness factor, obtaining a unique solution which should be as smooth as possible. In practice, this roughness factor has been designed to minimize the difference between the gradient of the implicit function evaluated on adjacent elements [4], [8], which is dependent on the mesh.

A version of DSI based on the Cartesian grid is introduced in [14]. The roughness factor is designed to minimize second directional derivatives of the implicit function using a finite difference approximation, and the implicit function is interpolated on the grid cells using a piecewise bilinear interpolation. The discontinuities are handled by deleting the grid cells they affect, putting those cells out of bounds of the model, which implies that the grid resolution is fine enough to include stratigraphic data in-bounds. The global grid resolution

WIT Transactions on Engineering Sciences, Vol 122, © 2019 WIT Press
www.witpress.com, ISSN 1743-3533 (on-line)
doi:10.2495/BE410171

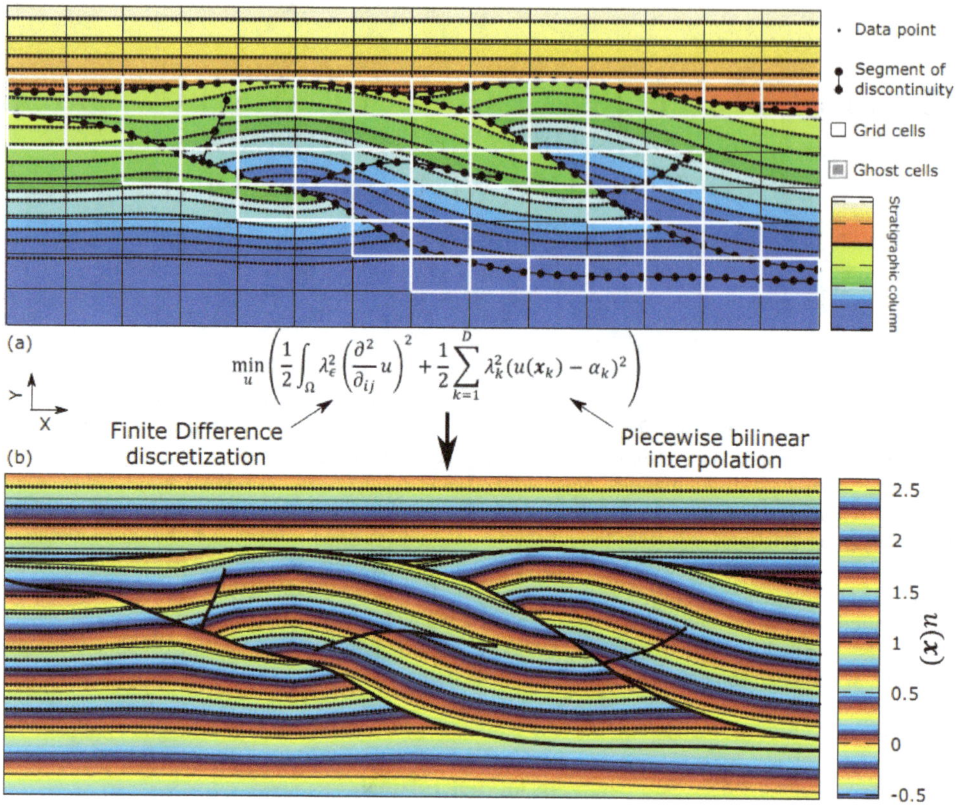

Figure 1: The proposed method applied on a synthetic cross section in two dimensions with folds, intersected faults and an erosion (figure modified from [13] with the present method). (a) The input data and the used interpolation grid; (b) The output structural model with extracted iso-values (fine lines).

is therefore determined based on the smallest distance of input data points to discontinuities, instead of being controlled by the complexity of the studied structures.

The roughness factor in DSI is defined as a discrete operator [10]. In practice, this factor may take different forms depending on the chosen domain discretisation scheme [4], [8], [14]. Instead, another approach is to find a continuous operator based on physical arguments, and then discretise it accordingly [13], [15]. An advantage of the latter is that it naturally provides appropriate weights to use during the least square's optimization stage, making the solution less sensitive to the resolution of the discretisation scheme.

In this paper, the proposed implicit method penalizes data constraints by the continuous bending energy, as in [13] and [15]. This problem is discretised on the Cartesian grid, using a piecewise bilinear evaluation on data terms and a finite difference approximation of energy terms, which gives a final linear system to solve close to [14], but with volumetric weights obtained from the discretisation. The discontinuities are handled with a ghost-cell technique, resulting in an implicit function defined everywhere in the domain of study. The proposed method has no requirement on the grid resolution other than to capture the studied geological structures. Fig. 1 summarizes the method on a synthetic cross section in two dimensions.

The first section of this article presents the proposed modelling method, and the second gives additional practical information on the way ghost nodes are handled and the related limits.

2 GRID BASED IMPLICIT STRUCTURAL MODELLING

The implicit function $u(\boldsymbol{x})$ is defined in the vector space V as

$$V = \left\{ u(\boldsymbol{x}) = \sum_{l=1}^{N} \Phi_l(\boldsymbol{x}) u_l = \boldsymbol{\Phi}(\boldsymbol{x})^T \cdot \boldsymbol{U} \;\middle|\; \boldsymbol{x}(x,y) \in \Omega \right\}, \tag{1}$$

where $\Omega \subset \mathbb{R}^2$ is the domain of study, $\boldsymbol{\Phi}^T = [\Phi_1, \dots, \Phi_N]$ is a set of linearly independent functions forming the basis of functions, and $\boldsymbol{U}^T = [u_1, \dots, u_N]$ is a set of constants.

Implicit structural modelling consists in finding the unknown coefficients u_l to construct the implicit function $u(\boldsymbol{x})$ (eqn (1)) that represents geological strata while honouring the input data. To achieve this, the proposed method performs a spatial regression of data points with a bending energy penalization (or thin plate energy) [16], [17] by minimizing

$$J(u) = \frac{1}{2} \int_{\Omega} \lambda_\epsilon^2 \left(\left(\frac{\partial^2}{\partial_{xx}} u \right)^2 + \left(\frac{\partial^2}{\partial_{yy}} u \right)^2 + 2 \left(\frac{\partial^2}{\partial_{xy}} u \right)^2 \right) + \frac{1}{2} \sum_{k=1}^{D} \lambda_k^2 (u(\boldsymbol{x}_k) - \alpha_k)^2, \tag{2}$$

with λ_ϵ the energy penalization weight, λ_k and α_k the weight and the expected value at a data point position $\boldsymbol{x}_k$, and D the number of data points.

2.1 Construction of the implicit function

2.1.1 Grid discretisation

The domain Ω is discretised on a regular grid of N nodes $\boldsymbol{x}_l$ (i.e., the grid corner points). Each coefficient u_l is associated to a node $\boldsymbol{x}_l$, and the implicit function $u(\boldsymbol{x})$ is uniquely defined by these coefficients u_l as the shape functions $\Phi_l(\boldsymbol{x})$ are linearly independent. Such a grid is illustrated in Fig. 1.

2.1.2 Bilinear quadrilateral interpolation functions

The implicit function coefficients u_l are interpolated within the grid elements with bilinear functions $\Phi_l(\boldsymbol{x})$. These are created with the isoparametric quadrilateral elements with four nodes (i.e., the corners only) [18]. For any position $\boldsymbol{x}$ in Ω, the containing cell c is found. The position $\boldsymbol{x}(x,y)$ and the four cell nodes $\boldsymbol{x}_l \in c$ are transformed in the cell space as the position $\boldsymbol{\xi}(\xi,\eta)$ and the four corners $\boldsymbol{\xi}_1^c(-1, -1)$, $\boldsymbol{\xi}_2^c(1, -1)$, $\boldsymbol{\xi}_3^c(1, 1)$ and $\boldsymbol{\xi}_4^c(-1, 1)$. The interpolation functions $N_1^c(\boldsymbol{\xi}), N_2^c(\boldsymbol{\xi}), N_3^c(\boldsymbol{\xi})$ and $N_4^c(\boldsymbol{\xi})$ in this space associated to the corresponding functions $\Phi_l(\boldsymbol{x})$ are defined as

$$\begin{aligned} N_1^c(\boldsymbol{\xi}) &= \frac{1}{4}(1-\xi)(1-\eta), & N_3^c(\boldsymbol{\xi}) &= \frac{1}{4}(1+\xi)(1+\eta), \\ N_2^c(\boldsymbol{\xi}) &= \frac{1}{4}(1+\xi)(1-\eta), & N_4^c(\boldsymbol{\xi}) &= \frac{1}{4}(1-\xi)(1+\eta). \end{aligned} \tag{3}$$

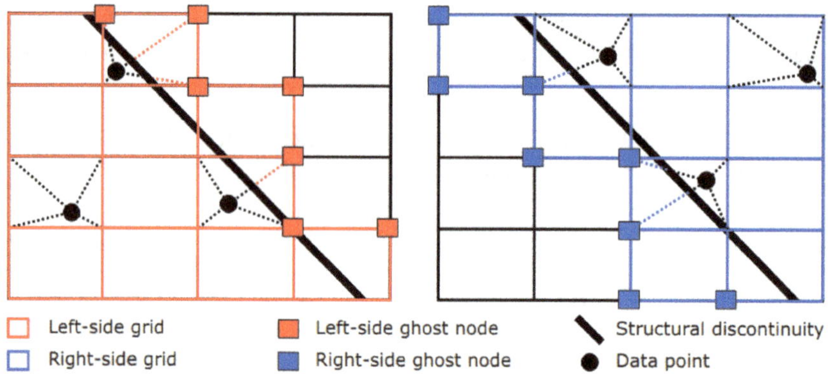

Figure 2: Illustration of the used ghost-cell technique by separating the evaluation on the left and on the right side of a discontinuity. Dotted segments represent the segment-discontinuity intersection tests for the use of ghost nodes.

2.1.3 Handling the structural discontinuities

Geological faults and unconformities are represented with a discontinuous jump in the implicit function $u(\boldsymbol{x})$. As a sum of continuous interpolation functions (eqn (1)), introducing a discontinuity in $u(\boldsymbol{x})$ calls for modifications either on the domain discretisation scheme, or on the interpolation functions. The presented method uses a ghost-cell technique [19], adding degrees of freedom on the grid cells crossed by discontinuities as illustrated in Figs 1 and 2. Points are added on either side of the discontinuities on top of already existing grid corners; in the following, the N_i grid corners $\boldsymbol{x}_{l_i}$ are called interpolation nodes and the N_g added points $\boldsymbol{x}_{l_g}$ are called ghost nodes.

The implicit function is augmented to

$$u(\boldsymbol{x}) = \sum_{l_i=1}^{N_i} \Phi_{l_i}(\boldsymbol{x}) u_{l_i} + \sum_{l_g=N_i+1}^{N_i+N_g} \Phi_{l_g}(\boldsymbol{x}) u_{l_g} = \sum_{l=1}^{N} \Phi_l(\boldsymbol{x}) u_l, \tag{4}$$

with N_i and N_g the number of interpolation and ghost nodes respectively. When evaluating $u(\boldsymbol{x})$, if the position $\boldsymbol{x}$ belongs to a cell crossed by discontinuities, the cell corners used for interpolation (Section 2.1.2) are found following Fig. 2: if the segment between $\boldsymbol{x}$ and a corner (dotted line) intersects the discontinuity, then the node used for eqn (3) is the corresponding ghost node $\boldsymbol{x}_{l_g}$, otherwise it is the interpolation node $\boldsymbol{x}_{l_i}$. With this technique, the implicit function is continuously evaluated on either side of the discontinuity, regardless its geometry.

2.2 Structural modelling problem

The modelling problem (eqn (2)) is solved by least squares optimization. Equations related to the energy term are written on all the nodes, while equations related to the data term are written on the data points.

2.2.1 Data points equations

Geological horizons are given by iso-values of the implicit function $u(\boldsymbol{x})$. An input data point is thus bound to its horizon h by specifying its expected iso-value α_h. A data equation of this type is written as

$$\lambda_k(u(\boldsymbol{x}_k) - \alpha_h) = 0, \quad \forall \boldsymbol{x}_k \in h. \tag{5}$$

There is no *a priori* knowledge on the expected iso-values α_h. The method approximates theses values during a pre-processing stage on input data which fixes the top and bottom horizons to a chosen value and then spatially interpolates the other horizons values linearly in between [20]. The data weights λ_k are generally taken constant between all data but can vary depending on the reliability of the data (e.g., borehole data are generally more reliable than seismic data).

2.2.2 Energy penalization

The bending energy term is regularly discretised on N_i cells Ω_{l_i} with the same volume v and centred on each interpolation node $\boldsymbol{x}_{l_i}$. This discretisation thus represents the dual-grid of the interpolation grid. On each node, energy equations are written as

$$\begin{aligned} \lambda_\epsilon\sqrt{v}(\frac{\partial^2}{\partial xx}u(\boldsymbol{x})) = 0, \\ \lambda_\epsilon\sqrt{v}(\frac{\partial^2}{\partial yy}u(\boldsymbol{x})) = 0, \\ \lambda_\epsilon\sqrt{2v}(\frac{\partial^2}{\partial xy}u(\boldsymbol{x})) = 0. \end{aligned} \tag{6}$$

We chose to extend this discretisation to the ghost nodes: energy equations are also written on ghost nodes with the same volume of integration v.

The interpolation functions $\Phi_l(\boldsymbol{x})$ chosen in eqn (3) are only C^0 at the nodes $\boldsymbol{x}_l$. The second derivatives of eqn (6) are therefore approximated by finite difference as

$$\frac{\partial^2}{\partial xx}u(\boldsymbol{x}_l) \approx \frac{u(x_l + \Delta_x, y_l) - 2\,u(x_l, y_l) + u(x_l - \Delta_x, y_l)}{\Delta_x^2}, \tag{7}$$

$$\frac{\partial^2}{\partial yy}u(\boldsymbol{x}_l) \approx \frac{u(x_l, y_l + \Delta_y) - 2\,u(x_l, y_l) + u(x_l, y_l - \Delta_y)}{\Delta_y^2}, \tag{8}$$

$$\frac{\partial^2}{\partial xy}u(\boldsymbol{x}_l) \approx \frac{u(x_l + \Delta_x, y_l + \Delta_y) - u(x_l + \Delta_x, y_l - \Delta_y) - u(x_l - \Delta_x, y_l + \Delta_y) + u(x_l - \Delta_x, y_l - \Delta_y)}{4\Delta_x\Delta_y}, \tag{9}$$

where Δ_x and Δ_y are the grid spacing in the x and y axes respectively. As the interpolation functions $\Phi_l(\boldsymbol{x})$ have the Kronecker delta property (i.e., $\Phi_l(x_k) = \delta_{kl}, \forall(k,l) \in N; u(\boldsymbol{x}_l) = u_l, \forall l \in N$), eqns (7), (8), and (9) reduce to a linear combination of the coefficients u_l. We define the combination coefficients $\delta_m^{xx}(\boldsymbol{x}_l)$, $\delta_m^{yy}(\boldsymbol{x}_l)$ and $\delta_m^{xy}(\boldsymbol{x}_l)$ associated to the nodes $\boldsymbol{x}_m (m \in N)$ as illustrated on Fig. 3(a) for the three equations respectively.

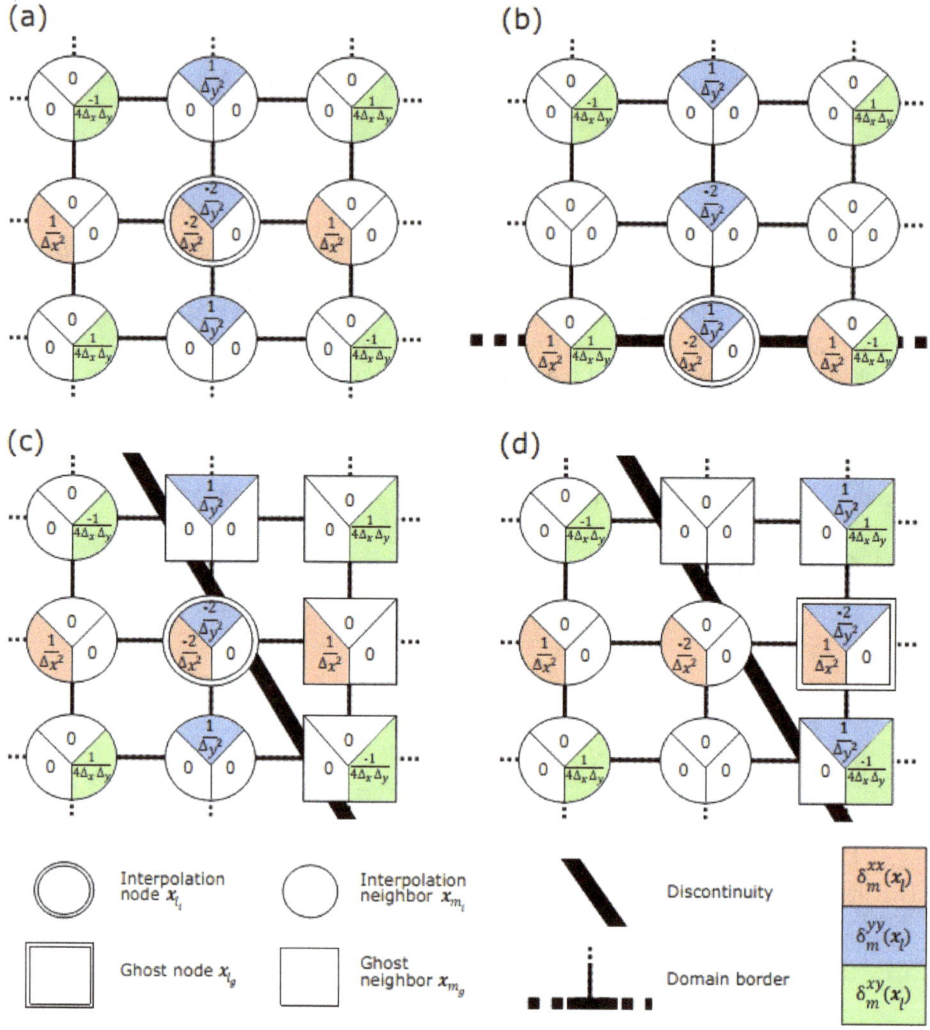

Figure 3: Finite difference schemes for second derivatives with the related coefficients on different types of nodes. (a) An interpolation node; (b) An interpolation node on a border; (c) An interpolation node close to a discontinuity; and (d) A ghost node.

2.2.3 System to solve

Combining eqn (5), written on each data point $\boldsymbol{x}_k$, and eqns (7), (8), and (9), written on each node $\boldsymbol{x}_l$, we define the least squares system to solve, of size (3N + D, N), as

$$\begin{bmatrix} \lambda_\epsilon\sqrt{v}\,\delta_1^{xx}(\boldsymbol{x}_1) & \cdots & \lambda_\epsilon\sqrt{v}\,\delta_N^{xx}(\boldsymbol{x}_1) \\ \lambda_\epsilon\sqrt{v}\,\delta_1^{yy}(\boldsymbol{x}_1) & \cdots & \lambda_\epsilon\sqrt{v}\,\delta_N^{yy}(\boldsymbol{x}_1) \\ \lambda_\epsilon\sqrt{2v}\,\delta_1^{xy}(\boldsymbol{x}_1) & \cdots & \lambda_\epsilon\sqrt{2v}\,\delta_N^{xy}(\boldsymbol{x}_1) \\ \vdots & & \vdots \\ \lambda_\epsilon\sqrt{v}\,\delta_1^{xx}(\boldsymbol{x}_N) & \cdots & \lambda_\epsilon\sqrt{v}\,\delta_N^{xx}(\boldsymbol{x}_N) \\ \lambda_\epsilon\sqrt{v}\,\delta_1^{yy}(\boldsymbol{x}_N) & \cdots & \lambda_\epsilon\sqrt{v}\,\delta_N^{yy}(\boldsymbol{x}_N) \\ \lambda_\epsilon\sqrt{2v}\delta_1^{xy}(\boldsymbol{x}_N) & \cdots & \lambda_\epsilon\sqrt{2v}\delta_N^{xy}(\boldsymbol{x}_N) \\ \lambda_1\phi_1(\boldsymbol{x}_1) & \cdots & \lambda_1\phi_N(\boldsymbol{x}_1) \\ \vdots & & \vdots \\ \lambda_D\phi_1(\boldsymbol{x}_D) & \cdots & \lambda_D\phi_N(\boldsymbol{x}_D) \end{bmatrix} \cdot \begin{bmatrix} u_1 \\ \vdots \\ u_N \end{bmatrix} = \begin{bmatrix} 0 \\ 0 \\ 0 \\ \vdots \\ 0 \\ 0 \\ 0 \\ \lambda_1\alpha_1 \\ \vdots \\ \lambda_D\alpha_D \end{bmatrix}. \tag{10}$$

After solving eqn (10), the implicit function $u(\boldsymbol{x})$ can be evaluated everywhere in the domain Ω using the coefficients u_l. If visualized on the grid of interpolation (i.e., the visualization points are the same as the corners of the grid), the implicit function values are given by the coefficients u_{l_i} $(l_i = \{1, \dots, N_i\})$. As our aim is to assess the quality of the interpolation also in the cells close to the discontinuities, the visualization grid is created finer than the interpolation grid and implicit values are evaluated using eqns (3) and (4). Fig. 1 was created using the proposed method on a (15x10) interpolation grid and a (200x200) visualization grid; energy weights (λ_ϵ) and data weights $\lambda_k (k = \{1, \dots, D\})$ were set to 1.

3 HANDLING GHOST NODES IN STRUCTURAL GEOLOGICAL SETTINGS

3.1 Structural discontinuities with complex geometries

3.1.1 Fault tips

Faults represent slip surfaces that can laterally vanish in space, where the displacement on each side of the fault becomes nil. The implicit function is supposed to be discontinuous on the fault and to become continuous at the fault tips. The chosen strategy to achieve the transition from an interpolation with ghost nodes to an interpolation without ghost nodes is illustrated at a fault tip in Fig. 4. As in Section 2.1.3, ghost nodes are added on either side of the fault and segment-discontinuity intersection tests are performed to define the nodes used in eqn (3) for the evaluation of the implicit function at a position $\boldsymbol{x}$.

3.1.2 Intersections of discontinuities

Several discontinuities (e.g., faults or erosions) can branch one onto another in space. Fig. 5 illustrates a simple case of two faults intersecting each other, defining three independent areas. The presented method creates as many ghost nodes as needed to interpolate on the cells close to the faults on each area. In this example, two ghost nodes are associated to each regular interpolation node, creating two additional coefficients u_{l_g} for each corner.

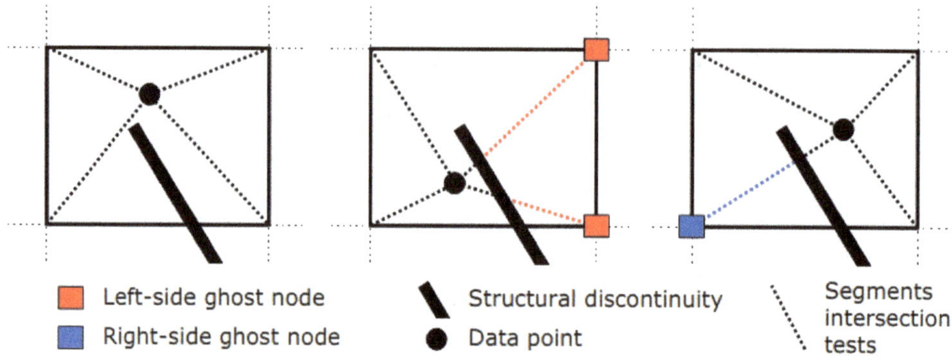

Figure 4: Illustration of the ghost-cell technique applied on a fault tip. A ghost node is used only if the segment between that node and a given data point intersects the discontinuity.

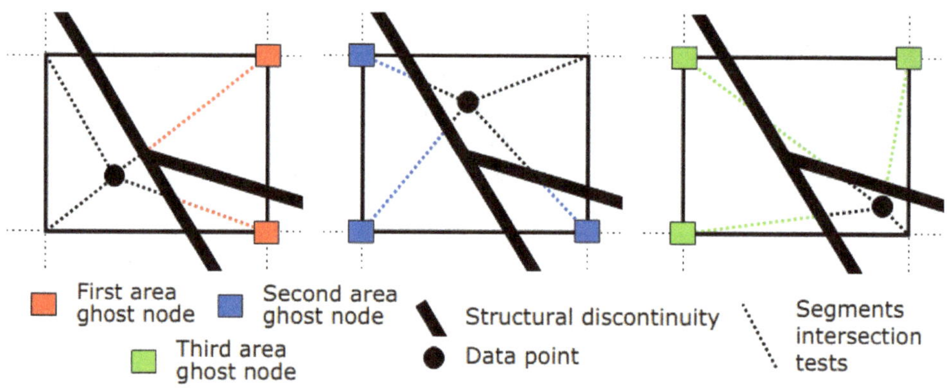

Figure 5: Illustration of the ghost-cell technique applied on several intersecting discontinuities. Ghost nodes are added for each area sealed by the discontinuities.

3.2 Finite difference with ghost nodes

The centred finite difference scheme presented in Fig. 3(a) can only apply to nodes far from the borders and the discontinuities. According to the discretisation of the energy term of eqn (2), the energy equations must be written on every interpolation node $\boldsymbol{x}_{l_i}$ and ghost node $\boldsymbol{x}_{l_g}$, which ensures homogeneous weights for all energy equations in the least squares system (10). Figs 3(b)–(d) illustrate the positions of the nodes $\boldsymbol{x}_m$ and their associated coefficients $\delta_m^{xx}(\boldsymbol{x}_l)$, $\delta_m^{yy}(\boldsymbol{x}_l)$ and $\delta_m^{xy}(\boldsymbol{x}_l)$ for three possible positions of a node $\boldsymbol{x}_l$ with respect to the domain borders or the discontinuities. These configurations are often combined, on a border interpolation node close to a discontinuity for instance.

As in Section 2.1.3, the nodes $\boldsymbol{x}_m$ are found between interpolation and ghost nodes with segment-discontinuity intersection tests.

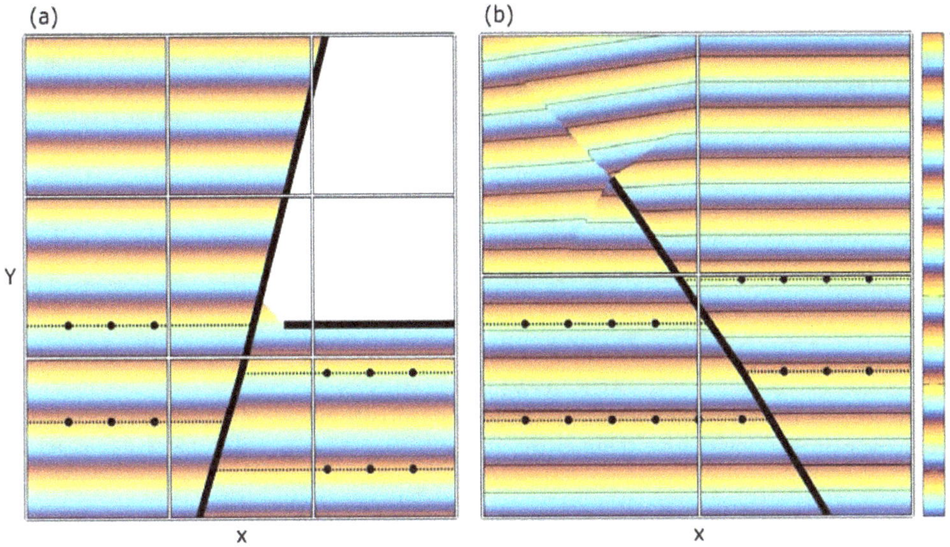

Figure 6: Resolution issues represented on two synthetic cross sections with faults and two parallel horizons. (a) Two disconnected faults treated as connected; (b) Rays of discontinuous jumps in the implicit function observed at the tip of a fault.

3.3 Grid resolution issues

The resolution of the grid can have a direct impact on the solution. In Fig. 6, the grid is chosen very coarse (i.e., (4x4) in Fig. 6(a) and (3x3) in Fig. 6(b)) to observe the influence on the resulting implicit function (observed on fine visualization grids of size (200x200) in both).

Fig. 6(a) shows how the method behaves when two faults are close to each other in the same cell but do not intersect: the connectivity between the two regions on the right side is not captured, the upper right area is left isolated without data, and hence has no solution. This is due to the way the nodes $\boldsymbol{x}_m$ are defined, using segment-discontinuity intersection tests (Sections 3.1.2 and 3.2). It shows that a resolution adapted to the scale of the geological structures being studied is expected with the proposed method.

Furthermore, the technique used at fault tips (Section 3.1.1) can generate discontinuous jumps in the implicit function within the containing cell (Fig. 6(b)). This is observed only because the visualization grid was taken way finer than the interpolation grid. This technique, based on segment-discontinuity intersection tests, is very close to the visibility criterion [21], which is known to generate the same artefacts [22]. Some other meshless techniques such as the diffraction and the transparency criteria [22], [23] could be tested.

4 CONCLUSION AND PERSPECTIVES

We proposed a grid based structural modelling method in two dimensions that handles discontinuities with a ghost-cell technique. The modelling problem is posed as a spatial regression of data points penalized by the bending energy approximated with finite difference schemes. The linear system to solve is thus sparse, and only the grid cells affected by discontinuities are stored in memory to handle the ghost nodes.

The ghost-cell technique represents well complex structural geometries, with jumps in the implicit function only at discontinuities. However, the method is dependent on the resolution of the grid relatively to the spacing between the discontinuities and to the curvature of the underlying surfaces: a coarse grid may overlook finer structures, and it may generate finer artefacts in the implicit function close to fault tips. These issues could be respectively reduced with other tests than segment-discontinuity intersections to indicate the use of ghost nodes or using meshless optic criteria for instance [22], [23].

The method is presented in two dimensions, but the used formalism could be extended to three dimensions as well: with the adapted version of the bending energy, the corresponding finite difference schemes, and a definition of tests for discontinuities and ghost nodes based on three-dimensional objects.

ACKNOWLEDGEMENTS

This work was funded by Schlumberger, the Association Nationale Recherche Technologie, the Research for Integrative Numerical Geology consortium, and the Laboratory of Excellence Ressources 21. These organisations are gratefully acknowledged.

REFERENCES

[1] Mallet, J.L., Three-dimensional graphic display of disconnected bodies. *Mathematical Geology*, **20**(8), pp. 977–990, 1988. DOI: 10.1007/bf00892974.

[2] Lajaunie, C., Courrioux, G. & Manuel, L., Foliation fields and 3D cartography in geology: Principles of a method based on potential interpolation. *Mathematical Geology*, **29**(4), pp. 571–584, 1997. DOI: 10.1007/bf02775087.

[3] Chilès, J.P., Aug, C., Guillen, A. & Lees, T., Modelling the Geometry of Geological Units and its Uncertainty in 3D From Structural Data: The Potential-Field Method. Orebody Modeling and Strategic Mine Planning—Spectrum 14, Jul. 22–24.

[4] Frank, T., Tertois, A.L., & Mallet, J.L., 3d-reconstruction of complex geological interfaces from irregularly distributed and noisy point data. *Computers & Geosciences*, **33**(7), pp. 932–943, 2007. DOI: 10.1016/j.cageo.2006.11.014.

[5] Calcagno, P., Chilès, J.P., Courrioux, G, & Guillen, A., Geological modelling from field data and geological knowledge. *Physics of the Earth and Planetary Interiors*, **171**(1–4), pp. 147–157, 2008. DOI: 10.1016/j.pepi.2008.06.013.

[6] Caumon, G., Gray, G., Antoine, C. & Titeux, M.O., Three-dimensional implicit stratigraphic model building from remote sensing data on tetrahedral meshes: theory and application to a regional model of La Popa Basin, NE Mexico. *IEEE Transactions on Geoscience and Remote Sensing,* **51**(3), pp. 1613–1621, 2013. DOI: 10.1109/tgrs.2012.2207727.

[7] Hillier, M.J., Schetselaar, E.M., de Kemp, E.A. & Perron, G., Three-Dimensional modelling of geological surfaces using generalized interpolation with radial basis functions. *Mathematical Geosciences*, **46**(8), pp. 931–953, 2014. DOI: 10.1007/s11004-014-9540-3.

[8] Souche, L., Iskenova, G., Lepage, F. & Desmarest, D., Construction of structurally and stratigraphically consistent structural models using the volume-based modelling technology: Applications to an Australian dataset. *International Petroleum Technology Conference,* 2014.

[9] Mallet, J.L., *Geomodeling*, Oxford University Press, Inc., 2002.

[10] Mallet, J.L., Discrete Smooth Interpolation. *Computer-aided Design,* **24**(4), pp. 178–191, 1992.

[11] Pellerin, J., Lévy, B., Caumon, G. & Botella, A., Automatic surface remeshing of 3d structural models at specified resolution: A method based on Voronoi diagrams. *Computers & Geosciences*, **62**, pp. 103–116, 2014. DOI: 10.1016/j.cageo.2013.09.008.
[12] Karimi-Fard, M. & Durlofsky, L.J., A general gridding, discretization, and coarsening methodology for modeling flow in porous formations with discrete geological features. *Advances in Water Resources,* **96**, pp. 354–372, 2016. DOI: 10.1016/j.advwatres.2016.07.019.
[13] Renaudeau, J., Malvesin, E., Maerten, F. & Caumon, G., Implicit structural modeling by minimization of the bending energy with local meshless techniques. In prep 2018.
[14] Irakarama, M., Laurent, G., Renaudeau, J. & Caumon, G., Finite difference implicit modeling of geological structures. *80th EAGE Conference and Exhibition*, Copenhagen: Denmark, 2018.
[15] Renaudeau, J., Malvesin, E., Maerten, F. & Caumon, G., Implicit structural modeling with local meshless functions. *80th EAGE Conference and Exhibition*, Copenhagen, Denmark, 2018.
[16] Dubrule, O., Comparing splines and kriging. *Computers and Geosciences*, **10**(2–3), pp. 327–338, 1984. DOI: 10.1016/0098-3004(84)90030-x.
[17] Wahba, G., Spline models for observational data. *Society for Industrial and Applied Mathematics*, **59**, 1990.
[18] Dhatt, G. & Touzot, G., Une présentation de la méthode des éléments finis, Presses Université Laval, 1981.
[19] Fedkiw, RP, Aslam, T, Merriman, B. & Osher, S., A non-oscillatory Eulerian approach to interfaces in multimaterial flows (the ghost fluid method). *Journal of Computational Physics,* **152**(2), pp. 457–492, 1999. DOI: 10.1006/jcph.1999.6236.
[20] Collon, P., Steckiewicz-Laurent, W., Pellerin, J., Laurent, G., Caumon, G., Reichart, G. & Vaute, L., 3D geomodelling combining implicit surfaces and Voronoi-based remeshing: A case study in the Lorraine coal basin (France). *Computers & Geosciences*, **77,** pp. 29–43, 2015. DOI: 10.1016/j.cageo.2015.01.009.
[21] Belytschko, T., Lu, Y.Y. & Gu, L., Element-Free Galerkin methods. *International Journal for Numerical Methods in Engineering,* **37**(2), pp. 229–256, 1994.
[22] Belytschko, T., Krongauz, Y., Fleming, M., Organ, D. & Liu, W.K.S., Smoothing and accelerated computations in the Element Free Galerkin method. *Journal of Computational and Applied Mathematics,* **74**(1–2), pp. 111–126, 1996. DOI: 10.1016/0377-0427(96)00020-9.
[23] Organ, D., Fleming, M., Terry, T. & Belytschko, T., Continuous meshless approximations for nonconvex bodies by diffraction and transparency. *Computational Mechanics*, **18**(3), pp. 225–235, 1996. DOI: 10.1007/s004660050142.

CAUCHY BVP FOR ELASTIC HALF-PLANE POSED IN DISPLACEMENT ORIENTATIONS

ALEXANDER N. GALYBIN
The Schmidt Institute of Physics of the Earth (IPE RAS), Russia

ABSTRACT

This study presents a Cauchy-type boundary value problem of plane elasticity in which the boundary conditions are posed in terms of the orientations of the displacement vector and its normal derivative. No magnitudes of the displacements are specified. The problem is reduced to a singular integral equation by using the well-known Muskhelishvili's theory based on the complex potentials. The solvability of the integral equation is analysed in accordance with the Gakhov's approach, which reveals that the problem has a finite number of linearly independent solutions depending on the index of the corresponding Riemann BVP. The index is defined through the orientations of the contour displacements. More detailed analysis is performed for the case of elastic half-plane since previously it has been shown that the shape of the domain does not influence the solvability. A numerical approach for solving the problem for the arbitrary domain is outlined.

Keywords: plane elasticity, boundary value problems, complex potentials, singular integral equations.

1 INTRODUCTION

This study continues the investigation of the incorrectly posed boundary value problems, BVP, of plane elasticity which formulations employ the boundary magnitudes of neither stresses nor displacements. Such formulations are found in geomechanics where the magnitudes of forces/displacements acting on the margins of tectonic plates forming the earth's crust are unknown. Instead the orientations of principal stresses and displacements are known not only near the margins but inside the plates as well. This type of data has become the main motivation for the development of non-standard models for the stress analysis by formulation of incorrectly posed BVPs [1]–[6]. The main feature of these problems is non-uniqueness of the solution, which is expressed as the presence of certain number of arbitrary parameters entered into the general solution for the stress (displacement) field. These, in principle, can be identified if the magnitudes of stresses (displacements) are measured at a number of discrete points.

This investigation is mainly based on the following approach used previously. Firstly, the boundary conditions in terms of the stress/displacement orientations are formulated through the boundary values of the complex potentials (holomorphic functions in the domain considered). This allows one to reduce the BVP to a system of singular integral equations, SIE, for the determination of unknown auxiliary functions that are usually proportional to the jump of the displacements across the contour (provided that the stress vector is continuous across the contour). The obtained system of SIE can be converted into a single complex SIE that contains the unknown complex valued function, its complex conjugate, the singular and regular integrals and their conjugates. Gakhov [7] has suggested the following approach suggested for analysing the solvability of the complex SIE. The first step assumes separation of the complex SIE in two parts. The first (dominant) part contains non-integral and singular terms while the second part contains all regular terms. The solvability of the full SIE is defined by its dominant part only. Therefore, as the second step, the dominant SIE is reduced to the Riemann BVP by applying a special integral operator specified by the coefficients of the dominant SIE (see Section 3 for more details).

WIT Transactions on Engineering Sciences, Vol 122, © 2019 WIT Press
www.witpress.com, ISSN 1743-3533 (on-line)
doi:10.2495/BE410181

Finally, the solvability of the Riemann BVP depends on the index determined by the coefficient of the Riemann problem. The latter is calculated as the increment of its argument (divided by 2π) after traversing the contour anticlockwise. If the index is positive then the Riemann problem has a finite number of solutions that is proportional to the index, otherwise the BVP does not have solutions in the class of holomorphic functions. This approach can also be adopted for numerical solving complete complex SIEs [6].

It is shown in [1]–[6] that the dominant SIE corresponds to the case of elastic half-plane regardless of the boundary condition used. Therefore we analyse the dominant SIE only, i.e. the case of half-plane, which makes it easy to operate with mathematical transformations but still keeps the necessary generality. In the case considered further on we arrive at one single complex SIE instead of a system of two real SIEs as e.g. in [1] or [5], [6] that can be solved successively. Therefore one modification of the previously reported technique is proposed to reduce the problem to the SIE suitable for Gakhov's approach. We also outline a numerical approach that is not limited to the dominant SIE but is applicable for the full SIE as well.

2 PROBLEM FORMULATION AND INTEGRAL EQUATIONS

2.1 Preliminaries

Let us introduce the following notations for the complex displacement vector

$$W \equiv 2G(u_1 + iu_2) = \omega \exp(i\beta), \tag{1}$$

and refer to the Kolosov–Muskhelishvili formula [8] for the displacements

$$W = \kappa\phi(z) - z\overline{\Phi(z)} - \overline{\psi(z)}. \tag{2}$$

Here G is the shear modulus, u_i (i=1,2) are the displacement components, $\omega \geq 0$ and β are respectively the modulus and the argument of the complex-valued function $W = W(z, \bar{z})$, where $z = x_1 + ix_2$, $\bar{z} = x_1 - ix_2$. For any domain Ω bounded by the contour Γ the complex potentials in (2) can be expressed through two auxiliary functions $g(t)$ and $h(t)$ as follows

$$\phi(z) = \frac{1}{2\pi i}\int_\Gamma \frac{g(t)\,dt}{t-z}, \quad \Phi(z) = \phi'(z) = \frac{1}{2\pi i}\int_\Gamma \frac{g'(t)\,dt}{t-z}, \quad \psi(z) = \frac{1}{2\pi i}\int_\Gamma \frac{h(t)\,dt}{t-z}, \quad z \in \Omega, \; z \,\overline{\in}\, \Gamma. \tag{3}$$

The boundary values of the displacement vector can be found by the Sokhotski–Plemelj formulas in the form

$$2W(\zeta) = \pm(\kappa g(\zeta) - \zeta\overline{g'(\zeta)} - \overline{h(\zeta)}) + \kappa\mathbf{S}(g) - \zeta\overline{\mathbf{S}(g')} - \overline{\mathbf{S}(h)}, \tag{4}$$

where +/- correspond to the interior/exterior domain respectively and the following notation for the singular integral is introduced

$$\mathbf{S}(g) = \frac{1}{\pi i}\int_\Gamma \frac{g(t)}{t-\zeta}dt, \quad \zeta \in \Gamma. \tag{5}$$

The functions $g(t)$ and $h(t)$ are not independent. Let us further select

$$h(\zeta) = \kappa\overline{g(\zeta)} - \bar{\zeta}g'(\zeta), \tag{6}$$

which provides the continuity of the displacements across the contour. These can be presented as follows

$$W(\zeta)=\kappa\mathbf{S}(g)-\tfrac{1}{2}\overline{\mathbf{S}\left((\overline{\zeta}-\overline{t})g'\right)}. \qquad (7)$$

It should be noted that the second term in the right hand side of (7) is not singular because

$$\lim_{t\to\zeta}\frac{\overline{t}-\overline{\zeta}}{t-\zeta}=\frac{d\overline{\zeta}}{d\zeta}=e^{-2i\theta(\zeta)}, \qquad (8)$$

where θ is the angle between the tangent to the contour and the real axis.

2.2 Boundary conditions

Let us assume that the orientation of the displacement vector, angle β=β(ζ), is known together with its normal derivative $\beta_n'(\zeta)$. The first condition means that arg$W$(ζ)=β(ζ) and leads to the boundary condition of the form

$$\mathrm{Im}\left(e^{-i\beta(\zeta)}W(\zeta)\right)=0 \quad\Leftrightarrow\quad e^{-i\beta(\zeta)}W(\zeta)-e^{i\beta(\zeta)}\overline{W(\zeta)},\quad \zeta\in\Gamma. \qquad (9)$$

This equation expresses the fact that $e^{-i\beta}W=\omega$ is real everywhere in the domain including its boundary. By differentiation of Im($e^{-i\beta}W$)=0 with respect to outward normal ***n*** one finds

$$e^{-i\beta(\zeta)}W_n'(\zeta)-e^{i\beta(\zeta)}\overline{W_n'(\zeta)}=i\beta_n'(\zeta)\left(e^{-i\beta(\zeta)}W(\zeta)+e^{i\beta(\zeta)}\overline{W(\zeta)}\right),\quad \zeta\in\Gamma. \qquad (10)$$

Taking (9) into account one can rewrite the right-hand side of (10) and present the second boundary condition in the form

$$e^{-i\beta(\zeta)}W_n'(\zeta)-e^{i\beta(\zeta)}\overline{W_n'(\zeta)}=2i\beta_n'(\zeta)e^{-i\beta(\zeta)}W(\zeta),\quad \zeta\in\Gamma. \qquad (11)$$

The derivatives of the complex-valued function $f=f(\zeta)$ on the contour Γ are defined as their limiting values. Respectively the tangent and normal derivatives are presented as follows

$$\frac{\partial f}{\partial \boldsymbol{l}}=\boldsymbol{l}\lim_{z\to\zeta}\frac{\partial f}{\partial z}+\overline{\boldsymbol{l}}\lim_{z\to\zeta}\frac{\partial f}{\partial \overline{z}}=\zeta'\lim_{z\to\zeta}\frac{\partial f}{\partial z}+\overline{\zeta}'\lim_{z\to\zeta}\frac{\partial f}{\partial \overline{z}}, \qquad (12)$$

$$\frac{\partial f}{\partial \boldsymbol{n}}=\boldsymbol{n}\lim_{z\to\zeta}\frac{\partial f}{\partial z}+\overline{\boldsymbol{n}}\lim_{z\to\zeta}\frac{\partial f}{\partial \overline{z}}=-i\zeta'\lim_{z\to\zeta}\frac{\partial f}{\partial z}+i\overline{\zeta}'\lim_{z\to\zeta}\frac{\partial f}{\partial \overline{z}}. \qquad (13)$$

Here one should notice that the derivatives in the left hand side are taken with respect to the arc length s, i.e. they can be expressed through the derivatives with respect to the complex variable ζ on the contour by the following relationship

$$\frac{\partial f}{\partial \boldsymbol{l}}=\frac{\partial f}{\partial s}=\frac{\partial f}{\partial \zeta}\zeta'=e^{i\theta}\frac{\partial f}{\partial \zeta}. \qquad (14)$$

By taking into account that $\zeta'=\exp(i\theta)$ one present the derivatives of the vector W on the contour Γ in the form

$$W_t'(\zeta) = e^{i\theta(\zeta)}\left(\kappa\Phi(\zeta) - \overline{\Phi(\zeta)}\right) - e^{-i\theta(\zeta)}\left(\zeta\overline{\Phi'(\zeta)} + \overline{\Psi(\zeta)}\right),, \tag{15}$$

$$W_n'(\zeta) = -ie^{i\theta(\zeta)}\left(\kappa\Phi(\zeta) - \overline{\Phi(\zeta)}\right) - ie^{-i\theta(\zeta)}\left(\zeta\overline{\Phi'(\zeta)} + \overline{\Psi(\zeta)}\right),, \tag{16}$$

where

$$\Psi(z) = \psi'(z) = \frac{1}{2\pi i}\int_\Gamma \frac{\kappa e^{-2i\theta(t)}\overline{g'(t)} - e^{-2i\theta(t)}g'(t) - \overline{t}g''(t)\,dt}{t-z}, \quad z\in\Omega,\ z\overline{\in}\Gamma. \tag{17}$$

It is evident from eqns (12), (13) that

$$W_t'(\zeta) + iW_n'(\zeta) = 2e^{i\theta(\zeta)}\left(\kappa\Phi(\zeta) - \overline{\Phi(\zeta)}\right), \tag{18}$$

$$W_t'(\zeta) - iW_n'(\zeta) = -2e^{-i\theta(\zeta)}\left(\zeta\overline{\Phi'(\zeta)} + \overline{\Psi(\zeta)}\right). \tag{19}$$

The latter formulas can be rewritten by using complex conjugation of (19) and by passing to the complex derivative using (14) that results in the following expressions

$$W_z'(\zeta) = \kappa\Phi(\zeta) - \overline{\Phi(\zeta)}, \tag{20}$$

$$\overline{W}_z'(\zeta) = -\overline{\zeta}\Phi'(\zeta) - \Psi(\zeta). \tag{21}$$

Note that these can be directly obtained by direct differentiation of (2). Eqns (18)–(21) provide the link between the tangential and normal derivatives and the boundary values of the derivatives with respect to z and its conjugate on the contour Γ. It is evident that if the boundary value of any complex-valued function is known than the tangential derivative is also known and if its normal derivative is known then one can find the boundary values of the derivatives with respect to z.

Let us combine two real conditions (9) and (11) into one complex by differentiation of $\mathrm{Im}(e^{-i\beta}W)=\mathrm{Im}(\omega)=0$ with respect to the variable z. Similarly to (11) one finds

$$e^{-i\beta(\zeta)}W_z'(\zeta) - e^{i\beta(\zeta)}\overline{W}_z'(\zeta) = 2i\beta_z'(\zeta)e^{-i\beta(\zeta)}W(\zeta), \quad \zeta\in\Gamma. \tag{22}$$

However in contrast to (11) this form is now complex and combines two real conditions equivalent to (9) and (11), therefore it can be used for obtaining a complex SIE.

2.3 Complex singular integral equation

Let us substitute (2) into the right hand side of (22) and (20) and (21) into the left hand side of (22) to obtain the following boundary value problem for the determination of the holomorphic functions φ, ψ and their derivatives

$$e^{-i\beta(\zeta)}\left(\kappa\Phi(\zeta) - \overline{\Phi(\zeta)}\right) + e^{i\beta(\zeta)}\left(\overline{\zeta}\Phi'(\zeta) + \Psi(\zeta)\right) = 2i\beta_z'(\zeta)e^{-i\beta(\zeta)}\left(\kappa\phi(\zeta) - z\overline{\Phi(\zeta)} - \overline{\psi(\zeta)}\right), \quad \zeta\in\Gamma. \tag{23}$$

Now it is possible to derive a SIE by substituting the boundary values of the complex potentials into (23) followed by the application of the Sokhotski–Plemelj formulas. However this would lead to a SIE that contains both unknown function $g(\zeta)$ and its derivative. No direct techniques for investigation of such SIEs are known, which necessitates further transformation in order to obtain a SIE of the Gakhov's type mentioned above.

Bearing in mind the relationship between the displacement vector and the unknown density of the Cauchy-type integral one can substitute (7) into (9) to obtain the following relationship

$$0 = e^{-i\beta(\zeta)}W(\zeta) - e^{i\beta(\zeta)}\overline{W(\zeta)} = e^{-i\beta(\zeta)}\kappa\mathbf{S}(g) - e^{i\beta(\zeta)}\kappa\overline{\mathbf{S}(g)} - i\,\mathrm{Im}\left[e^{i\beta(\zeta)}\mathbf{S}\left((\overline{\zeta} - \overline{t})g'\right)\right], \quad \zeta \in \Gamma. \tag{24}$$

Here the last term in the right-hand side is not singular.

By differentiation of (24) with respect to ζ one finds

$$2i\beta_{\zeta}'(\zeta)\left(e^{-i\beta(\zeta)}\kappa\mathbf{S}(g) + e^{i\beta(\zeta)}\kappa\overline{\mathbf{S}(g)}\right) =$$

$$= e^{-i\beta(\zeta)}\kappa\mathbf{S}(g') - e^{i\beta(\zeta)}\kappa\overline{\mathbf{S}\left(e^{-2i\theta(\zeta)}g'\right)} - i\frac{d}{d\zeta}\mathrm{Im}\left[e^{i\beta(\zeta)}\mathbf{S}\left((\overline{\zeta} - \overline{t})g'\right)\right], \quad \zeta \in \Gamma. \tag{25}$$

This provides the relationship between the unknown function and its contour derivative that is further considered as the only unknown function after transformation of the right-hand side of (22) and (23). The left hand sides of (23) can be also transformed in view of the Sokhotski–Plemelj formulas, which eventually leads to the following complex SIE

$$e^{-i\beta}\left[\pm(\kappa g' - \overline{g}') + \mathbf{S}(g') - \overline{\mathbf{S}(g')}\right] + e^{i\beta}\left[\pm e^{2i\theta}(\kappa g' - \overline{g}') + \overline{\mathbf{S}(e^{-2i\theta}(\kappa g' - \overline{g}'))} + \frac{d}{d\zeta}\mathbf{S}((\overline{\zeta} - \overline{t})g')\right]$$

$$-\kappa\frac{\beta_z'}{\beta_\zeta'}\left(e^{-i\beta}\mathbf{S}(g') - e^{i\beta}\overline{\mathbf{S}\left(e^{-2i\theta}g'\right)} - \frac{i}{\kappa}\frac{d}{d\zeta}\mathrm{Im}\left[e^{i\beta}\mathbf{S}\left((\overline{\zeta} - \overline{t})g'\right)\right]\right) = 0, \quad \zeta \in \Gamma. \tag{26}$$

Eqn (26) represents the full complex SIE for the problem considered. It should be complemented by the integral conditions expressing the fact that the function g returns to its initial value after the complete traverse of the contour,

$$\int_{\Gamma} g'(t)\,dt = 0. \tag{27}$$

3 THE CASE OF HALF-PLANE

3.1 The SIE for half-plane

It has been mentioned above that solvability of the BVP can be investigated for the case of half-plane, for which the formulas of the previous section can be readily simplified by assuming θ=0, Imζ=0, in particular the singular operator assumes the following form

$$\mathbf{S}(g) = \frac{1}{\pi i}\int_{-\infty}^{\infty}\frac{g(t)}{t - \zeta}dt, \quad \mathrm{Im}\,\zeta = 0. \tag{28}$$

It obeys the following properties

$$\mathbf{S}\left[\mathbf{S}(g)\right] = g, \tag{29}$$

$$\overline{\mathbf{S}(g)} = -\mathbf{S}(\overline{g}). \tag{30}$$

The former is valid for any closed contour, while the latter is valid only for the real axis.

It is also evident that due to (27)

$$\mathbf{S}\left((\overline{\zeta}-\overline{t})g'\right)=0\,. \tag{31}$$

We further consider the case of upper half-plane and therefore the SIE assumes the following form

$$(e^{-i\beta}+e^{i\beta})(\kappa g'-\overline{g}')+\kappa\left(e^{-i\beta}-e^{i\beta}-\frac{\beta_z'}{\beta_\zeta'}e^{-i\beta}\right)\mathbf{S}(g')-\left(e^{-i\beta}+e^{i\beta}-\kappa\frac{\beta_z'}{\beta_\zeta'}e^{i\beta}\right)\overline{\mathbf{S}(g')}=0,\ \zeta\in\Gamma. \tag{32}$$

This equation is the homogeneous dominant SIE, a part of the full SIE of the form investigated by Gakhov [7]. Namely

$$ag+b\overline{g}+c\mathbf{S}g+d\overline{\mathbf{S}g}+\mathbf{K}g+\overline{\mathbf{L}g}=f\,. \tag{33}$$

Here f=0 and the coefficients a,b,c,d are known continuous functions of the complex contour variable specified as follows

$$a=\kappa(e^{-i\beta}+e^{i\beta}),\ b=-e^{-i\beta}-e^{i\beta}),\ c=\kappa\left(e^{-i\beta}-e^{i\beta}-\frac{\beta_z'}{\beta_\zeta'}e^{-i\beta}\right),\ d=-e^{-i\beta}-e^{i\beta}+\kappa\frac{\beta_z'}{\beta_\zeta'}e^{i\beta}. \tag{34}$$

The regular operators **K**g and **L**g can be found from the full SIE (26), in the case of half-plane they vanished. It has been shown [7] that the solvability of SIE (33) is determined by the index calculated through the increment of the argument of the complex valued function G after the complete traverse of Γ in the counter-clockwise direction

$$2\mathrm{K}\equiv Index\left(G\right)=\tfrac{1}{2\pi i}\left[\ln(G)\right]\Big|_{\Gamma}=\tfrac{1}{2\pi}\left[\arg(G)\right]\Big|_{\Gamma}$$
$$G=\tfrac{A-iB}{A+iB},\quad A-iB=(\overline{a}+\overline{c})(a-c)+(b+d)(\overline{d}-\overline{b})\,. \tag{35}$$

If 2K >0 (the coefficient 2 is introduced for convenience), then the solution of the homogeneous equation SIE (f=0) depends on 2K complex constants or 4K real constants, otherwise the SIE does not have solutions, see Gakhov [7].

3.2 Special case of boundary conditions

Let us consider the case $\beta_n'(\zeta)=0$, which can be important for some applications in geomechanics. In this case the coefficients a,b,c,d become

$$a=\kappa(e^{-i\beta}+e^{i\beta}),\ b=-e^{-i\beta}-e^{i\beta},\ c=-\kappa e^{i\beta},\ d=(\kappa-1)e^{i\beta}-e^{-i\beta}\,. \tag{36}$$

Therefore

$$A-iB=4\kappa\cos\beta(\kappa e^{i\beta}-e^{-i\beta}) \tag{37}$$

and consequently

$$G=\frac{A-iB}{A+iB}=e^{2i\beta}\,\frac{\kappa e^{i\beta}-e^{-2i\beta}}{\kappa-e^{2i\beta}}\,. \tag{38}$$

Since κ>1 the fraction in (38) does not affect the index as its real part is always positive, therefore, the index is found as follows

$$2\mathrm{K} \equiv Index(G) = \tfrac{1}{2\pi i}\left[\ln(\exp(2i\beta))\right]\Big|_{\Gamma} = \tfrac{\beta}{\pi}\Big|_{\Gamma}\,. \tag{39}$$

This result indicates that the index depends on the displacement orientations only, i.e. on the angle β, whose increment (after the complete traverse of Γ in the positive direction) should be positive in order to provide a solution of the considered BVP. If 2K≤0, no bounded (non-trivial) solutions exist. For 2K>0 the solution of (32) for the special case of the coefficients (36) includes an arbitrary polynomial of 2K-1 degree, which means that the solution is controlled by up to 4K real parameters.

3.3 Numerical approach

Direct application of any boundary element technique to the homogeneous dominant SIE of the form (32) would produce only a trivial solution due to f=0. Therefore we should transform this SIE such that its right-hand side becomes non-homogeneous and none of the homogeneous solutions is lost.

It has already been shown in [5] that application of the following integral operator

$$\boldsymbol{\Theta}(.) = \overline{a}\mathbf{I}(.) - b\overline{\mathbf{I}(.)} - \overline{c}\mathbf{S}(.) - d\overline{\mathbf{S}(.)} \tag{40}$$

(**I** is the identical operator, i.e. **I**f=f) reduces the dominant eqn (32) (as well as the full SIE (26)) to the form

$$\begin{gathered} Ag' - iB\mathbf{S}g' + \mathbf{Reg}(g', \overline{g}') = 0, \\ A = a\overline{a} - b\overline{b} - c\overline{c} + d\overline{d}, \quad iB = \overline{a}c - a\overline{c} + \overline{b}d - b\overline{d}, \quad \mathrm{Im}(A) = \mathrm{Im}(B) = 0. \end{gathered} \tag{41}$$

We further seek the solution of (41) as the sum of two functions

$$g' = g_0' + g_1'. \tag{42}$$

Here g_0' is the general solution of the homogeneous dominant equation

$$Ag_0 - iB\mathbf{S}g_0 = 0, \tag{43}$$

and g_1' is any particular solution of the non-homogeneous SIE

$$Ag_1' - iB\mathbf{S}g_1' + \mathbf{Reg}(g_1', \overline{g}_1') = -\mathbf{Reg}(g_0', \overline{g}_0') \tag{44}$$

where the right hand side is expressed through the analytical solution of SIE (43); it includes all linearly independent solutions of (43) and hence (44) is not homogeneous anymore. Therefore, any proper boundary element method can be applied to find the numerical solution of the non-homogeneous SIE (44). Finally the sum in (42) determines the solution of the homogeneous dominant SIE, where the analytical part g_0' is found in Gakhov [7]. Further on the solution for the contour derivative of the displacement vector is found from (7).

It should be noted that this algorithm without any modifications can be applied for solving the full SIE as well.

4 CONCLUSIONS

In this paper we investigated the solvability of an incorrectly posed BVP of the Cauchy type for elastic half-plane formulated in terms of the orientations of the displacements vector and its normal derivative on the boundary. This study complements the method of investigation proposed in [1] by reducing the full SIE of the form that contains both unknown function $g(\zeta)$ and its derivative to the SIE of the form as in (26) and then to the SIE of the form as in (33) that has been investigated earlier. A numerical approach capable of dealing with both homogeneous and non-homogeneous SIEs has been proposed.

ACKNOWLEDGEMENT

The author is thankful to the Russian Foundation for Basic Research that supported this work by the Grant 16-01-00260.

REFERENCES

[1] Galybin, A.N. & Mukhamediev, Sh.A., Plane elastic boundary value problem posed on orientation of principal stresses. *Journal of the Mechanics and Physics of Solids*, **47**(11), pp. 2381–2409, 1999. DOI: 10.1016/s0022-5096(99)00032-0.

[2] Galybin, A.N. & Mukhamediev, Sh.A., Integral equations for elastic problems posed in principal directions: application for adjacent domains. *WIT Transactions on Modelling and Simulation*, vol. 42, WIT Press: Southampton and Boston, pp. 51–60, 2006.

[3] Galybin, A.N., Boundary value problems of plane elasticity involving orientations of displacements and tractions. *Journal of Elasticity*, **102**(1), pp. 15–30, 2011. DOI: 10.1007/s10659-010-9259-4.

[4] Galybin, A.N., Boundary integral equations for plane elastic problems posed in terms of stress orientations. *Proceedings of the 8th UK Conference on Boundary Integral Methods*, University of Leeds, UK, 2011.

[5] Galybin, A.N., On solvability of the BVP formulated in terms of displacement orientations on the interface between dissimilar elastic materials. *International Journal of Computational Methods and Experimental Measurements,* **5**(3), pp. 369–376, 2017. DOI: 10.2495/cmem-v5-n3-369-376.

[6] Galybin, A.N., Boundary value problems for elastic half-planes posed in terms of stress and displacement orientations. *International Journal of Computational Methods and Experimental Measurements*, **6**(6), pp. 1173–1181, 2018. DOI: 10.2495/cmem-v6-n6-1173-1181.

[7] Gakhov, F.D., *Boundary Value Problems*. Nauka (3rd Russian edition, see also the translation of the first edition by Pergamon Press, 1966: Moscow), 1977.

[8] Muskhelishvili, N.I., *Some Basic Problems of the Mathematical Theory of Elasticity*, The Netherlands, 1963.

SIMPLIFIED ANALYTICAL SOLUTION OF THE CONTACT PROBLEM ON INDENTATION OF A COATED HALF-SPACE BY A SPHERICAL PUNCH

EVGENIY V. SADYRIN[1,2], ANDREY S. VASILIEV[1,2], SERGEI S. VOLKOV[1,2], BORIS I. MITRIN[1] & SERGEI M. AIZIKOVICH[1]
[1]Research and Education Center "Materials", Don State Technical University, Russia
[2]Research Institute for Mechanics, National Research Lobachevsky State University of Nizhni Novgorod, Russia

ABSTRACT
This paper is devoted to construction of a mathematical model combining simplicity for practical usage and high accuracy. It is based on the solution of an axisymmetric contact problem on penetration of a rigid indenter into an elastic half-space with a functionally graded or homogeneous coating. The problem is reduced to solution of a dual integral equation. Asymptotically exact expressions for indentation force, depth, contact stiffness and distribution of contact pressures are obtained in simplified analytical form using one-parameter approximation of the integral equation kernel transform. Numerical calculations are provided for a number of homogeneous and functionally graded coatings. Accuracy of the solution is analyzed against ratio of Young's moduli of coating and substrate and the value of relative coating thickness.
Keywords: contact, penetration, spherical indenter, simple analytical solution, functionally graded coating.

1 INTRODUCTION

One of the ways to improve reliability and increase the service life of machine parts, elements of building structures, components of micro- and nanoelectronics as well as other products of modern industry is the modification of the initial surface of products by applying protective coatings. Such coatings can have a multilayer or functionally graded structure. In practice, nanoindentation is one of commonly used approaches to determine the mechanical characteristics of coatings.

To interpret the nanoindentation results, methods based on mathematical models that use solutions of classical contact problems for isotropic homogeneous materials are widely used. Oliver and Pharr proposed [1] and then modernized [2] a method for determining the hardness and Young's modulus of homogeneous isotropic materials, which is based on the Sneddon's solution for an axisymmetric indenter [3]. Field and Swain [4] for the interpretation of the spherical indentation results proposed a method that uses Hertz's solution of the problem of a spherical indenter penetration into a homogeneous half-space. Despite the fact that these methods do not take into account the influence of the substrate on the experimental results, under certain experimental conditions (indentation depth does not exceed 10% of the coating thickness [5]) and certain ratios of the mechanical properties of the coating and substrate [6]–[9] the properties of homogeneous coatings can be recovered. However, applying such methods in case of significant difference in Young's moduli of the coating and substrate can lead to significant underestimation or overestimation of reconstructed value of Young's modulus of the coating. In addition, this approach cannot be used to determine the Young's modulus of thin coatings, since the required indentation depth becomes comparable to the surface roughness height. In this situation, mathematical models based on contact problems for coated solids can be effectively used.

Plane and axisymmetric contact problems for elastic solids with coatings with arbitrarily varying elastic properties in depth were studied by Ke and Wang [10] and Liu et al. [11].

WIT Transactions on Engineering Sciences, Vol 122, © 2019 WIT Press
www.witpress.com, ISSN 1743-3533 (on-line)
doi:10.2495/BE410191

They used an approximation of the function describing the variation of the Young's modulus by a piecewise linear function. The singular integral equation of the problem was solved by the collocation method. Guler and Erdogan [12] constructed a solution of the plane contact problem on indentation considering the exponential variation of the elastic moduli by depth using similar collocation technique. Argatov et al. [13]–[16] developed mathematical models for nanoindentation of samples with a single-layer uniform thick coating using the indenters of various shapes.

In the papers by Aizikovich [21], Vasiliev et al. [17]–[19] and Volkov et al. [20] solution of axisymmetric and plane contact problems for elastic solids with a functionally graded coating were constructed using the bilateral asymptotic method. This approach has a few advantages. It effectively works for the entire range of values of the relative thickness of the coating; arbitrary variation of elastic properties in depth may be modeled. Also, the problem solutions are constructed in analytical form. However, the developed models are multiparameter, and it is rather difficult to use them in practice.

This paper is devoted to the construction of a simplified mathematical model of a spherical stamp indentation, allowing to take into account both the elastic properties of the coating and the contribution of the substrate. The model is based on the solution of an axisymmetric contact problem on indentation of an elastic half-space with functionally graded or homogeneous coating by a rigid punch. The solution is also constructed using the bilateral asymptotic method [21]. The main difference from the results obtained before is in the fact that the kernel transform is approximated by a ratio of two quadratic functions containing only one parameter. In this case, the scheme for construction of an approximated analytical solution is substantially simplified in comparison with the general case [17]–[20], in which the product of quadratic-fractional functions is used. The use of simple one-parameter approximation makes it possible to obtain the solution of the problem in an explicit analytical form, convenient in practice for engineering calculations.

2 PROBLEM STATEMENT

The mathematical model of the indentation of a coating-substrate system by a spherical indenter is based on the contact problem of the linear elasticity. The indenter is modeled by a rigid spherical punch of radius R (the results can be easily generalized to the case of a deformable elastic indenter [22]). The sample is modeled by an elastic half-space consisting of the functionally graded or homogeneous elastic layer (coating) of thickness H and a homogeneous half-space (substrate).

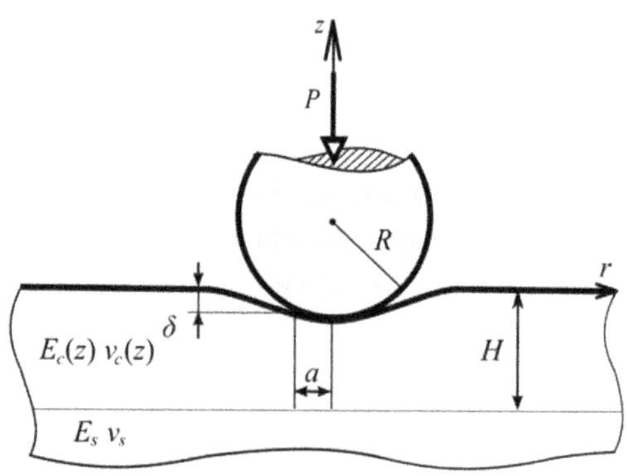

Figure 1: Statement of the contact problem.

Let us consider the cylindrical coordinate system (r, φ, z), the z axis is orthogonal to the surface of the half-space and passes through the center of the punch. Under the action of the centrally applied force P, the punch is displaced in the negative direction of the z axis by an amount δ. We consider the frictional forces between the base of the punch and the elastic layer to be negligibly small. The Young's modulus and Poisson's ratio of the half-space vary by depth according to the following laws:

$$\{E,\nu\} = \begin{cases} \{E_c(z)\,,\nu_c(z)\}\,, -H \le z \le 0; \\ \{E_s, \nu_s = \text{const}\}, -\infty < z < -H. \end{cases} \quad (1)$$

Here and after, index (c) corresponds to the coating while index (s) to the substrate.

Let the coordinate r be related to the radius of the contact zone a. In this case the boundary conditions on the surface of the half-space have the following form

$$z = 0: \quad \tau_{zr} = 0, \quad \begin{cases} \sigma_z = 0, \quad r > 1; \\ w = \; -\delta + a^2 r^2/2R \;, \; r \le 1. \end{cases} \quad (2)$$

To obtain simpler equations the spherical shape of a punch is approximated by a paraboloid. It can be done with high accuracy if $R >> a$.

Complete adhesion at the interface between the coating and the substrate is assumed:

$$z = -H: \quad \tau_{zr}^{(c)} = \tau_{zr}^{(s)}, \quad \sigma_z^{(c)} = \sigma_z^{(s)}, \quad w^{(c)} = w^{(s)}, \quad u^{(c)} = u^{(s)} \quad (3)$$

Here u and w are the displacements along r and z axes, respectively; $\sigma_z, \sigma_\phi, \sigma_r, \tau_{zr}$ are the components of the stress tensor. It is required to find the distribution of contact normal pressure:

$$\sigma_z \big|_{z=0} = -p(r), \quad r \le 1. \quad (4)$$

We assume that both the stresses and displacements vanish at $r \to \infty$ and $z \to -\infty$. To determine the size of the contact area, we will use the classical assumption of zero contact stresses at the contact boundary: $p(1) = 0$.

3 PROBLEM SOLUTION

Using Hankel integral transforms [23], the integral equation is obtained [17]:

$$\int_0^1 p(\rho)\rho d\rho \int_0^\infty L(u)\,\mathrm{J}_0(ur\lambda^{-1})\,\mathrm{J}_0(u\rho\lambda^{-1})du = \lambda\Theta\left(\delta - a^2 r^2/2R\right)a^{-1}, \quad r \le 1. \quad (5)$$

Following notations were used above: $\lambda = H/a$ is the relative thickness of the coating; $L(u)$ is the kernel transform of the integral equation which is generally determined numerically from the Cauchy problem solution for a system of ordinary differential equations with variable coefficients [17]. Let Θ and Θ_s be constants characterizing the effective elastic properties of the surfaces of the coating and the substrate:

$$\Theta = E_c(0)/\left(2(1-\nu_c^2(0))\right), \quad \Theta_s = E_s/\left(2(1-\nu_s^2)\right). \quad (6)$$

To solve eqn (5) we approximate the function $L(u)$ by the expression

$$L(u) \approx L_N(u) = \prod_{i=1}^{N}\left(u^2 + A_i^2\right)/\left(u^2 + B_i^2\right),\ A_i, B_i \in C. \tag{7}$$

Then the solution of the problem has the form analogous to [24]

$$p(r) = \frac{4a\Theta_s}{\pi R}\left[\sqrt{1-r^2} + \sum_{i=1}^{N} C_i \frac{A_i}{\lambda}\int_r^1 \frac{\sinh\left(A_i\lambda^{-1}t\right)}{\sqrt{t^2-r^2}}dt\right],\quad 0 \le r \le 1. \tag{8}$$

Constants $C_i\,(i = 1,\ldots,N)$ are determined from a system of linear algebraic equations:

$$\sum_{i=1}^{N} C_i A_i\left[\frac{B_k\sinh(A_i\lambda^{-1}) + A_i\cosh(A_i\lambda^{-1})}{B_k^2 - A_i^2}\right] = -\frac{\lambda(B_k+\lambda)}{B_k^2},\quad k = 1,2,\ldots,N. \tag{9}$$

To analyze the experiments on nanoindentation in practice, the indentation stiffness function is used $S = dP/d\delta$ [1]. Its value can be determined experimentally from the slope angle of tangent line to the unloading curve (of the graph demonstrating the dependence of indenter displacement from the indentation force) at the upper point of this curve (at the maximum depth of the indenter displacement).

Expressions for punch displacement, indentation force and stiffness have the form [24]:

$$\delta = \frac{a^2}{R}\left[1 + 2\sum_{i=1}^{N} C_i\cosh\left(\frac{A_i}{\lambda}\right) + 2\lambda^2\sum_{i=1}^{N}\left(\frac{1}{A_i^2} - \frac{1}{B_i^2}\right)\right], \tag{10}$$

$$P = \frac{8a^3\Theta_s}{3R}\left[1 + 3\sum_{i=1}^{N} C_i\left(\cosh\left(\frac{A_i}{\lambda}\right) - \frac{\lambda}{A_i}\sinh\left(\frac{A_i}{\lambda}\right)\right)\right], \tag{11}$$

$$S = \frac{dP}{d\delta} = 4a\Theta_s\frac{1 + 3\sum_{i=1}^{N} C_i\left(\cosh(A_i\lambda^{-1}) - A_i^{-1}\lambda\sinh(A_i\lambda^{-1})\right)}{1 + 2\sum_{i=1}^{N} C_i\cosh(A_i\lambda^{-1}) + 2\lambda^2\sum_{i=1}^{N}\left(A_i^{-2} - B_i^{-2}\right)}. \tag{12}$$

It is proved [24] that the obtained solution (8)–(12) is asymptotically exact for large and small values of the geometrical parameter of the problem λ. The error of the solution for intermediate values is related to the accuracy of the approximation of the kernel transform. On example of the torsion problem it was shown that these errors are of the same order of magnitude [25].

In order to simplify the solution as much as possible, let us assume that $N = 1$ in eqn (7). Then

$$L(u) \approx \left(u^2 + A^2\right)/\left(u^2 + B^2\right). \tag{13}$$

It is important to note that the approximation of the kernel transform is constructed in such a way that $L_N(0) = L(0)$ is satisfied. We introduce a parameter $\beta = \Theta_s/\Theta$ characterizing the "softness" of the coating. Coatings with $\beta > 1$ we call "soft", while with $\beta < 1$ we will call "hard" coatings. We also take into account that $L(0) = \beta^{-1}$ [24], hence $B = \sqrt{\beta}A$. Thus, eqns (8)–(12) may be simplified

$$p(r)=\frac{4a\Theta_s}{\pi R}\left[\sqrt{1-r^2}+\frac{(\sqrt{\beta}+\lambda A^{-1})(\beta^{-1}-1)}{(\sqrt{\beta}\sinh(A\lambda^{-1})+\cosh(A\lambda^{-1}))}\int_r^1\frac{\sinh(A\lambda^{-1}t)}{\sqrt{t^2-r^2}}dt\right],\quad 0\le r\le 1, \tag{14}$$

$$\delta=a^2\left(1+2\sqrt{\beta}\Delta_0\right)/R, \tag{15}$$

$$P=8a^3\Theta_s\left(1+3\left(\sqrt{\beta}+\lambda A^{-1}\right)\Delta_0\right)/(3R), \tag{16}$$

$$S=4a\Theta_s\left(1+\frac{\sqrt{\beta}+3\lambda A^{-1}}{1+2\sqrt{\beta}\Delta_0}\Delta_0\right), \tag{17}$$

$$\Delta_0=\frac{\lambda}{A^2}\left(\beta^{-1}-1\right)\left(\frac{A-\lambda\tanh(A\lambda^{-1})}{1+\sqrt{\beta}\tanh(A\lambda^{-1})}\right). \tag{18}$$

4 NUMERICAL RESULTS

To illustrate the simplified solutions obtained, we consider a set of homogeneous and functionally graded coatings. Let the Young's modulus of the substrate be E_0 = const, and the Poisson's ratio of both the coating and the substrate is 0.33. Let us consider a number of homogeneous coatings for which β = 0.1, 0.2, 0.5, 2, 5, 10 (we will call them coatings 1–6, respectively) and four functionally graded coatings:

$$\text{coatings 7 and 8: } E_c^{\text{lin}}=E_0\left(\beta-(1-\beta)zH^{-1}\right), \beta=0.5, 2.$$

$$\text{coatings 9 and 10: } E_c^{\text{exp}}=E_0\left(\beta e^k-1-(\beta-1)e^{-kzH^{-1}}\right)\left(e^k-1\right)^{-1}, k=-3; \beta=0.5, 2.$$

Graphs of kernel transforms of integral equations for coatings 1–10 are shown in Fig. 2. Kernel transforms for homogeneous coatings converge much faster to unity as $u\to\infty$. Kernel transforms for soft coatings 4–6, 8, 10 converge more slowly to the value of β^{-1} as $u\to 0$ than the kernel transforms for hard coatings 1–3, 7, 9. The smaller value of β, the slower convergence is.

Table 1 shows the values of error of kernel transforms approximation for coatings 1–10 using multiparameter approximation (7) and one-parameter approximation (13). As can be seen, the more the value of β differs from unity (upward or downward), the higher the error of one-parameter approximation. However, for many of the examples considered, even a one-parameter approximation allows to obtain a sufficiently high accuracy. The error of multiparameter approximation for all the examined coatings is not exceeding 0.20%, that indicates the high accuracy of the results.

To analyze the distribution of contact stresses, it is convenient to introduce dimensionless quantities. Let us denote $p_{\text{hom}}(r)$, P_{hom}, δ_{hom}, S_{hom} as the contact stresses, indentation force, punch displacement and indentation stiffness, respectively, in the Hertzian contact (for a homogeneous half-space with elastic properties that coincide with the substrate). We introduce dimensionless variables and functions:

$$p_0(r)=\frac{p(r)}{p_{\text{hom}}(0)}=\frac{\pi Rp(r)}{4a\Theta_s},\ \delta_0=\frac{\delta}{\delta_{\text{hom}}}=\frac{\delta R}{a^2},\ P_0=\frac{P}{P_{\text{hom}}}=\frac{3RP}{8a^3\Theta_s},\ S_0=\frac{S}{S_{\text{hom}}}=\frac{S}{4a\Theta_s}. \tag{19}$$

In order to analyze the difference in the distribution of contact stresses for a half-space with and without coating, we also consider the relative contact stresses:

$$p_{\text{rel}}(r)=p(r)/p_{\text{hom}}(r). \tag{20}$$

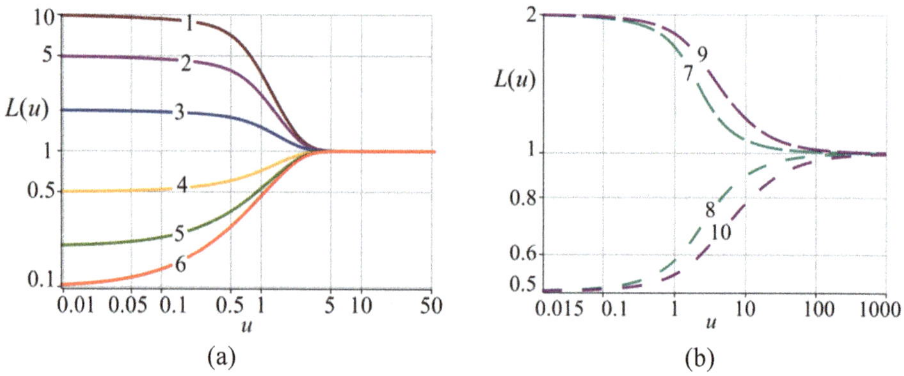

Figure 2: Kernel transforms for (a) homogeneous; and (b) functionally graded coatings.

Table 1: Values of error of kernel transforms approximation for coatings 1–10.

Coating no.	Type of inhomogeneity	β	Approximation error	
			$N = 1$	$N > 1$
1	Homogeneous	0.1	18.61%	0.15%
2	Homogeneous	0.2	11.78%	0.10%
3	Homogeneous	0.5	3.80%	0.08%
4	Homogeneous	2	4.30%	0.05%
5	Homogeneous	5	12.6%	0.11%
6	Homogeneous	10	20.70%	0.20%
7	Linear	0.5	3.80%	0.05%
8	Linear	2	6.80%	0.06%
9	Exponential	0.5	6.84%	0.08%
10	Exponential	2	9.34%	0.07%

Figs 3–5 demonstrate the graphs of indentation force, depth and stiffness for homogeneous coatings 1–6 and functionally graded coatings 7–10 for $N = 1$ and $N >> 1$. Logarithmic scale is used. For $\lambda \to 0$ the values of P_0, δ_0, S_0 tend to unit, which means that the force, depth and stiffness almost equal to the values corresponding to the Hertzian contact.

For $\lambda \to \infty$, values of P_0 and S_0 tend to β^{-1}, and δ_0 to 1. This means that these values coincide with the Hertzian ones for the elasticity moduli of the coating surface $E_c(0)$, $\nu_c(0)$. The behavior of P_0 and S_0 in a region of intermediate values of λ is very similar to the graph of the kernel transform (see Fig. 2). The graphs of P_0 and S_0 constructed using one-parameter and multiparameter approximations are sufficiently close for all the coatings considered, the graphs of δ_0 are less close to each other.

Figs 6 and 7 illustrate the distribution of contact pressure and relative contact pressure for coatings 4 and 10. One-parameter approximation allows to achieve a fairly good coincidence of the contact pressure distributions for small and large λ values for all the coatings considered, as well as for the intermediate λ in the case of a small ratio of the coating and substrate elastic moduli.

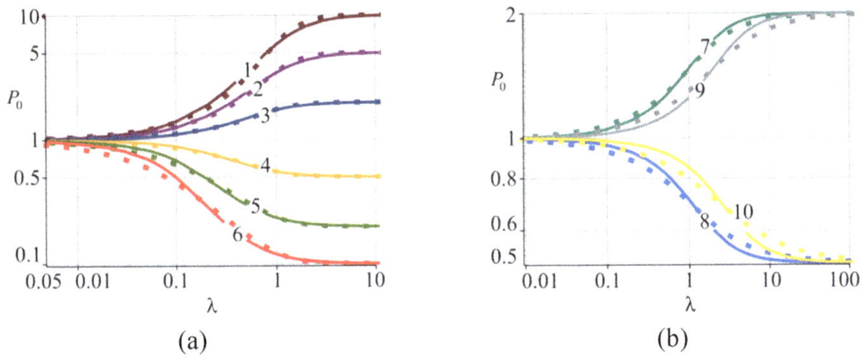

Figure 3: Indentation force for (a) homogeneous; and (b) functionally graded coatings for $N = 1$ (solid line) and $N >> 1$ (dotted line).

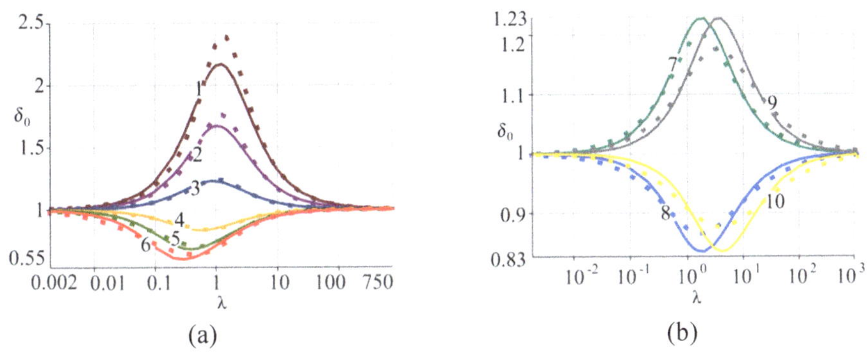

Figure 4: Indentation depth for (a) homogeneous; and (b) functionally graded coatings for $N = 1$ (solid line) and $N >> 1$ (dotted line).

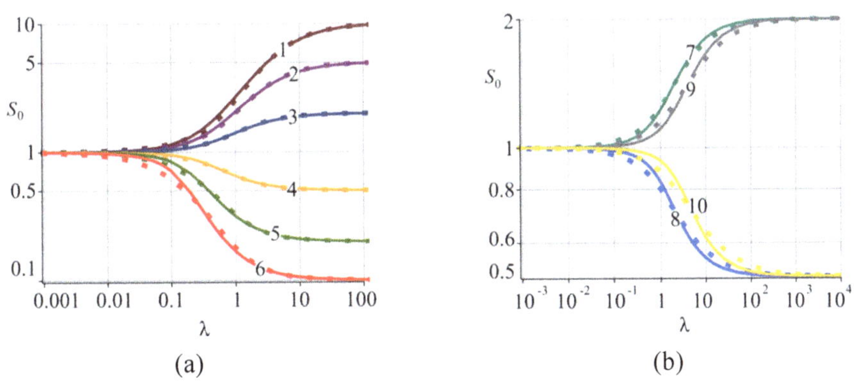

Figure 5: Indentation stiffness for (a) homogeneous; and (b) functionally graded coatings for $N = 1$ (solid line) and $N >> 1$ (dotted line).

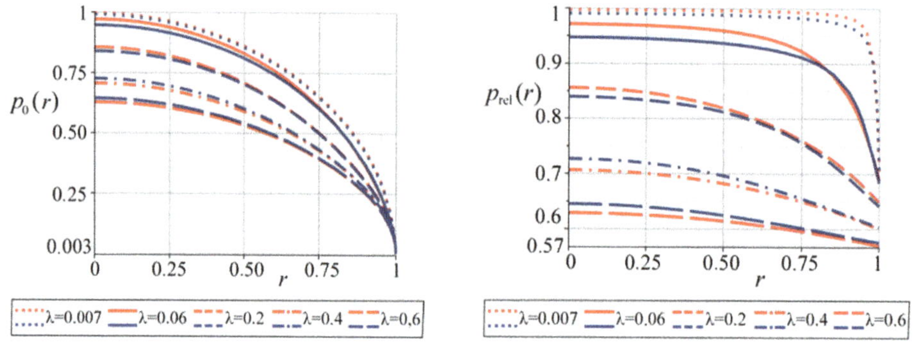

(a) (b)

Figure 6: Distribution of (a) dimensionless contact stresses; and (b) relative contact stresses for a homogeneous coating 4 ($\beta = 2$) for $N = 1$ (red lines) and $N >> 1$ (blue lines) for various λ.

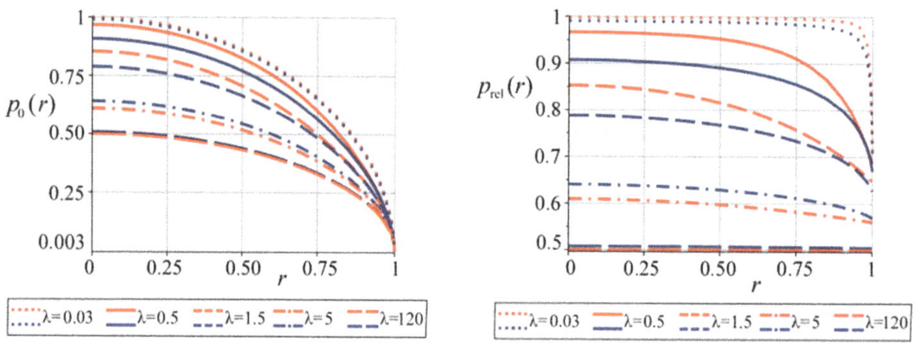

(a) (b)

Figure 7: Distribution of dimensionless contact stresses (a) and relative contact stresses (b) for a homogeneous coating 10 ($\beta = 2$) for $N = 1$ (red lines) and $N >> 1$ (blue lines) for various λ.

For a more detailed analysis of the accuracy of simplified solutions based on one-parameter approximation, we compare them with solutions obtained with multiparameter (much more accurate) approximation. Relative error of the simplified solution was calculated as:

$$\Delta_p = \left|1 - \frac{p_{N=1}(r)}{p_{N>1}(r)}\right| \cdot 100\%. \tag{21}$$

Fig. 8 demonstrates three-dimensional graphs of the relative error $\Delta_p(r)$ for coatings 7 and 1. For these coatings the one-parameter approximation of the kernel transform has the minimal value of 3.5% (for coatings 7 and 3) and maximal value of 29.30% (for coating 1), see Table 1. As can be seen, the maximum error in contact pressures is comparable to the error of approximation of the transform (3.80% for coating 7 and 18.61% for coating 1). In

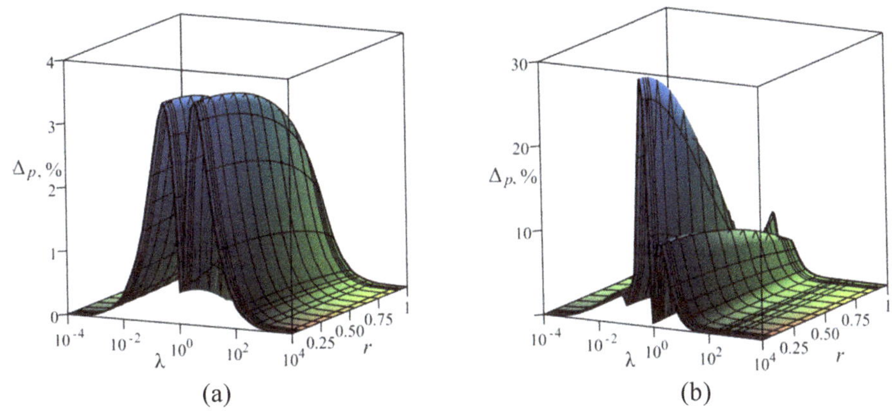

Figure 8: Relative error of the simplified solution for (a) coating 7; and (b) coating 1.

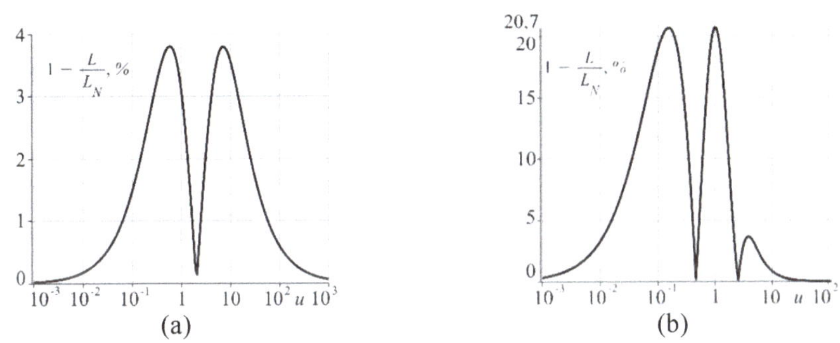

Figure 9: Relative error of simplified approximation of the kernel transform for (a) coating 3; and (b) coating 6.

the same way as the error graph of a one-parameter kernel transform approximation has two distinct hillocks (see Fig. 9), the error of contact pressures have them for any fixed value of $0 \leq r < 1$, with the exception of a small neighborhood of $r \approx 1$.

It can be seen from Fig. 8(b) that even with a relatively low accuracy of approximation (coating 1) for sufficiently large and small values of λ, the error of the solution is small. Let λ_0 be the value at which the maximum relative error of contact stresses is less than 1% for all $\lambda < \lambda_0$. Similarly, λ_∞ is such a value that for all $\lambda > \lambda_\infty$ the error is less than 1%. Table 2 lists the values of λ_0 and λ_∞ for all the 10 coatings. It is important to note that the maximum value of $\Delta_p(r)$ is calculated for $0 \leq r < r_0$, where r_0 is a certain value close to unit. In the vicinity of $r \approx 1$, for small values of λ, the relative error of contact pressures may be higher than 1%. This is due to the peculiarity of the behavior of the relative contact pressure distribution in this region. As was noted earlier [18], for any small value of λ there is a small neighborhood of $r \approx 1$ where $p_{\mathrm{rel}} \approx \beta^{-1/2}$. Since both the one-parameter and multiparameter approximations exactly coincide with the transform values for $u = 0$ and $u \to \infty$, then $\Delta_p(r) \to 0$ is fulfilled for $r \to 1$. However, a narrow area is observed to the left of $r = 1$ where the value of $\Delta_p(r)$ is quite high even for very small values of λ. It should also be noted that the absolute values of

Table 2: Values of λ_0 and λ_∞ for coatings 1–10.

Coating no.	β	Type of inhomogeneity	λ_0	λ_∞	$\max(\Delta_p)$
1	0.1	Homogeneous	0.003	15	29.30%
2	0.2	Homogeneous	0.005	10	14.52%
3	0.5	Homogeneous	0.03	50	3.51%
4	2	Homogeneous	0.01	4	4.22%
5	5	Homogeneous	0.002	5	12.36%
6	10	Homogeneous	0.001	6	23.14%
7	0.5	Linear	0.03	50	3.50%
8	2	Linear	0.02	100	6.51%
9	0.5	Exponential	0.03	120	6.33%
10	2	Exponential	0.03	300	9.13%

Table 3: Approximation error for P, S and δ for coatings 1–10.

Coating no.	β	Type of inhomogeneity	Δ_P	Δ_S	Δ_δ
1	0.1	Homogeneous	10.95%	7.33%	10.14%
2	0.2	Homogeneous	5.94%	3.65%	5.73%
3	0.5	Homogeneous	2.93%	2.71%	1.35%
4	2	Homogeneous	3.02%	2.78%	1.90%
5	5	Homogeneous	9.65%	9.11%	5.25%
6	10	Homogeneous	17.53%	16.57%	8.35%
7	0.5	Linear	3.01%	2.48%	2.44%
8	2	Linear	4.91%	4.64%	3.50%
9	0.5	Exponential	5.36%	4.03%	4.53%
10	2	Exponential	6.62%	6.28%	4.77%

the contact pressure are small near $r = 1$. That is why the value of relative error in this region is less important.

Similar to the contact stresses, let us analyze error of simplified expressions for the indentation force, depth and stiffness using following expressions:

$$\Delta_{\{P,S,\delta\}} = \left| 1 - \frac{\{P,S,\delta\}_{N=1}}{\{P,S,\delta\}_{N>1}} \right| \cdot 100\% . \tag{22}$$

Approximation error for P, S and δ are presented in the Table 3. For all considered coatings the values are even less than the error of approximation of the kernel transform. Graphs of relative errors of P, S and δ are illustrated for the coating 3 in Fig. 10.

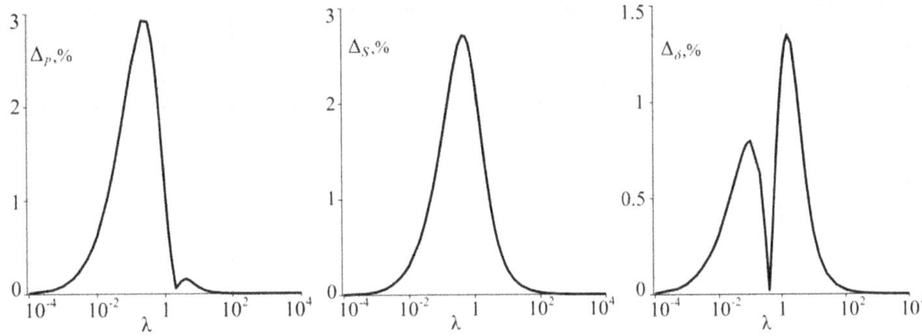

Figure 10: Relative error of the simplified solution for coating 3.

5 CONCLUSION

Simplified analytical expressions for the contact pressure, indentation force, depth and stiffness are obtained for the problem on indentation of a half-space with a homogeneous or functionally graded coating by a spherical punch. The accuracy of the results is studied in details. It is shown that the simplified formulas have sufficiently high accuracy for thin and thick coatings and for coatings of arbitrary thickness in the case of a relatively small difference in the elastic moduli of the coating and the substrate. For these cases, the obtained formulas can be used to recover coating characteristics from the indentation experiment. The approach can be applied to wide range of problems which can be reduced to a system of dual integral equations, including thermoelastic [26], electroelastic [27], fracture [28] problems.

ACKNOWLEDGEMENTS

Authors acknowledge the support of the Ministry of Education and Science of Russia in the framework of Governmental Assignment no. 9.1481.2017/4.6. E.V. Sadyrin was supported by the scholarship of the President of Russia no. SP-3672.2018.1.

REFERENCES

[1] Oliver, W.C. & Pharr, G.M., An improved technique for determining hardness and elastic modulus using load and displacement sensing indentation experiments. *Journal of Materials Research*, **7**(6), pp. 1564–1583, 1992. DOI: 10.1557/jmr.1992.1564.

[2] Oliver, W.C. & Pharr, G.M., Measurement of hardness and elastic modulus by instrumented indentation: Advances in understanding and refinements to methodology. *Journal of Materials Research*, **19**(1), pp. 3–20, 2004. DOI: 10.1557/jmr.2004.0002.

[3] Sneddon, I.N., The relation between load and penetration in the axisymmetric Boussinesq problem for a punch of arbitrary profile. *International Journal of Engineering Science*, **3**(1), pp. 47–57, 1965. DOI: 10.1016/0020-7225(65)90019-4.

[4] Field, J.S. & Swain, M.V., Determining the mechanical properties of small volumes of material from submicrometer spherical indentations. *Journal of Materials Research*, **10**(1), pp. 101–112, 1995. DOI: 10.1557/jmr.1995.0101.

[5] Bückle, H., *The Science of Hardness Testing and its Research Applications*, eds. J.H. Westbrook & H. Conrad, American Society For Metals: Ohio, pp. 453–494, 1973.

[6] Chen, J. & Bull, S.J., On the factors affecting the critical indenter penetration for measurement of coating hardness. *Vacuum*, **83**(6), pp. 911–920, 2009. DOI: 10.1016/j.vacuum.2008.11.007.

[7] Cai, X. & Bangert, H., Hardness measurements of thin films-determining the critical ratio of depth to thickness using FEM. *Thin Solid Films*, **264**(1), pp. 59–71, 1995. DOI: 10.1016/0040-6090(95)06569-5.

[8] Veprek-Heijman, M.G.J. & Veprek, S., The deformation of the substrate during indentation into superhard coatings: Bückle's rule revised. *Surface and Coatings Technology*, **284**, pp. 206–214, 2015. DOI: 10.1016/j.surfcoat.2015.10.064.

[9] Panich, N. & Sun, Y., Effect of penetration depth on indentation response of soft coatings on hard substrates: a finite element analysis. *Surface and Coatings Technology*, **182**(2–3) pp. 342–350, 2004. DOI: 10.1016/j.surfcoat.2003.07.002.

[10] Ke, L.L. & Wang, Y.S., Two-dimensional contact mechanics of functionally graded materials with arbitrary spatial variations of material properties. *International Journal of Solids and Structures*, **43**(18–19), pp. 5779–5798, 2006. DOI: 10.1016/j.ijsolstr.2005.06.081.

[11] Liu, T.J., Wang, Y.S. & Zhang, C., Axisymmetric frictionless contact of functionally graded materials. *Archive of Applied Mechanics*, **78**(4), pp. 267–282, 2008. DOI: 10.1007/s00419-007-0160-y.

[12] Guler, M.A. & Erdogan, F., Contact mechanics of graded coatings. *International Journal of Solids and Structures*, **41**(14), pp. 3865–3889, 2004. DOI: 10.1016/j.ijsolstr.2004.02.025.

[13] Argatov, I., Frictionless and adhesive nanoindentation: Asymptotic modeling of size effects. *Mechanics of Materials*, **42**(8), pp. 807–815, 2010. DOI: 10.1016/j.mechmat.2010.04.002.

[14] Argatov, I.I. & Sabina, F.J., Spherical indentation of a transversely isotropic elastic half-space reinforced with a thin layer. *International Journal of Engineering Science*, **50**(1), pp. 132–143, 2012. DOI: 10.1016/j.ijengsci.2011.08.009.

[15] Argatov, I.I. & Sabina, F.J., Asymptotic analysis of the substrate effect for an arbitrary indenter. *The Quarterly Journal of Mechanics and Applied Mathematics*, **66**(1), pp. 75–95, 2013. DOI: 10.1093/qjmam/hbs020.

[16] Argatov, I.I. & Sabina, F.J., Small-scale indentation of an elastic coated half-space: the effect of compliant substrate. *International Journal of Engineering Science*, **104**, pp. 87–96, 2016. DOI: 10.1016/j.ijengsci.2016.04.008.

[17] Vasiliev, A.S., Volkov, S.S. & Aizikovich, S.M., Indentation of an axisymmetric punch into an elastic transversely-isotropic half-space with functionally graded transversely-isotropic coating. *Materials Physics and Mechanics*, **28**, pp. 11–15, 2016.

[18] Vasiliev, A.S., Volkov, S.S., Belov, A.A., Litvinchuk, S.Y. & Aizikovich, S.M., Indentation of a hard transversely isotropic functionally graded coating by a conical indenter. *International Journal of Engineering Science*, **112**, pp. 63–75, 2017. DOI: 10.1016/j.ijengsci.2016.12.002.

[19] Vasiliev, A., Volkov, S., Aizikovich, S. & Jeng, Y.R., Axisymmetric contact problems of the theory of elasticity for inhomogeneous layers. *ZAMM-Journal of Applied Mathematics and Mechanics/Zeitschrift für Angewandte Mathematik und Mechanik*, **94**(9), pp. 705–712, 2014. DOI: 10.1002/zamm.201300067.

[20] Volkov, S., Aizikovich, S., Wang, Y.S. & Fedotov, I., Analytical solution of axisymmetric contact problem about indentation of a circular indenter into a soft functionally graded elastic layer. *Acta Mechanica Sinica*, **29**(2), pp. 196–201, 2013. DOI: 10.1007/s10409-013-0022-5.

[21] Aizikovich, S.M., Asymptotic solutions of contact problems of elasticity theory for media inhomogeneous in depth. *Journal of Applied Mathematics and Mechanics*, **46**(1), pp. 116–124, 1982. DOI: 10.1016/0021-8928(82)90091-0.

[22] Kudish, I.I., Volkov, S.S., Vasiliev, A.S. & Aizikovich, S.M., Lubricated point heavily loaded contacts of functionally graded materials. Part 1. Dry contacts. *Mathematics and Mechanics of Solids*, **23**(7), pp. 1061–1080, 2017. DOI: 10.1177/1081286517704689.

[23] Uflyand, Ya.S., *Integral Transformations In Problems of the Theory of Elasticity*, Nauka: Leningrad, 1967.

[24] Volkov, S.S., Vasiliev, A.S., Aizikovich, S.M., Seleznev, N.M. & Leontieva, A.V., Stress-strain state of an elastic soft functionally-graded coating subjected to indentation by a spherical punch. *PNRPU Mechanics Bulletin,* (4), pp. 20–34, 2016. DOI: 10.15593/perm.mech/2016.4.02.

[25] Aizikovich, S.M. & Vasiliev, A.S., A bilateral asymptotic method of solving the integral equation of the contact problem of the torsion of an elastic half-space inhomogeneous in depth. *Journal of Applied Mathematics and Mechanics*, **77**(1), pp. 91–97, 2013. DOI: 10.1016/j.jappmathmech.2013.04.011.

[26] Krenev, L.I., Aizikovich, S.M., Tokovyy, Y.V. & Wang, Y.-C., Axisymmetric problem on the indentation of a hot circular punch into an arbitrarily nonhomogeneous half-space. *International Journal of Solids and Structures*, **59**, pp. 18–28, 2015. DOI: 10.1016/j.ijsolstr.2014.12.017.

[27] Volkov, S.S., Vasiliev, A.S., Aizikovich, S.M. & Mitrin, B.I., Axisymmetric indentation of an electroelastic piezoelectric half-space with functionally graded piezoelectric coating by a circular punch. *Acta Mechanica*, pp. 1–14, 2017. DOI: 10.1007/s00707-017-2026-x.

[28] Aizikovich, S.M., Galybin, A.N. & Krenev, L.I., Semi-analytical solution for mode I penny-shaped crack in a soft inhomogeneous layer. *International Journal of Solids and Structures,* **53**, pp. 129–137, 2015. DOI: 10.1016/j.ijsolstr.2014.10.010.

Author index

Multiphase Flow: Theory and Applications

Edited by: ***P. VOROBIEFF****, Organisation University of New Mexico, USA and* ***C.A. BREBBIA,*** *Wessex Institute, UK*

The selected papers contained in this book present the latest research in one of the most challenging, yet most universally applicable areas of technology. Multiphase flows are found in all areas of technology and the range of related problems of interest is vast, including many areas of science and engineering.

Recently multiphase fluid dynamics have generated a great deal of attention, leading to many notable advances in experimental, analytical and numerical studies. It is perhaps, however, work on numerical solutions which is the most noticeable owing to the continuing improvements in computer software tools. Progress in numerical methods has permitted the solution of many practical problems, helping to improve our understanding of the physics involved.

The presented papers illustrate the close interaction between numerical modellers and researchers working to gradually resolve the many outstanding issues in our understanding of multiphase flow.

ISBN: 978-1-78466-311-7 **eISBN: 978-1-78466-312-4**
Published 2018 / 466pp

www.ingramcontent.com/pod-product-compliance
Lightning Source LLC
LaVergne TN
LVHW062236200926
848484LV00006B/1550
9781784662950